About Island Press

Island Press is the only nonprofit organization in the United States whose principal purpose is the publication of books on environmental issues and natural resource management. We provide solutions-oriented information to professionals, public officials, business and community leaders, and concerned citizens who are shaping responses to environmental problems.

In 2003, Island Press celebrates its nineteenth anniversary as the leading provider of timely and practical books that take a multidisciplinary approach to critical environmental concerns. Our growing list of titles reflects our commitment to bringing the best of an expanding body of literature to the environmental community throughout North America and the world.

Support for Island Press is provided by The Nathan Cummings Foundation, Geraldine R. Dodge Foundation, Doris Duke Charitable Foundation, Educational Foundation of America, The Charles Engelhard Foundation, The Ford Foundation, The George Gund Foundation, The Vira I. Heinz Endowment, The William and Flora Hewlett Foundation, Henry Luce Foundation, The John D. and Catherine T. MacArthur Foundation, The Andrew W. Mellon Foundation, The Moriah Fund, The Curtis and Edith Munson Foundation, National Fish and Wildlife Foundation, The New-Land Foundation, Oak Foundation, The Overbrook Foundation, The David and Lucile Packard Foundation, The Pew Charitable Trusts, The Rockefeller Foundation, The Winslow Foundation, and other generous donors.

The opinions expressed in this book are those of the author(s) and do not necessarily reflect the views of these foundations.

INVASIVE SPECIES

INVASIVE SPECIES

Vectors and Management Strategies

EDITED BY

GREGORY M. RUIZ AND JAMES T. CARLTON

ISLAND PRESS

Washington • Covelo • London

Cover caption: Between 1890 and 1914 Britain built two-thirds of the world's ships and transported 50 percent of the world's seaborne trade. The cover figure shows principal steamer routes and coaling stations as plotted on an Admiralty map of 1889, suggesting the scale and complexity of the global movement of ships—and of the organisms they transported. Modified from A. N. Porter, ed., 1991, *Atlas of British Overseas Expansion.* (Simon & Schuster).

ISLAND PRESS is a trademark of The Center for Resource Economics.

No copyright claim is made in the work of Richard Orr, Mary Palm, and Amy Rossman, employees of the federal government.

Library of Congress Cataloging-in-Publication Data
Invasive species : vectors and management strategies / Gregory M. Ruiz and James T. Carlton, Editors.
    p. cm.
Includes bibliographical references and index.
  ISBN 1-55963-902-4 (hardcover : alk. paper) -- ISBN 1-55963-903-2 (pbk. : alk. paper)
  1. Biological invasions. 2. Nonindigenous pests. 3. Biological diversity conservation. 4. Nonindigenous pests--Control. I. Ruiz, Gregory M. II. Carlton, James T.
  QH353.I62 2003
  577'.18--dc21
                            2003007223

British Cataloguing-in-Publication Data available.

Book design by: Brighid Willson

Printed on recycled, acid-free paper

Manufactured in the United States of America
09 08 07 06 05 04 03 02   8 7 6 5 4 3 2 1

Contents

Preface

Interest in the science and management of biological invasions has expanded rapidly in the past decade. Most invasions result from transport of organisms by human activities from one place to another, whether intentional or unintentional, allowing species to become established in new geographic regions. A steep rise in the number and impact of invasions has been observed for virtually all major habitats on Earth (Baskin 2002), propelling public concern, scientific interest, and calls for policy and management actions. Effective policy and management strategies—aimed at reducing the considerable ecological, economic, and human-health risks of invasions—depend upon a solid understanding of transfer mechanisms (vectors). In short, a scientific understanding of why, how, when, and where species are transported provides a critical foundation for vector management.

This volume explores the current knowledge base and policies surrounding invasion vectors, presenting updated contributions from a conference held November 8-11, 1999, at the Smithsonian Environmental Research Center (SERC) in Edgewater, Maryland. The "Conference on Pathways of Nonindigenous Species" was part of a series of workshops of the first phase of the Global Invasive Species Programme (GISP). GISP Phase I was developed in 1997 as a collaborative, international effort to address concerns about alien species invasions by initiating a scientifically based global strategy to understand invasions and limit their unwanted impacts. Phase I (Mooney 1999; http://globalecology.stanford.edu/DGE/Gisp) consisted of 11 program components and involved a diverse group of international science and policy experts. The goals of Phase I included (a) assembling information on patterns, prevention, and management of invasions across geographic regions, ecosystems, and taxonomic groups, and (b) distributing this information through a variety of avenues to governments, communities, and scientists.

A key component of GISP focuses on understanding invasion vectors and identifying management strategies to prevent invasions. To advance this component, we organized a conference to assess invasion vectors and vector management in terrestrial, freshwater, and marine ecosystems for major taxonomic groups in a variety of regions around the world. Information presented at this conference was further developed and expanded to create this volume.

STRUCTURE OF THE BOOK

This book is divided into three sections, each with a different purpose. The first section, comprised of nine chapters, highlights our present understanding about the operation of vectors and forms the book's data foundation. Examined are invasion causes, routes, and vectors across a diverse range of habitats and taxonomic groups. The major vectors that have led to the homogenization of terrestrial plants (especially "weeds"), fungi, insects, and vertebrates are examined in the first four chapters. Bridging both land and freshwater is an examination of the invasion histories of terrestrial and aquatic gastropod mollusks (snails and slugs), while a separate chapter focuses on freshwater fish, amphibians, reptiles, and mammal invasions. Two chapters concern the marine environment, considering invertebrate and plant invasions. The last chapter in this section focuses on an analysis of shipping vectors that serve to transport species into the world's largest freshwater ecosystems, the Laurentian Great Lakes of North America.

This first section does not attempt a comprehensive analysis that includes all guilds or taxonomic groups (e.g., freshwater macrophytes, phytoplankton, and various freshwater and terrestrial invertebrates are not covered here), but the data provided among chapters offer a detailed look at how literally thousands of species were transferred—and often are still being moved—around the world. The groups not treated here are important; fortunately, review essays are now available for many of them.

The second section consists of eight chapters that examine different approaches to management and policy, geared at reducing the likelihood of invasions and invasion impacts associated with vectors. In the first five chapters, contributors examine how vectors have been, are, or could be intercepted and managed. These chapters are focused upon specific countries as case histories, including Australia, New Zealand, South Africa, and the United States. An important bridging chapter that links the book's first and second sections is a view from South Africa, where vectors (such as forestry, horticulture, and shipping, and intentional releases of birds, mammals, and fishes) form the basis for modern perspectives of legislation and management action. Each chapter in this section takes a somewhat different approach—based upon the taxonomic array of species being tackled and upon national experience—but form a fascinating overview of both divergent and complementary strategies. The final three chapters address the emerging field of risk analysis, and its two major components, risk assessment and risk management. These three chapters each take a different focus, using a species-based approach, a pathways-based approach, and a "generic" approach that takes into account both species and pathways.

The final section consists of a single chapter as an overview and synthesis of vector ecology, drawing on the previous chapters and additional material. The goal of this last section is to integrate much of the information across previous chapters to develop a broad-based, conceptual framework and next steps for vector management.

TERMINOLOGY

While invasion terminology was discussed at the conference, we did not attempt to standardize authors' usages across chapters of specific terms for the transport of organisms through human action. Elsewhere (Carlton and Ruiz 2004) we have suggested a potential framework to standardize terms and definitions associated with vector science. This framework focuses on the terms used for causes (why a species is transported, whether accidental or deliberate), purposes (why a species is deliberately introduced), routes (the geographic path over which a species is transported), corridors (the physical conduit over or through which the vector moves within the route, such as footpaths, roads, and railroad beds), and vectors (how a species is transported—that is, the physical means or agents). Not all of our authors use these terms in the same way, but within the context of any one chapter, definitions and applications of words and concepts are clearly provided.

In similar fashion, the terminology used to characterize non-native species is extensive. Common terms include aliens, exotics, invaders, nonindigenous species, introduced species, immigrants, translocated species, naturalized species, colonists, adventives, neophytes, weeds, imports, and invasive species. The usage and particular meaning vary among taxonomic groups, geographic regions, as well as among individual scientists, managers, and policy-makers. This variation is perhaps most evident in the usage of the term "invasive," which often refers to species that spread or cause significant impacts (for discussion see reviews by Pyšek, 1995, Eser 1998, Richardson et al. 2000, Carlton 2002). For example, invasive plant species are considered by many to be those that establish self-sustaining populations and then undergo spread. For other taxonomic groups, and in the political arena, invasive species are frequently considered to be those organisms that cause significant and unwanted ecological, economic, or human-health impacts, having a distinctly negative connotation. As with transport activities, we have not attempted to standardize terminology used among chapters to describe non-native species.

We have used the term "invasive species" in the book title, in large part to be consistent with this focus and usage of the Global Invasive Species Programme. This term has the advantage of widespread recognition, both by the public and scientists, as being associated generally with colonization by non-native species. However, precise definitions of "invasive" remain in flux (as above; see Carlton 2002), as attributes of established populations— in terms of spread and impact—fall along a continuum (instead of binary bins). Moreover, the potential or realized attributes of many, if not most, non-native species are simply not known—within their native or newly colonized territories—further confounding such classification.

ACKNOWLEDGMENTS

We are most grateful to Harold Mooney for his encouragement and guidance throughout Phase I. We thank the many colleagues who reviewed chapters, and particularly Richard Mack, who both performed yeoman work as a reviewer of many chapters, and provided motivation and inspiration in assembling this book. We are in addition thankful to members of the Marine Invasion Research Laboratory at SERC, and especially Kelly Lion, who were instrumental in assisting us with all aspects of organizing the Conference. We also benefited greatly from editorial assistance by Monaca Noble and Kelly Lyles in producing the present volume. Barbara Dean, Barbara Youngblood and Meg Weaver, Island Press, patiently escorted us through the labyrinth of the editorial process and final book production.

Finally, we wish to thank the conference participants and sponsors for the opportunity to explore the many, complex facets of vector ecology and management. The conference was sponsored and funded by the Global Environment Facility, Scientific Committee on Problems of the Environment (SCOPE), Smithsonian Institution, and the United States Geological Survey (USGS).

—GREGORY M. RUIZ AND JAMES T. CARLTON

REFERENCES

Baskin, Y. 2002. *A plague of rats and rubbervines. The growing threat of species invasions.* Washington, DC: Island Press.

Carlton, J. T. 2002. Bioinvasion ecology: assessing invasion impact and scale. In *Invasive aquatic species of Europe. Distribution, impacts, and management,* E. Leppäkoski, S.Gollasch, and S. Olenin, eds., pp. 7–19. Dordrecht: Kluwer Academic Publishers.

Carlton, J. T. and G. M. Ruiz. 2004. Vector Science and Integrated Vector Management in Bioinvasion Ecology: Conceptual Frameworks. In *Best Practices for the Prevention and Management of Alien Invasive Species,* H. A. Mooney et al., eds. Washington, DC: Island Press.

Eser, U. 1998. Assessment of plant invasions: theoretical and philosophical fundamentals. In *Plant invasions: ecological mechanisms and human responses,* U. Starfinger, K. Edwards, I. Kowarik, and M. Williamson, eds., pp. 95–107. Leiden: Backhuys Publishers.

Mooney, H. A. 1999. The Global Invasive Species Programme (GISP). *Biological Invasions* 1: 97–98.

Pyšek, P. 1995. On the terminology used in plant invasion studies. In *Plant invasions—general aspects and special problems,* P. Pyšek, K. Prach, M. Remánjek, and M. Wade, eds., p. 71–81. Amsterdam: Academic Publishing.

Richardson, D. M., P. Pyšek, M. Rejmánek, M. G. Barbour, F. D. Paneta, and C. J. West. 2000. Naturalization and invasion of alien plants: concepts and definitions. *Diversity and Distributions* 6: 93–107.

Part I

INVASION CAUSES, ROUTES, AND VECTORS:

Spatial and Temporal Patterns in Terrestrial, Freshwater, and Marine Ecosystems

Chapter 1

Global Plant Dispersal, Naturalization, and Invasion: Pathways, Modes, and Circumstances

Richard N. Mack

"He who defends everything, defends nothing." This quote from the philosopher-king and military strategist Frederick the Great may seem an oblique manner by which to begin a chapter on the global dispersal of plants. Its relevance here lies in the goal of predicting global plant dispersal as a means to curb, if not prevent altogether, the entry of species in new ranges in which they could be invasive (*sensu* Mack et al. 2000). Both military defense and the quarantine for nonindigenous species deal with the pathways, modes (or conveyances), and circumstances of the foreigners' arrival as well as their number and composition. More specific, the questions for the military commander and for his/her plant quarantine counterpart follow a similar track. By what route(s) or pathway(s) will the foreigners arrive? Where within the defended territory will they enter? How many will initially and subsequently arrive? How will they arrive, that is, by what mode? What circumstances and features of the new locale will foster or hamper their persistence, geographic spread, and the consequences of their actions? Equally important but largely outside the scope of topics considered here is the character of the aliens; that is, will they be alike and respond similarly to all factors in the defended territory?

Whether dealing with human intruders or potentially harmful plant immigrants, answering all these questions correctly is daunting, especially when the potential entry points are numerous and the threat of entry is long term. The global dispersal of plants is a huge topic with many components. I do not attempt here to cover this subject in the comprehensive manner that Ridley achieved in his 1930 classic work or the more focused accounts by Guppy (1906) and van der Pijl (1969). Instead, I outline (1) pathways of both historic and modern importance to plant dispersal that have led to naturalization and invasion. (2) Drawing on the myriad modes by which plants are moved long distances, I illustrate several modes that have long attracted biologists (e.g., solid ballast, commercial seed lots) and then (3) outline the historic and modern consequences of the chief mode by which nonindigenous plants enter new ranges—deliberate introductions.

PATHWAYS OF PLANT DISPERSAL

I use the term pathway or route in a strict sense here: advance or progression in a particular direction, regardless of the mode (i.e., conveyance) that disperses plants along that pathway. By definition then, a pathway has a starting point and one or a series of destinations, as opposed to a probability distribution of destinations.

Natural Pathways

Atmospheric, oceanic, and river currents have always formed pathways or routes for plant dispersal. For instance, the Gulf Stream has carried seeds and plant propagules not only through the Caribbean but also as far as the British Isles. One consequence of this long-term dispersal has been the arrival of many subtropical species in the Scilly Isles, southwest of the main islands of Britain (Lousley 1971). Similar long-distance dispersal has facilitated the spread of mangrove species through Oceania and elsewhere (Murray 1986). Such movement can be highly directional: off the western coast of North America, the California Current carries plant flotsam in a distinctly southerly direction each summer. Alternatively, species are not as likely to reach a new range along these natural pathways, if the range lies outside the path of prevailing currents.

Ocean and air currents, of course, continue to affect plant distribution; their influence on the dynamic composition of any flora is a direct function of the frequency with which they carry plants to new ranges (Ridley 1930). Not surprisingly, strand and shoreline species commonly immigrate in this manner (Smith 1999). The sprawling morning glory, *Ipomoea pescaprae*, is a cosmopolitan subtropical/tropical shoreline species (Ridley 1930); its huge geographic distribution is a direct consequence of its floating seed and tolerance for sandy, salt-spray environments. Even among species not confined to shorelines, natural forces continue to transport species across impressive distances. Natural forces have apparently dispersed plants from Australia to New Zealand in modern times (Mack and Lonsdale 2001); these immigrations have produced adventives and possibly some naturalizations but no invasions. The role that the natural forces of moving air and water play today in changing plant distributions is small, compared with the role of humans. These natural forces are not feeble, just infrequent, in their global impact.

Pathways Developed by Humans

The human dispersal of plants has long followed many natural pathways, especially when human transport was substantially dependent on wind and water currents. The Gulf Stream along with the weaker Canary Current off North Africa and the westerly South Equatorial Current formed a great triangular route for sailing ships between Europe, West Africa, and the Caribbean that was in full swing by the early eighteenth century (Viola and Margolis 1991). Thus, ships augmented flotsam and other living rafts as conveyors of plants from Africa to South America and the Caribbean. The list of locations to which plants could be carried by ships was greatly expanded with the well-known advances in ship construction and navigation, beginning in the late fifteenth century and the later advent of steam-powered ships. It is hard to overestimate the increased likelihood of transoceanic plant dispersal as a result of these innovations. In effect, any two anchorages now potentially share a connecting pathway. As a result, species have been introduced to new ranges that they would not have reached by ocean currents alone; for example, species native to temperate Britain reached temperate New Zealand and vice versa (Good 1964).

With the enormous versatility of steamship travel, webs of transoceanic pathways soon became well established. These routes were shaped by the desire to speed commerce between trading partners, which were often European (and later American) nations and their overseas colonies. Even before the advent of steamships, a path for plant dispersal had developed between western Europe and the North American eastern seaboard and soon thereafter between the Netherlands and Britain and their colonies at the Cape of Good Hope, in the South China Sea, and Australia (Mack 1999 and references therein). These routes initially relied on ocean and wind currents; they multiplied as coaling stations were established. For instance, by 1889 Britain alone had established with its colonies and other trading partners 156 coaling stations that spanned the world (Porter

1991) (Fig. 1.1). France, Germany, Portugal, and Spain also maintained ports and connecting routes with their overseas colonies (Emmer and Gaastra 1996).

With steamships, landfalls that were exceptionally remote and unlikely to receive plant immigrants by natural forces (e.g., Ascension and St. Helena Islands in the southern Atlantic) became the recipients of many deliberately and accidentally introduced plants (Cronk 1989). Furthermore, along these webs of routes thousands of plant species were efficiently moved within colonial empires. By the late nineteenth century Britain had established botanical gardens at key locations worldwide (e.g., Calcutta, Cape Town, Hobart, Port of Spain, Singapore) (McCracken 1997). These facilities became bases for plant collection from which newly discovered species could be shuttled within the British Empire via London's Kew Gardens—an early hub-and-spoke transportation network. Never before had plant immigration operated on such a massive scale, replete with test gardens in potential new ranges (McCracken 1997). As discussed below, scores of species that were transported along pathways developed in the nineteenth century among Europe, North America, and Europe's colonies later became naturalized; some have become invasive (e.g., the woody plants, *Rhododendron ponticum, Acacia ieucocephala, Leucaena nilotica,* and *Lantana camara*). The tradition of massive worldwide plant exchanges can be traced to the development of these colonial programs, beginning as early as 1600 (Kloot 1987, Mack 1991, 1999, McCracken 1997).

The spread of plants was extended inland with the growth of canals (Mills et al. 2000) and, later, railroads (Dewey 1896, Mack 1991). Canals were seen as essential routes for eighteenth-century commerce in the United States, a view that continued until they were largely supplanted by railroads at the end of the nineteenth century. In the meantime, navigable canal networks laced the interior of the eastern United States and provided a mode for both deliberate and inadvertent spread of nonindigenous plants. Reconstructing the nineteenth-century spread for accidental introductions along canals is necessarily circumstantial and relies on the collection history of these species alongside or near old canal routes. Invasion by the aquatic herb *Lythrum salicaria* (purple loosestrife) has been documented in this manner. Thompson et al. (1987) contend that with the exception of the canal system in interior Pennsylvania, the early-nineteenth-century inland spread of *L. salicaria* in the northeastern United States was tied closely to canal traffic moving from navigable East Coast estuaries. And most sites of purple loosestrife's pre-1880 establishment in the region were along the Erie Canal, which bisected New York and the Delaware & Raritan Canal, which bisected New Jersey (Thompson et al.

COMMUNICATIONS : PRINCIPAL STEAMER ROUTES
AND COALING STATIONS, 1889

FIGURE 1.1. Principal steamer routes and coaling stations (holding at least 500 tons) within the British maritime system in 1889 (Porter 1991). Britain's colonies and trading partners spanned the globe and made the rapid transport of plants a routine procedure.

1987). Canals in the United States still provide avenues along which non-indigenous plants spread. Dispersal of highly invasive species (e.g., *Elodea canadensis* [waterweed], *Hydrilla verticillata* [hydrilla], *Pistia stratiotes* [water lettuce]) that can inhabit waterways in Florida has been facilitated by recreational boats moving along the state's canals (Schardt and Schmitz 1991). Clearly, canals in the United States continue to provide a pathway for the inland movement of nonindigenous species (Mills et al. 2000).

Rail systems were built with remarkable speed in the nineteenth century. Rail lines in the eastern half of the United States expanded from a few hundred kilometers in 1832 to more than 48,000 km by 1860; they expanded much more after 1865, especially west of the Mississippi (Meinig 1986). Reliable rail transport did much to facilitate the growth of a truly national market for seed merchants, transforming their trade from a local cottage industry to a national mercantile system (Mack 1991). Establishment of even a single U.S. transcontinental rail line in 1869 ensured that living plants could be carried in days, instead of weeks or months, across the continent. This possibility was soon realized even for accidentally introduced species. *Sisymbrium altissimum* was given the derisive name "Jim Hill mustard" under the widespread belief in the late nineteenth century that it had spread along tracks of the Great Northern Railway, which was controlled by railroad magnate James J. Hill. Spread of many other species was certainly facilitated by rail lines and rolling stock (Muhlenbach 1979), including such aggressive invaders as *Bromus tectorum* (cheatgrass) and *Salsola iberica* (Russian thistle) (Mack 1986). More important was the opportunity railroads and a reliable transcontinental postal service provided to the deliberate dispersal of plants. Pathways that had previously operated intermittently, or not at all, became major avenues for seed dispersal (Mack 1991).

Air transport in the twentieth century expanded the opportunity for plant movement among traditional seaport destinations and also created opportunities for direct movement of nonindigenous plants between the interiors of continents. Plant materials now move directly by air not only between Liverpool and Cape Town, or Auckland and Yokohama, but also between Atlanta, Charlotte, Dallas/Ft. Worth, Denver, and Orlando and cities in central Europe. None of these U.S. cities can be reached from the sea by navigable rivers or canals. Thus, the opportunity for plants from central Europe to rapidly reach suitable habitats in the interior United States and vice versa has increased enormously with the constant movement of airfreight. Similar long-distance links between land-locked cities occur worldwide: Frankfurt–Johannesburg, Moscow–Beijing, Frankfurt–New

Delhi, Mexico City–São Paulo. Rapid transit also means that species can be transported as actively growing vegetative plants, rather than exclusively as seeds or other dormant life-forms, thereby expanding the list of transportable species. Predicting possible pathways of plant movement has been made much more difficult since the advent of commercial aviation. Certainly the opportunity for plant entry to some locales remains small (e.g., much of Central Asia), but for an increasing number of locales, predicting pathways has largely been reduced to asking when a pathway or route will develop, not whether it will develop.

MODES OF PLANT DISPERSAL

Mode refers to the manner or conveyance, or both, by which plants are carried along a pathway. Plant dispersal by animals has been reported extensively (e.g., Ridley 1930, van der Pijl 1969). Consequently, I will not discuss those modes here. Although animals, such as migratory marine birds, may be modes (i.e., vectors) of transport over great distances (Ridley 1930), most plant dispersal by animals is local. Major exceptions involve the role that livestock have played. Spread of nonindigenous plants has often been attributed to the movement of livestock regionally or even transcontinentally (Mack 1986 and references therein), although the historic role of livestock as modes of plant transport is difficult to evaluate retrospectively.

Human-Mediated Modes of Dispersal

Accidental, compared with deliberate, plant dispersal by humans is a recurring distinction in this chapter. In accidental dispersal, the plants—whether seeds, vegetative shoots, or other perennating structures—were not the objects of transport. The likelihood that these hitchhikers will survive the voyage is dependent on their tolerance of the environment during transit and the voyage's duration. Steps are usually taken in deliberate introductions to ensure plant survival in transit, protection that continues with post-immigration cultivation (Mack 2000).

Four Extraordinary Examples

Viable plant materials, especially seeds, are capable of being carried on or within an amazing array of modes. Fortunately, the most bizarre modes are also the least likely to result in plant naturalization, if only because they occur so infrequently. Such "long shots" in the plant dispersal lottery have long fascinated biologists, and a rich tradition of assembling these anecdotes continues (Grenfell 1987). Ever mindful of Frederick the Great's admonition, the limited resources devoted to thwart plant entry should not be fractionated to anticipate every conceivable scenario. Nevertheless, the four examples below collectively illustrate the breadth of inadvertent and deliberate plant transport by humans.

Xanthium spinosum (Bathurst burr) and its congener, *X. pungens* (Noogoora burr), are serious nonindigenous pests in Australian range-lands (Cuthbertson and Parsons 1992). Bathurst burr reputedly arrived in Australia in the 1840s in the tails of horses imported from Chile. Maiden (1920) maintained that its spread in Australia was facilitated by an advertising stunt. He was indignant that advertisers in 1918 used the plant's burr (i.e., the hooked spines on the fruit) as the body of a paper-winged butterfly that "practical jokers" hurled at passersby. Once the burr was attached, the pedestrian became the unwitting disperser of the winged handbill and the fruit. Many Bathurst burr-bodied paper butterflies must have been discarded along paths once they were detached from clothing, potentially setting the stage for the plant's establishment in a new locale.

As this first example illustrates, clothing is well known informally as a seed collector. But its direct role in seed dispersal is almost totally unevaluated. One rare exception is the account of the Reverend Woodruffe-Peacock, vicar of Cadney-cum-Howsham, North Lincolnshire, who identified the seeds attached to his hunting partner's clothing during a day's traverse of coverts (planted thickets used as animal cover) and stubble fields. His examination yielded nine species on his friend's jacket and another nine species elsewhere on his clothing. He maintained that the repeated action of such seed dispersal by humans could readily have increased the ruderal flora in an English covert in the 120 years since its establishment (Woodruffe-Peacock 1918).

International gatherings, including fairs, conventions, and professional meetings, provide an ideal opportunity for the dispersal of organisms, including plants. The delegates may disperse after the meeting to the furthermost reaches of the globe. Living plants, including seeds and cuttings, are sometimes distributed as souvenirs at these gatherings. This seemingly innocuous practice can have disastrous results. As often cited, *Eichhornia crassipes* (water hyacinth) was reputedly first brought into the United States by Japanese delegates to a cotton exposition in New Orleans in 1884 (Barrett 1989). Water hyacinth is not native to Japan; the Japanese delegates had arrived recently in New Orleans from Venezuela and brought along the attractive plant to share with others! In similar fashion, I have received seeds at several international scientific meetings, most recently in

1998 at the VII International Congress of Ecology in Florence, Italy. In the registration packet each delegate received a small vial containing unidentified seeds. The clear implication was that these seeds could be carried home and sown.

The fourth example is the most bizarre form of seed dispersal of which I am aware. By the 1890s, plant invaders were becoming increasingly prevalent in agriculture in the western United States (Mack 1986). Part of the task for the newly established state and federal agricultural experiment stations was to alert landowners to weeds, presumably before these nonindigenous species became so prevalent that control was futile. Hillman (1893a, 1893b), a Nevada experiment station official, provided pamphlets to landowners that described and illustrated troublesome species that were not yet in the state. He went one extraordinary step further in his desire to ensure that his clientele had the maximum information available for plant identification. Each of his pamphlets also had a mature specimen of each species, including seeds! These were dried specimens pasted onto the pamphlet pages; no provision was made to encapsulate the seeds, and I do not know whether he took the precaution to sterilize the seeds before he distributed the pamphlets. It is apparent, however, that with time, either the seeds would have been shaken loose or the whole pamphlet would have been discarded. Hillman's pamphlets are unique: the announcements of the impending arrival of these invaders could also have been the same modes that delivered them to far-flung locales in their new range (Mack 1986).

Ballast: Past, Present, and Future

One mode of accidental dispersal, solid ship ballast, has a venerable history. Victorian era biologists were fascinated with nonindigenous species appearing "spontaneously" on the wharves, piers, and ballast dumping grounds in and around seaports. The ballast consisted of the rubble, gravel, stones, or any other dense debris that could be used to stabilize ships at sea. Once in port, the debris was unceremoniously dumped along the shore before outbound cargo was loaded. (Local ordinances prudently forbade dumping in the harbor itself.) This highly heterogeneous material was gathered so indiscriminately at the port of embarkation that it often incorporated plants, insects, and other soilborne organisms (Lindroth 1957). After the ballast was unloaded, plants were often seen growing atop the ballast heaps (Martindale 1877).

Many of the earliest and sole collections for species in a new range were made on nineteenth-century ballast heaps. At least eighty-one species that are reputedly in the Pennsylvania flora have been found only on ballast around the Philadelphia harbor, and none of these species have been collected in the state in more than a hundred years (Rhoads and Klein 1993) (Table 1.1). The native origins of these herbaceous species have a decidedly European, especially Mediterranean, character: *Centaurea melitensis,*

TABLE 1.1. Adventive species found in the years indicated on solid ballast around Philadelphia harbor. In each case the collection from ballast remains the only Pennsylvania records for the species. Grasses are listed separately from Cyperaceae under Monocots (records from Rhoads & and Klein 1993).

Dicots

Agalinus fasciculata 1864	*Alternanthera paronychioides* 1863
Amaranthus pumilus 1865	*Ammi visnaga* 1879
Amsinckia intermedia 1877	*Amsinckia lycopsoides* 1877
Anacyclus clavatus 1882	*Arnoseris minima* 1877–78
Asperula arvensis 1866	*Aster tripolium* 1867–80
Atriplex tatarica 1874	*Bidens pilosa* var. *bimucronata* 1879
Bidens pilosa var. *radiata* 1868–79	*Cakile maritima* 1878–80
Calamintha nepeta 1865	*Carduus tenuiflorus* 1877–79
Centaurea melitensis 1877	*Chenopodium rubrum* 1865–77
Chrysanthemum segetum 1877–78	*Clarkia purpurea* 1877
Corrigiola littoralis pre-1900	*Ecballium elaterium* ca. 1864
Eupatorium cannabinum 1879	*Fumaria parviflora* 1877
Glaucium flavum 1890	*Heliotropium supinum* 1879
Heliotropium undulatum 1879	*Ipomoea imperati* 1865
Iva frutescens ssp. *oraria* 1865	*Lappula marginata* 1878
Lepidium heterophyllum 1870s	*Linaria supina* 1879–80
Ludwigia leptocarpa 1865	*Merremia dissecta* ca. 1865
Mesembryanthemum crystallinum 1879	*Mollugo cerviana* 1879
Ornithopus perpusillus 1878 & 1879	*Parietaria judaica* 1879–1921
Plantago coronopus 1878	*Plantago heterophylla* 1864–65
Polypremum procumbens 1864–65	*Richardia brasiliensis* pre-1900
Salvia verbenacea 1879	*Scandix pecten-veneris* 1878–80
Scolymus hispanicus 1878–80	*Scrophularia aquatica* 1878–83
Senecio erucifolius 1865–79	*Senecio sylvaticus* 1878
Sesbania exaltata 1864–94	*Sesuvium portulacastrum* 1865
Sida linifolia 1879	*Solanum luteum* 1866
Stachys annua 1865–1920	*Stachys arvensis* 1878–83
Stachys sylvatica 1879–1921	*Teesdalia nudicaulis* 1877–80
Trifolium carolinianum 1865	*Trifolium lappaceum* 1879
Turgenia latifolia 1870s	*Verbascum nigrum* ca. 1865
Zygophyllum fabago 1880	

Monocots

Carex distans 1865	*Cyperus croceus* 1864–65
Fimbristylis vahlii 1864–79	*Schoenoplectus maritimus* 1865–77
Alopecurus creticus 1877	*Alopecurus rendlei* 1880
Briza minor 1879–89	*Bromus arenarius* 1879
Corynephorus canescens 1878–79	*Crypsis alopecuroides* 1879
Digitaria serotina 1865	*Eustachys petraea* 1865
Festuca ciliata 1878	*Lophochloa cristata* ca. 1878
Parapholis incurva ca. 1878	*Paspalum paspalodes* 1864–79
Phleum subulatum 1879	*Spartina patens* pre-1900
Sporobolus indicus ca. 1865	*Sporobolus pyramidatus* pre-1900

Ecballium elaterium, Linaria supina, Teesdalia nudicaulis. The use of rubble from walls as ballast is also implied by the appearance of *Parietaria judaica,* a common wall-inhabiting species in Britain (Mabberley 1997). Species from Africa (*Mesembryanthemum crystallinum, Sesuvium portulacastrum*) were also found, as well as the South American native *Richardia brasiliensis.* Unknown is whether these plants arrived directly from their native ranges or from other donor areas. Nevertheless, the list is testimony to the distant and varied locales from which ballast plants were being loaded onto ships bound for Philadelphia.

Given the chance circumstances that characterize immigrations in solid ballast, it is not surprising that few naturalized species can be attributed to introduction by this mode. Not only could the environment at the point of disembarkation be unsuitable, but the likely small size of the founder population (Panetta and Randall 1994) and the rarity of even inadvertent cultivation on ballast would also rule against naturalization (Mack 2000). Early destruction would have been the almost universal fate for plants conveyed in this manner.

Despite the precarious environment, solid ballast has been a mode, if rarely a spectacular one, for plant naturalization. *Spartina alterniflora* (saltwater cordgrass), one parent of the aggressive hybrid *Spartina anglica* (common cordgrass), was carried to Britain as seed in solid ballast (Thompson 1991). Accidental plant dispersal in ballast may also have played an important role in an early misconception about the taxonomic affinities among tropical floras. Linnaeus erroneously concluded that tropical floras worldwide are strikingly similar, based on plant collections he received from his correspondents at far-flung tropical stations. This remarkable error, for otherwise so astute an observer, arose apparently because his collectors were either unable or unwilling to venture into the tropical interior and confined their plant collecting to seaports. Even by the mid-eighteenth century these widely separated coastal locales shared a common flora that had been carried pantropically by humans (Stearn 1958). Ballast and other debris are among the likely modes for the dispersal of these species.

Although solid ballast is rarely used today, ore, crushed rock, sand, and gravel are environmentally equivalent and can contain living plants. Movement by ships and barges of rock, sand, soil, and gravel is primarily a domestic enterprise in which material is shuttled along rivers or a coastline. The distances involved can be substantial: more than 3,000 km of the Amazon River are navigable by barge, and the route bisects a broad array of physical environments. Ores of iron, bauxite, vermiculite, and other minerals are also international maritime commodities and may contain viable plants, incorporated from the overlying soil at the mine or the loading pier. Through repeated plant surveys at large ore piles in coastal Maryland and Virginia in the 1950s, Reed (1964) found an astounding 550 angiosperm species; he reported that 80 percent of these species were not recorded in the regional flora. Although many species survived only briefly as herbaceous adventives (*Amaranthus lividus, Anchusa italica, Echium italicum,* and *Heliotropium procumbens*), others, such as *Cleome viscosa,* were repeatedly detected on these ore heaps.

Modern use of ballast employs seawater, which all too commonly is jettisoned into the brackish water of a seaport or estuary. This dispersal mode is exceedingly important in the global transferal of marine invertebrates and protozoans (Carlton and Geller 1993). Unfortunately, its role in the dispersal of higher plants, especially land plants, has yet to be evaluated comprehensively. Unknown is the extent to which ballast water carries small viable seeds of aquatic vascular plants, such as *Carex, Juncus, Scirpus, Thalassia,* and *Zostera.* Marine algae could also be readily transported in this brine. A restriction for the viable transfer of these organisms is their ability to tolerate the environment in a closed tank of seawater for weeks at a time. For example, the survival of seeds without water-impervious coats would be unlikely (Murray 1986).

Straw and Hay: Bulky Conveyances for Plant Dispersal

Straw, which is often (but by no means confined to) the dried remains of tall grasses, has long been used as packing material (Muenscher 1955). Whatever seeds were still attached to the straw at harvest could be readily incorporated within the packed goods. It is impossible to know the extent to which seeds spread in this manner have become founder populations in a new range, although circumstantial evidence has long been cited. For example, Dewey (1896) reported that the grasses *Bromus tectorum* and

Bromus sterilis were first noticed in Denver, Colorado, in the vicinity of a crockery store—the apparent descendants of seeds in discarded packing boxes.

Hay, straw used as animal forage, has long been considered a significant source of plant introduction (Maiden 1914). Here straw is itself the cargo, although the concern once again deals with seeds left adhering to the straw. A common motivator for the long-distance transport of hay is regional drought in which local hay production falls, and livestock survival is threatened. The 1980–81 drought in parts of New South Wales, Australia, mirrors a circumstance that has been repeated in many countries. Massive shipments of hay were transported into the drought-stricken region from locations as far away as 900 km. A sampling of hay bales (26 kg each) was examined in detail for extraneous seeds. Each bale contained an average of 68,700 seeds with a range from 104 to 364,000 seeds. An astonishing 105 extraneous species were detected among just thirty-eight bales. Although most of these species were considered innocuous, all but one bale contained at least one restricted or prohibited species, such as *Polygonum aviculare* (wireweed), *Avena fatua* (wild oats), *Rumex acetosella* (sorrel) (Thomas et al. 1984). For some species, the number of seeds in a bale could have been large enough to form a persistent population (Panetta and Randall 1994).

Given its bulk, hay would seem an unlikely transoceanic commodity. But many nations, including Japan, Korea, Malaysia, Saudi Arabia, Singapore, and Taiwan, import hay. As a result, hay is a potential mode for international plant dispersal. Japan is perhaps the international leader in hay inspection for noxious, nonindigenous species. The Japanese Plant Protection and Quarantine Service imposes a zero tolerance standard on hay contaminated with seeds of *Agropyron repens* (quack grass). If quack grass seeds are found in the single bale that Japanese inspectors draw from the cargo, the whole shipment is blocked from importation (W. Ford, pers. comm.). Such a practice does reduce the likelihood of entry by quack grass but overlooks other equally noxious species. Furthermore, by sampling only a single bale per shipload, undetected quack grass could obviously still enter the country.

Samplings similar to those reported by Thomas et al. (1984) need to be conducted repeatedly in many countries to replace supposition with quantified results on the role of hay transport in plant dispersal. In addition, the fate of extraneous seeds transported in hay needs to be followed through repeated censuses (e.g., Mack and Pyke 1983) in paddocks and rangelands. As with other modes of plant dispersal, the likelihood of naturalizations or invasions sparked from hay contaminants is a function of the size of their founder populations and their post-immigration environment, including fertilization and irrigation (Mack 2000).

Weeds in War

War involves the abrogation of social and political norms, including quarantine restrictions. Plants that may have been prohibited from crossing a political boundary, or even unlikely to be dispersed at all, can be transported great distances with the movement of troops and their machinery. We will probably never know the extent to which great human invasions across Eurasia extended the range of vascular plants. Such transport of plants is nevertheless plausible, especially as seeds attached to the horses and other livestock that were maintained by invading armies or the debris dumped on foreign shores (Kornas 1990). We do know, however, that plant dispersal has been recorded coincident with war for as long as it has been purposely investigated.

The American Civil War (1861–65) was perhaps the first in which plants were transported rapidly on land, thanks to the extensive use of railroads. Steamships also played a similar role along the U.S. East Coast and the Gulf Coast. Dewey (1896) reported that the southern range limit for the noxious *Cirsium arvense* (Canada thistle) in the eastern United States (ca. 1890) coincided with the location of a Union Army supply station at Remington, Virginia; hay contaminated with thistle seeds was the alleged mode of introduction. Mohr (1878) provided a retrospective assessment of the introduction of *Lespedeza striata* (Japanese clover) near the seaport of Charleston, South Carolina, and its spread coincident with troop movements in the Gulf Coast states. He maintained that the legume was likely distributed in dung by livestock that were driven behind advancing armies. It apparently continued to spread in the region after the war (Mohr 1878).

Plant dispersal in war materiel was also assessed upon conclusion of the Franco-Prussian War (1870–71). Forage imported from Algeria contained many nonindigenous species that became adventive in central France, mainly at Blois, Cheverny, Vendôme, and Orléans. More than 150 species were detected in several surveys in 1872; two-thirds were annuals or biennials (A. B. M. 1872). The fate of most of the descendants of this diverse adventive flora is unknown, although at least three species (*Medicago murex, M. polymorpha,* and *Trifolium nigrescens*) are now members of the French flora (Tutin et al. 1980).

No other war has matched World War II for the area of the planet that was enveloped in conflict. As a consequence, many remote archipelagos in the Pacific Ocean suddenly became the disembarkation points for troops and materiel. A diverse array of conveyances—caterpillar-treaded vehicles, aircraft, ship cargo, and troops' clothing—were all modes for plant dispersal. The biota of small, isolated oceanic islands experienced perhaps the most devastation through the inadvertent release of nonindigenous plants. For example, Johnston Island is an isolated speck of land in the Central Pacific. A 1923 expedition recorded only three native species on the island. By 1946, the island's flora had risen to 27 species; the 24 new immigrants were all the result of deliberate (e.g., *Casuarina equisetifolia, Messerschmidia argentea, Terminalia catappa*) and accidental (e.g., *Portulaca oleracea, Sonchus oleraceus*) introduction by the military (Fosberg 1949). Fosberg (1957) tallied 135 nonindigenous species that arrived among forty islands in Micronesia during World War II. Most had been introduced accidentally. A few, however, were spread deliberately, such as *Cynodon dactylon* (Bermuda grass), which was sown to revegetate islands devastated by fighting. As unlikely as it seems, some entry even resulted from seeds attached to clothing. Fosberg (1957) cites records of the grasses *Paspalum conjugatum* and *Cenchrus echinatus* that made inter-island transfers as hitchhikers on clothing. Plant dispersal to oceanic islands has been particularly serious because the native floras are miniscule and entry of nonindigenous species has commonly been associated with substantial disturbances (roads, supply depots, even bomb and shell craters) that facilitate the establishment of alien ruderals (Fosberg 1957).

World War II also brought about the spread of plants in Europe. In northern Finland, extensive and recurring maneuvers by the combatants inadvertently introduced approximately 140 nonindigenous species—a large number, given the size of the native ruderal and pre-war nonindigenous flora in this isolated boreal region. Hay imported from central Europe, France, the Netherlands, Belgium, and Scandinavia included an array of herbaceous species (e.g., *Alchemilla acutiloba, A. glabra, Cardaminopsis arenosa, Centaurea scabiosa*) that can be distinguished from species brought from the adjacent Soviet Union (e.g., *Alchemilla baltica, Briza media, Centaurea phrygia*). Regardless of source, these species characterize ruderal or frequently disturbed sites, including hay fields and pastures (e.g., *Achillea millefolium, Achillea ptarmica, Briza media, Erodium cicutarium, Plantago lanceolata, Poa compressa, Rumex crispus*) as well as species deliberately cultivated for hay (e.g., *Dactylis glomerata, Festuca pratensis, Melilotus albus, Phleum pratense*). Few of these adventives survived until 1970, and the remainder (principally *Cardaminopsis arenosa* and *Alchemilla* spp.) continued to survive only as small colonies (Ahti and Hamet-Ahti 1971). Not surprisingly, none of the species that Hamet-Ahti

(1984) gently terms "polemochores" (literally, the seeds distributed by disputes) has become established in the mature northern boreal forest.

Military conflicts continue, if intermittently, and so does the potential for the spread of nonindigenous species in this mode. In this regard, two recent military actions, the Gulf War in 1991 and the more recent incursion by Australian troops into East Timor, are instructive. Transferring large caterpillar-treaded vehicles from central Europe to Saudi Arabia was unlikely to disperse species that could tolerate desert conditions, even briefly. In contrast, Australia prudently recognized that moving vehicles from its tropical base at Darwin, Northern Territory, to the similar environment of East Timor and back could provide a much more likely opportunity for plant dispersal and naturalization (Anon. 2000). Its military took steps to clean vehicles before transport. Although current military conflict is thankfully not at the scale of the World Wars, it does remain a potential mode for plant dispersal around the globe. In a twist on Frederick the Great's dictum, prevention of plant dispersal becomes the best defense against unwanted plant entry. Preventing nonindigenous plant dispersal is unlikely to appear in the operational instructions for most combatants.

Seed Contaminants in Crop Seeds

Despite the concerted efforts of modern agriculture, crop fields are not monocultures; they routinely contain weeds that coevolved with crops (Harlan 1965). Seeds of these weedy species are often harvested with the crop, unless elaborate steps are taken to exclude them (Wolff 1951). And because seeds are the most environmentally resistant and transportable of any life stage for an angiosperm, crop seeds and associated weed seeds can be stored for long periods and transported great distances.

Hundreds of species are suspected to have been transported by agriculture to new ranges where they became naturalized, particularly in Europe (Kornas 1990). Three eleventh-century granaries that were excavated in Poland reveal the character of this dispersal mode. The seeds of thirty-nine extraneous species were mixed with the crop seeds and ranged from 6.9 to 9.4 percent of the seed lots. Not all were nonindigenous species; some were natives that may have been inadvertently harvested from adjacent riverine communities. Nevertheless, extraneous species included such common crop weeds as *Agrostemma githago* (corn-cockle), *Bromus secalinus* (chess), *Bromus arvensis, Rumex acetosella,* and *Polygonum convolvulus* (black bindweed) (Kosina 1978). These species could readily have been spread across Europe in trade.

Effective efforts to remove extraneous seeds mixed in crop seeds did not arise until the last half of the nineteenth century. Seed merchants first in Germany and then elsewhere in Europe were required to clean seeds for domestic use, although not necessarily for export. The continuing practice of indiscriminately exporting contaminated seeds abroad had serious implications for the United States. As late as 1895, the United States imported the majority of seeds sown for many vegetable crops; the list included alfalfa, beet, brussels sprouts, cress, endive, radishes, spinach, turnip, and even some forage species, such as *Anthoxanthum odoratum* (sweet vernal grass) (Hicks 1895). All undoubtedly contained extraneous seeds. Once alerted to the hazard of this uninspected mode of plant introduction, states rapidly established at least rudimentary inspection for commercial seeds. In large measure, however, the contribution of seed contaminants to the naturalized and even invasive flora of the United States had already occurred. Seed analyses from states as widely separated from each other as New York and Washington reveal that as recently as 1913–16, seed contaminants in commercial seed lots were both diverse and often extensive. In New York, seed lots of *Trifolium pratense*, *Trifolium repens*, *Poa pratensis*, and lawn grass (unidentified species) contained 6.40, 5.58, 3.99, and 15.42 percent extraneous species (Munn 1914). During the same period in Washington,

TABLE 1.2. Number of occurrences of the seeds of nonindigenous species among thirty-three samples of red clover seed (*Trifolium pratense*) (Department of Agriculture of the state of Washington, July 1, 1914 to June 30, 1916).

Amaranthus graecizans 1	*Amaranthus retroflexus* 2
Anthemis cotula 4	*Brassica arvensis* 2
Brassica nigra 2	*Bromus secalinus* 1
Capsella bursa-pastoris 1	*Carduus lanceolatus* 16
Chaetochloa viridis 10	*Chenopodium album* 9
Chenopodium botrys 1	*Cuscuta* spp. 2
Digitaria sanguinalis 1	*Geranium dissectum* 1
Gramineae (unident.) sp. 7	*Helianthus petiolaris* 1
Lactuca virosa 1	*Lolium* spp. 6
Lotus corniculatus 1	*Panicum capillare* 5
Panicum crus-galli 10	*Phleum pratense* 5
Plantago aristata 1	*Plantago lanceolata* 22
Poa compressa 1	*Prunella vulgaris* 3
Rumex acetosa 1	*Rumex acetosella* 25
Rumex crispus 5	*Salsola tragus* 4
Secale cereale 1	*Stellaria media* 1
Trifolium hybridum 5	*Trifolium repens* 5

seed lots of red clover collectively contained as many as 34 extraneous seed species (Washington Department of Agriculture 1914, 1916) (Table 1.2). At least four of the species detected as seed contaminants in these samples from Washington—*Bromus secalinus, Chenopodium album* (lamb's quarters), *Rumex crispus* (yellow dock), *Stellaria media* (common chickweed)— were also found in the aforementioned samples from eleventh-century Polish granaries, testimony to these species' intimate and extraordinarily long association with crops. Such association would have greatly facilitated their naturalization (Mack 2000).

Nonindigenous species continued to enter new ranges as seed contaminants, despite improvements in seed-cleaning procedures. Imported grain that was sown, rather than milled for consumption, would have contributed to this spread. Based on extensive records collected from the nineteenth century until 1975, Suominen (1979) claimed that approximately 370 nonindigenous species entered Finland in this mode. Russia, central and western Europe, and North America have all been donors to the Finnish adventive flora. South America, Turkey, and Algeria have also contributed species (Suominen 1979), further evidence of the geographic breadth of the international grain market.

Despite the enactment of seed purity laws in many nations, extraneous nonindigenous seeds are still common contaminants in the international seed trade. For example, at least 47 species were detected in shipments of soybean seeds from the United States to Japan in 1994–95 alone, and 26 species were detected among commercial legume seeds from Australia in the same period (Enomoto 1999). Although these seed lots involve different crops and were transported from different donor countries, they share common species (e.g., *Amaranthus patulus, Ambrosia artemisiifolia* var. *elatior* [common ragweed], *Chenopodium album, Echinochloa colona, Persicaria scabra*); others are unique to the crop or donor region. These species point to the potential that the seed trade still has to contribute to the nonindigenous flora of nations actively involved in international trade. Although the prevalence of seed contaminants in crop seeds has been reduced markedly in the last 100 years, it remains a source of accidental plant introduction and cannot be ignored completely.

MAJOR MODES OF DISPERSAL: THE PRIMACY OF DELIBERATE INTRODUCTIONS

Identifying the main mode by which nonindigenous plants arrive in new ranges is straightforward and unequivocal; deliberate introductions are the chief culprit. Humans have deliberately distributed plants worldwide for millennia: the Romans introduced many species, such as *Alliaria petiolata* (garlic mustard), *Agrostemma githago,* and *Nepeta cataria* (catnip), into Britain (Godwin 1975). But these efforts pale by comparison to plant dispersal in the post-Columbian era in which Europeans strove to introduce their economic plants into their new colonies, as well as export newly discovered species to Europe (Viola and Margolis 1991, Mack 1999).

Many aspects of this global plant exchange are remarkable, including the rapidity with which this commerce was established. Even before 1500, Columbus was introducing plants to his ill-fated Caribbean colony, Navidad (Mack 2001). In large measure, agriculture in Europe's overseas colonies was based on introduced species: wheat, rice, barley, and oats for temperate regions; sugarcane, rice, and indigo for subtropical and tropical regions. Maize, a native of the New World, was rapidly spread elsewhere (Viola and Margolis 1991 and references therein). Imported plants included the whole range considered necessary to sustain European life, including medicinal plants (Mack 1999).

This traffic in plants was not confined to seeds. The HMS *Bounty* was carrying potted breadfruit from the South Pacific when the voyage was aborted in 1789; the ship's master, the much-maligned Captain Bligh, was following accepted practice by transporting plants in large pots on the open deck. Plant mortality was high in such exposed conditions, even when the crew did not mutiny! Development in the early nineteenth century of Wardian boxes, basically glass-enclosed growth chambers, meant that many more species could be transported worldwide in a vegetative state (McCracken 1997). As the size and geographic extent of Europe's colonial holdings grew, the volume and diversity of plant exchanges grew proportionally. Not only crops, but species for forage, soil stabilization, seasonings, and medicine were imported in large numbers (Shaughnessy 1986). Among these early plant introductions were some species that have since become naturalized, and some of these have in turn become invaders (Shaughnessy 1986, Wells et al. 1986, Mack 1999).

Decisions based on human preferences and biases have always played the principal role in shaping the character of deliberate plant dispersal, from the selection of the export species, to the new ranges in which species were introduced. In this regard, human preference has reached its greatest expression in the choices of nonindigenous species as ornamentals. More species have been transported to new ranges for this purpose than for any other (Kloot 1987, Mack and Erneberg, 2002). By the late eighteenth century, and probably much earlier, species were being imported into what would become the United States expressly for aesthetic reasons, and among these species were some that have since become naturalized (e.g.,

Lysimachia longifolia, Rosa eglanteria [sweetbrier], *Salix babylonica* [white willow]) (Mack and Erneberg, 2002).

The frequency of introducing ornamental species into new ranges has increased enormously in the last 200 years and will increase further as global commerce grows. There are approximately 250,000 angiosperm species. Estimates vary as to the number of these (and their subspecies and cultivars) that are available today in the world market, but the records for New Zealand mirror a worldwide phenomenon. At least 25,000 taxa are commercially available to New Zealand gardeners (Gaddums 1999). The likelihood of naturalization, much less invasion, by most of these species is low. Thousands have been cultivated in New Zealand for a century and show no signs of establishing outside cultivation. Nevertheless, the New Zealand naturalized and invasive floras do include many one-time ornamentals, deliberately introduced species that have proliferated, spread, and brought about much environmental damage (e.g., *Cytisus scoparius* [Scotch broom], *Hedychium gardnerianum* [Kahili ginger], *Lemna minor* [duckweed], *Lonicera japonica* [Japanese honeysuckle], *Ulex europaeus* [gorse]) (Roy et al. 1998). Unknown is how many of those now growing in cultivation in New Zealand will become invasive. Plant invaders commonly undergo a long, deceptively innocuous lag phase in their new range before they proliferate and spread (Kowarik 1995, Mack et al. 2000), so more invaders will almost certainly emerge.

The important role played by the importation and sale of nonindigenous species is becoming better understood. Hundreds of species sold in the nineteenth century in Australia, Europe, and the United States are now naturalized, and a significant number have become invasive (Clement and Foster 1994, Kloot 1987, Mack 1991, Mack and Erneberg, 2002). But the deliberate, and largely unregulated, global transfer of species continues. A sampling of species currently available from just two small seed catalogs—one U.S., the other Canadian—illustrates the problem (Table 1.3). A revival in the popularity of herbal remedies has sparked the appearance of catalogs largely devoted to species with putative healing properties. Many of these species have long been naturalized in North America and cause little or no harm (e.g., *Arctium lappa* [great burdock], *Capsella bursa-pastoris* [shepherd's purse], *Chenopodium album, Galium aparine* [bedstraw], *Nepeta cataria, Taraxacum officinale* [dandelion], *Tragopogon pratensis* [goat's beard]), but there is no compelling argument to increase their range through commerce. In contrast, some commercial species are exceedingly damaging to agriculture or natural environments, or both (*Cynoglossum officinale* [common hound's tongue], *Hypericum perforatum* [St. John's wort], *Isatis tinctoria* [dyer's woad], *Schinus terebinthifolius* [Brazilian pepper]) and a few present risks to human health (*Atropa*

TABLE 1.3. Nonindigenous species available as seasonings or for medicinal purposes from two seed catalogs in 1999. Each species is either naturalized or even invasive in the United States. Although these species vary greatly in the severity of the damage they cause in agriculture and natural communities, none should be available commercially.

Arctium lappa[1]	*Atropa belladonna*[1,2]
Capsella bursa-pastoris[1]	*Chenopodium album*[2]
Cynoglossum officinale[1,2]	*Cytisus scoparius*[2]
Datura stramonium[1,2]	*Dipsacus sativus*[2]
Echium vulgare[1,2]	*Eucalyptus globulus*[2]
Galium aparine[2]	*Hyoscyamus niger*[1,2]
Hypericum perforatum[1,2]	*Imperata cylindrica*[2]
Isatis tinctoria[1,2]	*Lythrum salicaria*[2]
Marrubium vulgare[1]	*Nepeta cataria*[1]
Paulownia tomentosa[1]	*Plantago lanceolata*[2]
Pueraria lobata[1,2]	*Rhamnus cathartica*[2]
Ricinus communis[1]	*Rumex acetosella*[1]
Rumex crispus[1,2]	*Schinus terebinthifolius*[2]
Silybum marianum[1]	*Tanacetum vulgare*[1,2]
Taraxacum officinale[1,2]	*Tragopogon pratensis*[1]
Tribulus terrestris[1,2]	*Urtica dioica*[1,2]
Verbascum thapsus[1,2]	

[1]*Horizon Herbs, Strictly Medicinal Growing Guide and 1999 Catalog.* Williams, Oregon.
[2]*Richters Herb Catalog.* 1999. Goodwood, Ontario.

belladonna [belladonna], *Ricinus communis* [castor bean], *Urtica dioica* [stinging nettle]). In some cases, these species are sold with at least some recognition by the seller that the species can be detrimental. One firm warns that the *Imperata cylindrica* it sells "can be invasive"; that is an understatement. *Imperata cylindrica* (alang alang) is a scourge in much of the tropics (Brook 1989) and is becoming invasive in southern Florida (Lippincott 2000).

Self-imposed restrictions by some nurseries—that is, they refuse to mail some species to vulnerable locales—offer little help. For example, the refusal by one merchant to sell *Datura stramonium* (jimsonweed), a narcotic-producing species, in Canada is rendered useless by U.S. seed merchants who will readily ship the same species to Canadian addresses. The refusal by other merchants to sell *Lythrum salicaria* in North America is an expost facto statement that this species is already invasive across a wide swath of North America (Thompson et al. 1987); little market probably remains for it. But simultaneously offering this species for export simply opens the possibility it will become invasive elsewhere.

The global merchandising of living plants is a multibillion-dollar business in which numerous pathways and modes of dispersal are well developed (U.S. Department of Agriculture 1998). Currently, almost any species can be imported into the United States, provided it is disease-free and does not appear on the federal noxious seed or weed lists (Westbrooks 1993). These are low hurdles for plant entry and do not differ markedly from the rules in many other countries (Anon. 1994).

Potential for the importation into the United States of thousands of previously unavailable native Chinese species represents a particularly serious problem. China and the United States overlap extensively in latitude, and the amplitude of their physical environments is roughly similar (Hou 1983). Furthermore, the Chinese flora is significantly more diverse than the U.S. flora (Qian and Ricklefs 1999) and includes many potential ecological counterparts to U.S. native species. As the Chinese horticultural trade with the United States burgeons, there will be much opportunity for the deliberate introduction of species that could become naturalized or even invasive (Mack 2001). Maximum effort should be mustered to evaluate the potential of these species to survive outside cultivation in the United States.

Not surprisingly, the emergence of e-commerce has enhanced domestic and international plant trade. Catalogs for a growing number of commercial plant nurseries can now be accessed through Web sites, such as http://www.seedquest.com and http://www.seedman.com/index/medicine.htm. The Internet is neither a pathway nor a mode for plant dispersal; it does, however, have the potential to greatly expand globally the clientele for plants and to facilitate customers' use of existing pathways for the delivery of viable plant material. Anyone with Internet access can instantly order nonindigenous plants, no matter how distant the vendor. This communication network obviously can bring about much good; but it also has the potential to exacerbate the movement of nonindigenous species into new ranges where they could cause great economic and environmental damage.

CONCLUSIONS

I began this chapter with a quote from Frederick the Great that serves as a powerful reminder for effectively intercepting detrimental nonindigenous plants. Attempting to defend against every possible mode of dispersal, arriving by every possible pathway, at any point in the defended territory, is impossible and invites disaster. Instead, priorities must be erected. Minimal or even no resources should be allocated to some pathways and modes. Such a decision was made, for example, in deciding not to remove the seeds from vehicles entering one Australian national park; the likelihood of plant naturalization via this mode was determined to be insignificant (Lonsdale and Lane 1994). The decision frees resources to intercept species moving in more important modes or to control invasive species already in the park. Unlike the task for intercepting alien insects and marine invertebrates, identifying the quarantine priorities for nonindigenous vascular plants is straightforward, even if implementing them remains daunting. Most future plant naturalizations and invaders will stem from deliberate plant importations. Thanks to modern seed purity restrictions, seed contaminants will remain only a secondary contributor.

Emphasis on carefully evaluating the consequences of deliberate introductions is justified, based on the outcomes of plant introductions. Species chosen for ornamentation do not have greater intrinsic ability to become naturalized than other species. But the sheer number of introduced ornamentals means that some fraction will become naturalized and even invasive. Maximum effort then needs to be expended on evaluating these species, preferably before they enter a new range (e.g., Steinke and Walton 1999). Once nonindigenous species are introduced, their ability to persist without cultivation should be monitored carefully. Attention to other modes—for example, importation in hay, bulk cargo, or animals—is of tertiary importance. Ironically, the majority of our environmental wounds from introduced plants are self-inflicted; that is, humans have deliberately introduced most of the species that now cause harm in new ranges. We can, however, use that knowledge effectively to prevent this mode of dispersal from wreaking even more damage.

Acknowledgments

I thank Jim Carlton and Greg Ruiz for the excellent workshop they organized in November 1999 that led to this publication. I also thank R. A. Black, W. Ford, R. H. Groves, A. Kelly, W. M. Lonsdale, R. Old, S. Reichard, and B. M. Waterhouse for useful discussions and assistance. J. T. Carlton, C. L. Kinter, and D. Simberloff provided very helpful reviews of the manuscript.

References

A. B. M. 1872. Spontaneous appearance of exotic forage plants in France after the late war. *Nature* 6: 263.

Ahti, T. and L. Hamet-Ahti. 1971. Hemerophilous flora of the Kuusamo district, northeast Finland, and the adjacent part of Karella, and its origin. *Annales Botanica Fennica* 8: 1–91.

Anonymous. 1994. Agreement on the application of sanitary and phytosanitary measures. Annex 1. Geneva: World Trade Organization.

Anonymous. 2000. Evaluation of the quarantine risk associated with military and humanitarian movements between East Timor and Australia. *Northern Australia Quarantine Strategy.* Canberra: Australian Quarantine and Inspection Service.

Barrett, S. C. H. 1989. Waterweed invasions. *Scientific American* 260: 90–97.

Brook, R. M. 1989. Review of literature on *Imperata cylindrica* (L.) Raeuschel with particular reference to South East Asia. *Tropical Pest Management* 35: 12–25.

Carlton, J. T. and J. B. Geller. 1993. Ecological roulette: biological invasions and the global transport of nonindigenous marine organisms. *Science* 261: 78–82.

Clement, E. J. and M. C. Foster. 1994. *Alien Plants of the British Isles.* London: Botanical Society of the British Isles.

Cronk, Q. C. B. 1989. The past and present vegetation of St. Helena. *Journal of Biogeography* 16: 47–64.

Cuthbertson, E. G. and W. T. Parsons. 1992. *Noxious Weeds of Australia.* Melbourne: Inkata.

Dewey, L. H. 1896. Migration of weeds. In *United States Department of Agriculture Yearbook*, pp. 263–286.Washington, DC: Government Printing Office.

Emmer, P. and F. Gaastra, eds. 1996. *The Organization of the Interoceanic Trade in European Expansion, 1450–1800.* Aldershot: Variorum.

Enomoto, T. 1999. Naturalized weeds from foreign countries into Japan. In *Biological Invasions of Ecosystem Pests and Beneficial Organisms*, E. Yano, K. Matsuo, M. Shiyomi, and D. A. Andow, eds., pp. 1–14. Tsukuba: National Institute of Agro-Environmental Sciences.

Fosberg, F. R. 1949. Flora of Johnston Island, Central Pacific. *Pacific Science* 3: 338–339.

Fosberg, F. R. 1957. The naturalized flora of Micronesia and World War II. *Eighth Pacific Science Congress* 4: 229–234.

Gaddums, M. 1999. *Gaddum's Plant Finder 2000.* Gisborne, New Zealand: New Zealand Plant Finder.

Godwin, H. 1975. *The History of the British Flora.* 2nd edition. Cambridge: Cambridge University Press.

Good, R. 1964. *The Geography of Flowering Plants.* 3rd edition. New York: Wiley.

Grenfell, A. L. 1987. *Myagrum perfoliatum* L. in Cambridgeshire. *Adventive News* 37. In *Botanical Society of the British Isles News* 47.

Guppy, H. B. 1906. Observations of a naturalist in the Pacific between 1896 and 1899. *Plant-Dispersal.* Volume 2. London: Macmillan.

Hamet-Ahti, L. 1984. Changes of the northern boreal vegetation and flora in Finland after the Second World War. *Phytocoenologia* 12(2/3): 359–361.

Harlan, J. R. 1965. The possible role of weed races in the evolution of cultivated plants. *Euphytica* 14: 173–176.

Hicks, G. H. 1895. Pure seed investigation. *United States Department of Agriculture Yearbook*, pp. 389–404. Washington, DC: Government Printing Office.

Hillman, F. H. 1893a. Nevada weeds. I. *Nevada Agricultural Experiment Station Bulletin* 21.

Hillman, F. H. 1893b. Nevada weeds. II. *Nevada Agricultural Experiment Station Bulletin* 22.

Hou, H. Y. 1983. Vegetation of China with reference to its geographical distribution. *Annals of the Missouri Botanical Garden* 70: 509–549.

Kloot, P. M. 1987. The naturalised flora of South Australia 3. Its origin, introduction, distribution, growth forms and significance. *Journal of the Adelaide Botanical Garden* 10: 99–111.

Kornas, J. 1990. Plant invasions in central Europe: historical and ecological aspects. *Biological Invasions in Europe and the Mediterranean Basin*, F. Di Castri, A. J. Hansen, and M. Debussche, eds., pp. 19–36. Dordrecht: Kluwer.

Kosina, R. 1978. The cultivated and wild plants from the XIth century granaries on the cathedral-island in Wroclaw. *Berichte Deutsche Gesellschaft Botanische* 91: 121–127.

Kowarik, I. 1995. Time lags in biological invasions with regard to the success and failure of alien species. In *Plant Invasions*, P. Pyšek, K. Prach, M. Rejmanek, and M.Wade, eds., pp. 15–38. Amsterdam: SPB Academic Publishing.

Lindroth, C. H. 1957. *The Faunal Connections between Europe and North America*. New York: Wiley.

Lippincott, C. L. 2000. Effects of *Imperata cylindrica* (L.) Beauv. (Cogongrass) invasion on fire regime in Florida sandhill (USA). *Natural Areas Journal* 20: 140–149.

Lonsdale, W. M. and A. M. Lane. 1994. Tourist vehicles as vectors of weed seeds in Kakadu National Park, Northern Australia. *Biological Conservation* 69: 277–283.

Lousley, J. E. 1971. *Flora of the Isles of Scilly*. Newton Abbot: David and Charles.

Mabberley, D. J. 1997. *The Plant-Book: A Portable Dictionary of the Vascular Plants*. 2nd edition. New York: Cambridge University Press.

Mack, R. N. 1986. Alien plant invasion into the Intermountain West: a case history. In *Ecology of Biological Invasions of North America and Hawaii*, H. A. Mooney and J. Drake, eds., pp. 191–193. New York: Springer-Verlag.

Mack, R. N. 1991. The commercial seed trade: an early disperser of weeds. *Economic Botany* 45: 257–273.

Mack, R. N. 1999. The motivation for importing potentially invasive plant species: a primal urge? In *Proceedings of the VI International Rangeland Congress*, pp. 557–562. Townsville, Australia.

Mack, R. N. 2000. Cultivation fosters plant naturalization by reducing environmental stochasticity. *Biological Invasions* 2: 111–122.

Mack, R. N. 2001. Motivations and consequences of the human dispersal of plants. In *The Great Reshuffling: Human Dimensions in Invasive Alien Species*, J. A. McNeely, ed., pp. 23–34. Gland: International Union for the Conservation of Nature.

Mack, R. N. and M. Erneberg. 2002. The United States naturalized flora: largely the product of delibrate introductions. *Annals of the Missouri Botanical Garden.* 89: 176–189.

Mack, R. N. and W. M. Lonsdale. 2001. Humans as global plant dispersers: getting more than we bargained for. *BioScience* 51: 689–710.

Mack, R. N. and D. A. Pyke. 1983. The demography of *Bromus tectorum* L.: variation in time and space. *Journal of Ecology* 71: 69–93.

Mack, R. N., D. Simberloff, W. M. Lonsdale, H. Evans, M. Clout, and F. A. Bazzaz. 2000. Biotic invasions: causes, epidemiology, global consequences and control. *Ecological Applications* 10: 689–710.

Maiden, J. H. 1914. Australian vegetation. In *Federal Handbook: prepared in connection with the 84th meeting of the British Association for the Advancement of Science,* G. H. Knibbs, ed., pp. 163–209. Melbourne: A. J. Mullett, Government Printer.

Maiden, J. H. 1920. *The Weeds of New South Wales.* Part I. Sydney: William Applegate Gullick.

Martindale, I. C. 1877. More about ballast plants. *Botanical Gazette* 2: 127–128.

McCracken, D. P. 1997. *Gardens of Empire: Botanical Institutions of the Victorian British Empire.* New York: Leicester University Press.

Meinig, D. W. 1986. *The Shaping of America: A Geographical Perspective on 500 Years of History.* Vol. 2. New Haven: Yale University Press.

Mills, E. L., J. R. Chrisman, and K. T. Holeck. 2000. The role of canals in the spread of nonindigenous species in North America. In *Nonindigenous Freshwater Organisms: Vectors, Biology and Impacts,* R. Claudi and J. H. Leach, eds., pp. 347–379. Boca Raton: Lewis Publishers.

Mohr, C. 1878. Foreign plants introduced into the Gulf States. *Botanical Gazette* 3: 42–46.

Muenscher, W. C. L. 1955. *Weeds.* 2nd edition. New York: Macmillan.

Muhlenbach, V. 1979. Contributions to the synanthropic (adventive) flora of the railroads in St. Louis, Missouri, U.S.A. *Annals of the Missouri Botanical Garden* 66: 1–108.

Munn, M. T. 1914. Seed tests made at the station during 1913. *New York Agricultural Experiment Station Bulletin,* p. 378. Geneva, New York.

Murray, D. R. 1986. Seed dispersal by water. In *Seed Dispersal,* D. R. Murray, ed., pp. 49–79. Sydney: Academic Press.

Panetta, F. D. and R. P. Randall. 1994. An assessment of the colonizing ability of *Emex australis. Australian Journal of Ecology* 19: 76–82.

Pijl, L. van der. 1969. *Principles of Dispersal in Higher Plants.* Berlin: Springer-Verlag.

Porter, A. N., ed. 1991. *Atlas of British Overseas Expansion.* London: Routledge.

Qian, H. and R. E. Ricklefs. 1999. A comparison of the taxonomic richness of vascular plants in China and the United States. *American Naturalist* 154: 160–181.

Reed, C. F. 1964. A flora of the chrome and manganese ore piles at Canton, in the port of Baltimore, Maryland, and at Newport News, Virginia, with descriptions of genera and species new to the flora of eastern United States. *Phytologia* 10: 321–327.

Rhoads, A. F. and W. M. Klein. 1993. *The Vascular Flora of Pennsylvania: Annotated Checklist and Atlas.* Philadelphia: American Philosophical Society.

Ridley, H. N. 1930. *The Dispersal of Plants throughout the World.* Kent: L. Reeve.

Roy, B., I. Popay, P. Champion, T. James, and A. Rahman. 1998. *An Illustrated Guide to Common Weeds of New Zealand.* Canterbury: New Zealand Plant Protection Society.

Schardt, J. D. and D. C. Schmitz. 1991. 1990 Florida aquatic plant survey, the exotic aquatic plants. *Technical Report 91-CGA.* Tallahassee: Florida Department of Natural Resources.

Shaughnessy, G. L. 1986. A case study of some woody plant introductions to the Cape Town area. In *The Ecology and Management of Biological Invasions,* I. A. W. Macdonald, F. J. Kruger, and A. A. Ferrar, eds., pp. 37–43. Cape Town: Oxford University Press.

Smith, J. 1999. *Australian Driftseeds: A Compendium of Seeds and Fruits Commonly Found on Australian Beaches.* Armidale: University of New England.

Stearn, W. T. 1958. Botanical exploration to the time of Linnaeus. *Proceedings of the Linnean Society of London* 169: 173–196.

Steinke, E. and Walton, C. 1999. Weed risk assessment of plants imports to Australia: policy and process. *Australian Journal of Environmental Management* 6: 157–163.

Suominen, J. 1979. The grain immigrant flora of Finland. *Acta Botannica Fennica* 111: 1–108.

Thomas, A. G., A. M. Gill, P. H. R. Moore, and F. Forcella. 1984. Drought feeding and the dispersal of weeds. *Journal of the Australian Institute of Agricultural Science* 50: 103–107.

Thompson, D. Q., R. L. Stuckey, and E. B. Thompson. 1987. Spread, impact, and control of purple loosestrife (*Lythrum salicaria*) in North American wetlands. *U.S. Dept. of Interior/Fish and Wildlife Service Research* 2. Washington, DC.

Thompson, J. D. 1991. The biology of a successful invader: what makes *Spartina anglica* so good? *BioScience* 41: 393–401.

Tutin, T. G., P. W. Ball, and A. O. Chater, eds. 1980. *Flora Europaea.* Cambridge: Cambridge University Press.

U.S. Department of Agriculture. 1998. Census of Horticultural Specialties. Washington, DC: National Agricultural Statistics Service.

Viola, H. J. and C. Margolis. 1991. *Seeds of Change: Five Hundred Years Since Columbus.* Washington, DC: Smithsonian.

Washington Department of Agriculture. 1914. First Annual Report 1913–1914. *Seed Laboratory Report,* pp. 101–111.

Washington Department of Agriculture. 1916. First Annual Report 1914–1916. *Seed Laboratory Report,* pp. 164–178.

Wells, M. J., R. J. Poynton, A. A. Balsinhas, K. J. Musil, H. Joffe, E. van Hoepen, and S. K. Abbott. 1986. The history of introduction of invasive plant species to southern Africa. In *The Ecology and Management of Biological Invasions in Southern Africa.* I. A. W. Macdonald, F. J. Kruger, and A. A. Ferrar, eds., pp. 21–35. Cape Town: Oxford University Press.

Westbrooks, R. G. 1993. Exclusion and eradication of foreign weeds from the United States by USDA APHIS. In *Biological Pollution,* B. N. McKnight, ed., pp. 225–241. Indianapolis: Indiana Academy of Science.

Wolff, S. E. 1951. Harvesting and cleaning grass and legume seed in the Western Gulf Region. *Agriculture Handbook No. 24.* Washington, DC: United States Department of Agriculture.

Woodruffe-Peacock, E. A. 1918. A fox-covert study. *Journal of Ecology* 6: 110–125.

Chapter 2

Invasion Pathways of Terrestrial Plant-Inhabiting Fungi

Mary E. Palm and Amy Y. Rossman

The terrestrial plant-inhabiting fungi present many challenges when evaluating invasion pathways. One reason is that knowledge of the biodiversity of fungi is grossly incomplete. It is estimated that less than 10 percent of the fungi have been discovered and described (Hawksworth 1991, Hawksworth and Rossman 1997). Of the fungi that have been discovered, few have been studied extensively regarding their biology such as pathogenic potential, host range, and potential for genetic variability. Even defining a "species" is difficult in many groups of fungi. For these reasons it is not easy to predict precisely the risk posed by the many fungi that could be transported and introduced into new environments.

One means of obtaining insight into potential invasion pathways of plant-inhabiting fungi is to review past introductions, particularly how these fungi were moved from country to country. This chapter includes some basic information on fungal biology to provide an understanding of the characteristics of fungi, thereby explaining why it has been difficult to determine and avert pathways for the movement of fungi. Specific pathways by which fungi have been introduced into new environments, primarily the United States, are discussed. Emphasis is placed on the risks posed by each pathway. Two assumptions about these pathways are examined to determine their validity. Finally, we discuss knowledge gaps that must be filled in order to prevent future movement and inadvertent introductions of disease-causing, plant-inhabiting fungi.

BIOLOGY OF FUNGI

The body of a fungus is a microscopic, threadlike structure that lives and grows within a substrate that acts as a food source. Fungi invade their food source mainly through the production of enzymes that allow the fungus to grow within the substrate and absorb nutrients. Because of this absorptive mode of nutrition, fungi can occur on or in all living organisms as well as in dead plant material—often inconspicuously. Fungi occur on and in insects, nematodes, and other nonplant organisms, but this discussion is limited to fungi that exist on and in terrestrial plants.

Fungi survive over time, and are distributed to new substrates naturally, by means of sexual or asexual spores or both. The asexually produced spores often serve to disseminate the fungus short distances during one disease cycle, as in the case of apple scab and wheat stem rust. These abundant, microscopic spores are frequently wind disseminated, waterborne, or carried by insects or other vectors to new, susceptible plant hosts. The sexually produced spores can serve as survival structures, initiating disease at the beginning of a cycle, as in the spring in temperate regions. Fungi also can survive as mycelium within their living or decaying hosts, or as specialized survival structures, such as sclerotia, produced in the host or soil.

All plants, whether diseased or healthy, are host to a variety of fungi. Some fungi, such as rusts, smuts, downy mildews, and powdery mildews, are obligate parasites attacking living host plants. These fungi usually occur on only one host plant species, a group of related plant species, or one host plant genus. Other fungi are saprotrophs and decay already-dead organic matter. Still other fungi cause diseases of living plants but can also survive as saprotrophs in dead plant material; these are called facultative saprotrophs.

Results of research on fungal biology during the past two decades have revealed that the situation is more complex than realized previously. Researchers have discovered that nearly all woody and herbaceous plants are colonized by a group of fungi that do not fit easily into those general biological categories—these endophytes are generally defined as fungi that exist within a plant but cause no apparent harm to the plant (see Redlin and Carris 1996). The distinction between an endophyte that causes no disease and a latent plant pathogen that elicits disease symptoms is not clear. It has been suggested that the endophyte and plant are in a balanced antagonism, whereas the pathogen and plant are in an unbalanced antagonism, in which the pathogen can suppress the plant's defense reaction and cause disease (Rodriguez and Redman 1997, Schulz et al. 1999). Can the same fungus have both lifestyles, or are they mutually exclusive? Are some of these apparent nonpathogenic endophytes actually latent pathogens—that is, not causing disease at the time, but able to cause disease under certain conditions? Or, are there two distinct biological groups: one that is always nonpathogenic and another that is latent but can be pathogenic under the right conditions? If the latter is true, how can we distinguish the nonpathogenic endophytes from the latent pathogens? Research is needed to answer these questions, and new methods for detecting pathogens not visibly causing disease must be developed in order to prevent their introduction.

Defining species and thus applying biologically meaningful scientific names for fungi has become increasingly difficult. Obligate parasites such as the rusts, smuts, powdery mildews, and downy mildews have been assumed to be host-specific; this is usually reflected in their scientific names. Plant pathogenic fungi that function as facultative saprobes also have been considered to be host-specific; many species are described and identified on the basis of their host. Recent molecular studies of genera such as *Phomopsis* and *Phyllosticta* have shown that such assumed host-specificity does not have a sound genetic basis (Baayen et al. 2002, Farr et al. 1999, Rehner and Uecker 1994). Strains of *Phomopsis* from a single plant host species have been found to be genetically diverse, suggesting that species cannot be defined by their plant host alone. Conversely, strains of *Phomopsis* that are genetically identical should be considered one species and may infect a number of different plant hosts. For this reason, accurate identification based on morphology alone has become difficult for some groups of fungi.

The issues concerning biology of the fungi that are especially relevant in examining invasion pathways include the biology of endophytes, the difficulty of detecting fungi on plant hosts, the increasingly questionable host-specificity of fungi, the need for biologically meaningful scientific names, and the difficulty of defining fungal species. These factors must be considered in making accurate and scientifically sound plant quarantine decisions and setting policies that regulate the movement of plants while minimizing the risk of introducing plant pathogenic fungi.

PATHWAYS FOR THE MOVEMENT OF PLANT-INHABITING FUNGI

Plant-inhabiting fungi are transported by humans, primarily in association with their plant hosts. These hosts may be propagative or nonpropagative plant material, including wood and wood products. Soil, fungi in pure culture, and natural pathways are also means of transport. These pathways are defined here and their risks discussed. Examples are pre-sented of diseases caused by fungi introduced by each pathway from one part of the world into new areas where their damage may be even worse than in the native habitat. These introductions have affected all types of plants—native and cultivated, annual and perennial, herbaceous and woody. With increasingly sophisticated means of genetically identifying specific strains and populations, it is possible to determine if a disease-causing fungus was introduced and how fungi are being moved around the world. In most cases, the introduction of plant-disease-causing fungi can be traced to the introduction of their host plant.

Propagative Plant Material

Propagative plant material poses the greatest risk as a pathway for success-fully introducing invasive fungi into a new environment. Such living plant material includes seeds, woody and herbaceous cuttings, bare-rooted plants such as nursery stock, and plants in growing media. These substrates are host to many fungi that may or may not be detectable by visual means. Propagative plant material may be annuals, perennials, or tree species. Cer-tain types of propagative plant material, such as woody nursery stock from larger landmasses transported among areas with comparable climates, are considered to pose the greatest risk, particularly from pathogenic fungi (Wilson 1987). However, germplasm of annual plants, especially those that are propagated vegetatively, also pose a risk, as suggested by the examples below. The greatest volume of propagative plant material is brought into the United States for commercial purposes; but plants for propagation may also be introduced by plant hobbyists, tourists with souvenirs, or immi-grants who merely want to bring something with them from their home country.

Woody Nursery Stock

An example of a fungus that was introduced along with its native host plant seedlings, and was devastating to a related but different host plant, is chestnut blight. Japanese chestnut trees (*Castanea crenata*), native to Asia, were brought to the northeastern United States beginning in 1876 (Pow-ell 1900) from their native Asia. It was at this time that the disease was likely inadvertently introduced along with the host propagative material. Diseased American chestnut trees (*Castanea dentata*) were observed in the early part of the twentieth century (Anagnostakis 1987). The native American chestnut trees were particularly susceptible to this fungus, and the disease spread quickly. Within three decades, the native American chestnut tree was eliminated as the dominant forest species in the northeastern United States, and, to this day, it remains a minor understory shrub. Later, the fungus was transported back to Europe, probably on chestnut wood, where many European chestnut groves were also destroyed. Chestnut blight, along with citrus canker and white pine blister rust, two other tree diseases, was one of the main reasons that U.S. Plant Quarantine legislation was passed in 1912 (Foster 1991). Chestnut blight exemplifies the devastating effect an introduced fungus can have on native plant species.

Some fungi that cause little or no disease on their native host can clearly cause severe damage on related plant hosts when introduced into a new environment. This was the situation with chestnut blight and also may be the case for dogwood anthracnose in the United States. Dogwood anthracnose, a relatively new disease of U.S. native flowering dogwoods (*Cornus florida* and *C. nuttallii*), was first observed in the Pacific Northwest and eastern United States in the 1970s (Daughtrey et al. 1996). The disease is caused by a previously unknown fungus, *Discula destructiva* (Redlin 1991). The low genetic diversity observed between isolates of *D. destructiva* obtained from trees of both coasts of the United States (Trigiano et al. 1995), as well as the nearly simultaneous appearance of the disease on both coasts and its rapid spread, suggests that the fungus causing dogwood anthracnose was introduced recently (Daughtrey et al. 1996). One possible pathway for the introduction of this fungus is asymptomatic seeds or plants of kousa dogwood (*Cornus kousa*) from Asia, where the fungus may be a weak pathogen or endophyte of kousa dogwood, because that host has been determined to be relatively resistant to dogwood anthracnose.

Beech bark disease is caused by the beech scale insect and several similar species of *Neonectria*. Beech scale was introduced into a Canadian public garden on European beech trees (see Houston 1994). It has been suggested that *Neonectria coccinea* var. *faginata*, a taxon that is more aggressive to beech than native pathogens, also was introduced. Results of a study by Mahoney et al. (1999), in which they compared isolates using RFLP analyses, support the suggestion that *N. coccinea* var. *faginata* was indeed introduced into North America from Europe (Mahoney et al. 1999).

Herbaceous Plant Germplasm

Disease-causing fungi, introduced via propagative plant material, may severely impact perennials (especially woody plants, as mentioned in the examples above), but may also have a negative impact on annuals. Crops grown in monocultures are particularly susceptible to disease as are horticultural crops grown and maintained in greenhouses.

Potatoes were introduced to other parts of the world from the Andes region of South America several centuries ago. In the early to mid–nineteenth century *Phytophthora infestans*, which causes potato late blight, was transported from its native Mexico into many parts of the world with cultivated potato germplasm (Fry et al. 1993). Due to weather conditions conducive to the disease and the reliance of the Irish on the potato as a staple, the effects of this disease were devastating to the population and changed the course of history with the death or emigration of more than 3 million Irish people. Interestingly, this disease has reemerged as a major problem worldwide, most recently in the United States, during the past decade. This reemergence is due to a recent introduction of a more virulent, fungicide-resistant strain of the pathogen (Fry and Goodwin 1997). Because of the ability to characterize strains of this fungus using molecular sequences, it has been possible to determine the specific strain causing the recent epiphytotic in the United States. Potato late blight exemplifies a disease that can have a major effect on an annual crop. Once the disease-causing fungus is introduced, the severity of the problem is related to the monoculture of a crop, the occurrence of suitable environmental conditions, and the reliance on one crop for income or food.

The ornamentals industry constantly strives to develop new products and therefore continually introduces new germplasm for the U.S. industry. Recently, many new powdery mildew diseases have become serious problems in this industry, including powdery mildews of poinsettia, cascading petunias, New Guinea impatiens, sedum, and many others. The appearance of so many new powdery mildew diseases, in a relatively short time period, is puzzling. Margery Daughtrey (Cornell University, pers. comm.) suggests that some of these fungi may have been introduced on plant cuttings used for propagation grown in new areas of the world. Transporting propagative plant material around the world increases the chance of introducing new powdery mildews. Powdery mildews could be introduced on cuttings of plant varieties that are fairly resistant to the disease and therefore without obvious visible effects. Additionally, initial powdery mildew infections can be inconspicuous (Celio and Hausbeck 1998), especially on cuttings, and therefore may be overlooked during phytosanitary inspections. When those cuttings are brought into greenhouse production, the fungus is able to spread and thrive, so that disease becomes a visible problem. For example, holiday season poinsettias used to be grown from cuttings to full-sized plants entirely in the United States. In the past years, poinsettias have been propagated in several other countries from where the unrooted cuttings are shipped to the United States for distribution. Poinsettia powdery mildew, now prevalent in U.S. greenhouses, may have been introduced inadvertently this way. Daughtrey (pers. comm.) also suggests that some powdery mildews were introduced many years ago but did not become established in production because of the post–World War II grower practice of regularly scheduled fungicide applications. Frequent use of a partially systemic fungicide may have suppressed the diseases caused by previously introduced but latent or inconspicuous powdery mildews. With the current practice of integrated pest management, fungicides are sprayed only after a disease problem is detected, thus allowing the powdery mildew fungi to develop and sporulate conspicuously.

Seeds

Seeds serve as an efficient pathway for the introduction of plant diseases (McCubbin 1954, McGee 1997, Neergaard 1977). Seeds are often not visibly infected, and measures other than inspection would be needed to detect associated pathogens. A number of diseases have been introduced into various countries throughout the world by means of infected seed. *Gloeotinia temulenta*, originally described from Europe and one of the most serious diseases of ryegrass (*Lolium* sp.) in the northwestern United States, was introduced in the 1940s on seed (see Neergaard 1977). *Sphaeropsis sapinea*, a disease of *Pinus*, has been repeatedly introduced into pine plantations in South Africa via seed, resulting in a high level of genetic diversity and an increased burden in disease management (Wingfield et al. 2001).

Nonpropagative Plant Material

Considerable quantities of nonpropagative plant material, often called "plant material for consumption," are transported throughout the world. Such goods include fruits, vegetables, cut flowers, dried herbs, wood, and wood products. The same factors influencing the introduction of invasive fungi on propagative material influence this pathway. However, most of the associated fungi are destroyed along with the plant material following its intended use, which considerably lessens the probability that the associated fungi will survive, become established, and spread into a new environment. Although still posing some risk, especially if material will be near a susceptible host or if all plant parts are not totally destroyed relatively quickly, nonpropagative plant material generally poses less risk as a pathway for the introduction of fungi as compared to propagative material.

Raw wood and wood products are a major exception to the relatively low risk of nonpropagative plant material for consumption. Wood and wood products pose a considerably greater risk because they harbor large numbers of diverse fungi and may be stored and used over a long period of time, often outdoors. In addition, knowledge of what fungi are present in wood and wood products is incomplete and the biology of many wood-inhabiting pathogens is not well understood. As mentioned earlier, wood-inhabiting fungi, harmless to the trees in their native habitats (where the fungus and plant host coevolved), may cause severe problems when introduced into a new environment. Usually, fungi associated with wood and wood products cannot be observed through inspection. Many of the fungi on wood and wood products are vectored by insects, adding to the difficulty of transporting wood and wood products safely; once introduced, insect-borne fungi can move long distances very rapidly, especially if the vector is already present or is introduced at the same time.

Such was the case of Dutch elm disease, introduced in veneer logs from Europe along with the bark beetle vector (see Yarwood 1983). Dutch elm disease, caused by *Ophiostoma ulmi*, has destroyed American elms across North America as well as the native elms in much of Europe and Asia. Brasier (2001) delineated the rapid evolution among the fungi that cause this disease. As it turns out, there were two pandemics of this disease in Europe and North America that were caused by two related fungal species, which are nearly identical morphologically but differ biologically. *Ophiostoma ulmi* was introduced into North America around 1927 and spread across the continent. The second pandemic began in the 1940s and was caused by *O. novo-ulmi* (then undescribed), which is more fit and has replaced the original species in North America. Additionally, the two fungal species have been able to hybridize. Using molecular techniques, Brasier (2001) determined that these fungi evolved through interspecific gene transfer. The more virulent pathogen has been reintroduced into Europe through the shipment of logs from North America. This example points out the possibilities of introducing a more virulent form of a pathogen, even though the disease is already known from a region.

This is not an isolated example of rapid evolution among introduced fungal pathogens. In the Netherlands a new *Phytophthora* pathogen on *Primula* and *Spathyphyllum* has been shown to be a hybrid of an endemic species and an introduced species (Man in't Veldt et al. 1998). A different aggressive *Phytophthora* hybrid has developed that is pathogenic on alder trees in Europe. Neither parent species was a known pathogen of alder (Brasier et al. 1999). In New Zealand and North America, newly evolved hybrids between introduced species of the rust fungus *Melampsora* were able to infect poplar trees, which the parent fungi could not infect (Newcombe et al. 2000, Spiers 1998). These examples indicate that hybrid fungal pathogens can develop, and become more aggressive or infect otherwise resistant hosts, when new fungal germplasm is repeatedly introduced.

Soil

Soil harbors numerous propagules of pathogenic fungi, especially those that cause root and wilt diseases. These propagules are easily transported with soil; for this reason, many countries regulate soil. Wingfield et al. (2001) discuss the likelihood that *Rhizina undulata*, the cause of a conifer root disease, was introduced into South Africa in soil, associated with pine trees introduced when Europeans colonized the area. He also points out the fact that in South Africa "exotic pine and eucalypt plantation forestry would not have been possible . . . without the accidental introduction of mycorrhizae" (Wingfield et al. 2001, p. 135). In this case, these fungi (considered beneficial) probably were introduced by way of infected roots or through mycorrhizal propagules in the soil or both.

Fungi in Pure Culture

Fungi in pure culture are often exchanged between scientists or purchased from culture collections for laboratory research purposes. If fungi are used solely in research laboratories and disposed of properly, this pathway poses very little or no risk. Research that involves greenhouse or field-testing poses a greater risk, and each situation must be carefully evaluated.

ASSUMPTIONS

One assumption about invasion pathways of plant-inhabiting fungi is that no risk is posed by the repeated introduction of the same species once a fungus has been introduced. Numerous examples suggest that this is not true, as discussed above relative to the movement of the fungi associated with Dutch elm disease, which has been transported back and forth across the Atlantic Ocean. It is important to understand that pathogenic strains can be morphologically indistinguishable from those already established in a country and therefore difficult to differentiate without the use of molecular techniques.

A second assumption is that it is always possible to predict the risk of a fungus introduced into a new environment based on the biology, especially the pathogenicity, of that fungus on its native host in its native environment. As has already been pointed out, there are examples of fungi that had little effect on their native host but became serious pathogens when introduced into new areas. This is a well-documented phenomenon. In their native habitat, plant hosts and their pathogens evolved together, so there is generally some host resistance as well as other antagonists that keep a disease in balance. However, when a fungus is introduced into a new habitat, particularly into a monoculture of susceptible plant hosts under conditions conducive to the disease, the disease can develop and spread rapidly. Further, when there is little host resistance to the pathogen or no natural antagonists or neither, the disease may be devastating.

KNOWLEDGE GAPS

Considering that only 10 percent of the fungi that exist have been described, it is not surprising that new disease-causing fungi are being discovered regularly. Increased knowledge about the biological diversity of fungi will provide the basis for accurate predictions of the risk that those fungi pose. This knowledge is obtained through systematics and includes the discovery of novel fungi as well as the characterization of known fungal pathogens. Even in the United States, for which a comprehensive account exists of the 13,000 species of fungi associated with plants (Farr et al. 1989), many new species continue to be discovered and described. Systematics studies are needed to know which fungi occur on what plant hosts and where they occur. Each fungal species must be accurately and meaningfully defined, characterized, and named. Ideally, the pathogenicity, host range, breeding systems, genetic variability, and potential for genetic recombination would also be studied to provide scientific data for use in pest risk assessments (Palm 2001).

Fortunately, the tools for describing and characterizing fungi, using both morphological and molecular means, are rapidly becoming more sophisticated and more generally available. It is now possible to determine the genetic variability and relatedness of similar fungi, to identify hybrids of plant pathogenic fungi, and to track the movement of fungi around the globe. These tools make it possible to detect and control the movement of potentially invasive pathogens.

CONCLUSIONS

Disease-causing fungi will continue to pose a threat to new environments because global trade is increasing at a rapid rate, as are the diversity, volume, and rate of plant material being moved. Propagative and nonpropagative plants, especially wood and wood products, will continue to be high-risk invasion pathways for terrestrial plant-inhabiting fungi. Our knowledge of the most threatening fungi is still lacking. However, with the current interest and concern about invasive species and bioterrorism, there should be more funding to support efforts to begin to fill these gaps. Because of the huge diversity of fungi, it would be prudent to focus on the pathways that pose the most significant risk and the fungi that could be moved through those pathways.

Many new research tools exist that can be used to increase our knowledge of biodiversity of fungi in order to be able to more precisely predict the risk posed by individual fungi and individual pathways. In doing so, we will obtain the knowledge required to safely transport plant material from country to country while still safeguarding the agriculture and environment of individual countries and the world in general.

REFERENCES

Anagnostakis, S. L. 1987. Chestnut blight: the classical problem of an introduced pathogen. *Mycologia* 79: 23–37.

Baayen, R. P., P. J. M. Bonants, G. Verkley, G. C. Carroll, H. A. Van Der Aa, M. De Weerdt, I. R. van Brouwershaven, G. C. Schutte, W. Maccheroni Jr., G. C. de Blanco, and J. L. Azevedo. 2002. Nonpathogenic isolates of the citrus black spot fungus, *Guignardia citricarpa*, identified as a cosmopolitan endophyte of woody plants, *G. mangiferae (Phyllosticta capitalensis)*. *Phytopathology* 92: 464–477.

Brasier, C. M. 2001. Rapid evolution of introduced plant pathogens via interspecific hybridization. *BioScience* 51: 123–133.

Brasier, C. M., D. Cooke, and J. M. Duncan. 1999. Origin of a new *Phytophthora* pathogen through interspecific hybridization. *Proceedings of the National Academy of Sciences* 96: 5878–5883.

Celio, G. J. and M. K. Hausbeck. 1998. Conidial germination, infection structure formation, and early colony development of powdery mildew on poinsettia. *Phytopathology* 88: 105–113.

Daughtrey, M. L., C. R. Hibben, K. O. Britton, M. T. Windham, and S. C. Redlin. 1996. Dogwood anthracnose. Understanding a disease new to North America. *Plant Disease* 80: 349–358.

Farr, D. F., G. F. Bills, G. P. Chamuris, and A. Y. Rossman. 1989. *Fungi on Plants and Plant Products in the United States.* St. Paul: American Phytopathological Society Press.

Farr, D. F., L. A. Castlebury, and R. A. Pardo-Schultheiss. 1999. *Phomopsis amygdali* causes peach shoot blight of cultivated peach trees in the southeastern United States. *Mycologia* 91: 1008–1015.

Foster, J. A. 1991. Exclusion of plant pests by inspections, certifications and quarantines. In *CRC Handbook of Pest Management in Agriculture.* D. Pimentel, ed. Vol. I. pp. 311–338. Boca Raton: CRC Press, Inc.

Fry, W. E. and S. B. Goodwin. 1997. Resurgence of the Irish potato famine fungus. *BioScience* 47: 363–371.

Fry, W. E., S. B. Goodwin, A. T. Dyer, J. M. Matuszak, A. Drenth, P. W. Tooley, L. S. Sujkowski, Y. J. Koh, B. A. Cohen, L. M. Spielman, K. L. Deahl, D. A. Inglis, and K. P. Sandlan. 1993. Historical and recent migrations of *Phytophthora infestans*: chronology, pathways, and implications. *Plant Disease* 77: 653–661.

Hawksworth, D. H. 1991. The fungal dimension of biodiversity: magnitude, significance and conservation. *Mycological Research* 95: 641–655.

Hawksworth, D. H. and A. Y. Rossman. 1997. Where are all the undescribed fungi? *Phytopathology* 87: 888–891.

Houston, D. R. 1994. Major new tree disease epidemics: beech bark disease. *Annual Review of Phytopathology* 32: 75–87.

Mahoney, E. M., M. G. Milgroom, W. A. Sinclair, and D. R. Houston. 1999. Origin, genetic diversity, and population structure of *Nectria coccinea* var. *faginata* in North America. *Mycologia* 91: 583–592.

Man in't Veldt, W. A., W. J. Veenbaas-Rijks, E. Ilieva, A. W. A. M. de Cock, P. J. M. Bonants, and R. Pieters. 1998. Natural hybrids of *Phytophthora nicotianae* and *P. cactorum* demonstrated by isozyme analysis and random amplified polymorphic DNA. *Phytopathology* 88: 922–929.

McCubbin, W. A. 1954. *The Plant Quarantine Problem.* Copenhagen: E. Munksgaard.

McGee, D. C., ed. 1997. *Plant Pathogens and the Worldwide Movement of Seeds.* St. Paul: APS Press.

Neergaard, P. 1977. *Seed Pathology.* Vol. I. New York: John Wiley & Sons.

Newcombe, G., B. Stirling, S. McDonald, and H. D. Bradshaw, Jr. 2000. *Melampsora xcolumbiana,* a natural hybrid of *M. medusae* and *M. occidentalis. Mycological Research* 104: 261–274.

Palm, M. E. 2001. Systematics and the impact of accidentally introduced fungi on agriculture in the United States of America. *BioScience* 51: 141–147.

Powell, G. H. 1900. The European and Japanese chestnuts in the eastern United States. *11th Annual Report of the Delaware College Agricultural Experiment Station.* pp. 101–135.

Redlin, S. C. 1991. *Discula destructiva* sp. nov., cause of dogwood anthracnose. *Mycologia* 83: 633–642.

Redlin, S. C. and L. M. Carris. 1996. *Endophytic Fungi in Grasses and Woody Plants. Systematics, Ecology and Evolution.* St. Paul: APS Press.

Rehner, S. A. and F. A. Uecker. 1994. Nuclear ribosomal internal transcribed spacer phylogeny and host diversity in the coelomycete *Phomopsis. Canadian Journal of Botany* 72: 1666–1674.

Rodriguez, R. A. and R. S. Redman. 1997. Fungal life-styles and ecosystem dynamics: biological aspects of plant pathogens, plant endophytes and saprophytes. *Advances in Botanical Research* 24: 169–193.

Schulz, B., A. K. Rommert, U. Dammann, H. J. Aust, and D. Strack. 1999. The endophyte-host interaction: a balanced antagonism? *Mycological Research* 103: 1275–1283.

Spiers, A. G. 1998. *Melampsora* and *Marssonina* pathogens of poplars and willows in New Zealand. *European Journal of Forest Pathology* 28: 233–240.

Trigiano, R. N., G. Caetano-Anollés, B. J. Bassam, and M. T. Windham. 1995. DNA amplification fingerprinting provides evidence that *Discula destructiva,* the cause of dogwood anthracnose in North America, is an introduced pathogen. *Mycologia* 87: 490–500.

Wilson, C. L. 1987. Exotic plant pathogens—who's responsible? *Plant Disease* 71: 863.

Wingfield, M. J., B. Slippers, J. Roux, and B. D. Wingfield. 2001. Worldwide movement of forest fungi, especially in the tropical and southern hemisphere. *BioScience* 51: 134–140.

Yarwood, C. E. 1983. History of plant pathogen introductions. In *Exotic Plant Pests and North American Agriculture.* C. L. Wilson and C. L. Graham, eds. pp. 39–63. New York: Academic Press.

Chapter 3

Exotic Insects and
Their Pathways for Invasion

Keizi Kiritani and Kohji Yamamura

Lindroth (1957) conducted classical research on the pathway of introduced insects. Based on species for which the earliest dates could be established, it was found that 90 percent of exotic insect species present in 1840 in the United States were of the order Coleoptera (Sailer 1983). Lindroth (1957) was able to show that most of the coleopterous species were from southwestern England, where ships engaged in North American trade took soil ballast on board before sailing. The ballast contributed as a vector of European Coleoptera to North America.

Kiritani and Morimoto (1993) and Morimoto and Kiritani (1995) listed 324 exotic species of insects in Japan, including 85 species considered "possibly" exotic. Most of the species in the latter group had been introduced in Japan before the Meiji era (1868), when Japan had opened the door to foreign countries. Since Kiritani and Morimoto (1993), new records have been added, and a total of 415 species have been identified as known exotic species. Evidence suggests that 295 have invaded; the remaining 120 are "possible" exotic insects, which are believed to have invaded at some time in the past, but their occurrence has been poorly documented (Morimoto and Kiritani 1995, Kiritani 2002). Because an increasing number of newly invasive insects are reported each year, this figure gives a minimum value of exotic insect species established so far in Japan.

For the 415 known exotic species, we have provided a breakdown according to their taxonomic groups and major invasion pathways (Table 3.1). Further details of these invasions are discussed below. Only a subset

TABLE 3.1. Exotic insects in Japan belonging to different orders and their major ways for invasion.

Order	Exotic species in Japan (%)	Major ways for invasion
Coleoptera	158 (38)	accidental
Hemiptera	92 (22)	accidental
Lepidoptera	53 (13)	accidental and spontaneous
Hymenoptera	33 (8)	mostly intentional
Diptera	32 (8)	accidental and spontaneous
Thysanoptera	16 (4)	accidental
Blattaria	11 (2)	accidental
Others	20 (5)	accidental

TABLE 3.2. Number of exotic species that were available to analysis for various phases of invasion.

Total number of exotic insects	415	100%
Evidence of invasion is available	284	69%
Known probable time of invasion	185	45%
Known country of origin	168	40%
Known earliest occurrence in Japan	155	37%
Able to identify probable pathways	98	24%
Able to identify probable latency period	35	8%

of these exotic species was available for analysis of specific aspects of invasion (Table 3.2). For example, it was possible to identify probable pathways for invasion for only 98 (which account for 24%) of the known 415 exotic species.

DEFINITION OF TERMS

Introduction is used here to refer to any deliberate introduction mediated by humans. In a narrow sense, invasion refers to all of those cases not considered an introduction. In a broad sense, however, it refers to all cases of species entering a new geographic area. Accordingly, invasion (broad sense) can be divided into (1) introduction and (2) invasion (narrow sense). Invasion (narrow sense) can be further divided into (2a) spontaneous or natural invasion (i.e., natural dispersal) and (2b) human-assisted, accidental invasion (Fig. 3.1).

Pathway is defined rather broadly as the means that allows the entry or spread of a pest (FAO 1996). The relative importance of any pathway in carrying organisms from one place to another varies according to many

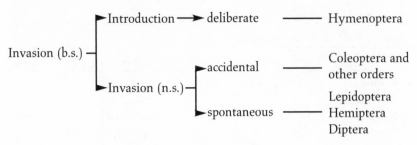

Figure 3.1. Classification of invasion.

factors, such as continent vs. island, inland vs. coastal area, importer vs. exporter, kinds of commodity, and transportation.

The most important pathways for accidental invasions of exotic species are related to international transport (i.e., trade and commerce and travel and tourism), whereas the most important vectors for deliberate introductions are related to biological control of agricultural pests and pollination of crops. In this chapter, we limit our analyses to cases of invasions in a narrow sense.

Exotic species are organisms that occur in places different from their area of natural distribution. Some exotic species become invasive, meaning that they threaten ecosystems, habitats, or species. Invasive species have in many cases caused disruption of ecological systems, homogenization of biota, and extinction (Mooney 1996).

Insects, which infest plant materials, are often the targets for plant quarantine. A quarantine pest can be defined as follows: a pest of potential economic importance to the area endangered thereby and not yet present there, or present but not widely distributed and being officially controlled (FAO 1996).

There are two categories of invasive insects of potential danger to agriculture. The first category includes a relatively small number of species that have a known high potential for damage, such as fruit flies and the codling moth in Japan. Ten insect species are designated high potential quarantine pests (Table 3.3). Of the 10 designated species, 2 were eradicated, 5 have not yet colonized, and 3 have become established in a limited area of Japan.

The second category of pests includes 28 species that are identified in Japan as potential quarantine pests. They are known to be destructive to crop plants grown in Japan. Three out of 28 species have recently become established (Table 3.3).

TABLE 3.3. List of quarantine insect pests identified as (A) designated high potential pests and (B) potential pests, and their population status in Japan.

(A) Designated high potential pests	
Mediterranean fruit fly (*Ceratitis capitata*)	Absent
Oriental fruit fly (*Bactrocera dorsalis*)	Eradicated in 1986
Melon fly (*B. cucurbitae*)	Eradicated in 1993
Queensland fruit fly (*B. tryoni*)	Absent
Colorado beetle (*Leptinotarsa decemlineata*)	Absent
Codling moth (*Cydia pomonella*)	Absent
Sweet potato vine borer (*Omphisa anastomosalis*)	Nansei Islands since 1941
Sweet potato weevil (*Cylas formicarius*)	Nansei Islands since 1903
West Indian sweet potato weevil (*Euscepes postfasciatus*)	Nansei Islands since 1947
Hessian fly (*Phytophaga destructor*)	Absent

(B) Potential pests	
Diptera (7 species)	All are absent
Lepidoptera (3 species)	Large white (*Pieris brassicae*) in 1996
Coleoptera (12 species)	Mexican bean beetle (*Epilachna varivestis*) in 1996
	Black vine weevil (*Otiorhynchus sulcatus*) in 1980
Hemiptera (6 species)	All are absent

SPONTANEOUS INVASIONS

Waterhouse (1991) stated that most arthropod immigration to oceanic islands was by aerial dispersal. The sudden infestation of *Leucaena* trees of Mimosaceae by *Heteropsylla cubana* around 1985 in Asian and Pacific regions, including Ogasawara (the Bonin Islands) and Nansei Islands (hereafter both are referred to as the southern islands), must have been triggered by moving air masses. Azuma (1986) stated that 7 species, or 20 percent, of 34 exotic lepidopterous species in Okinawa prefecture (in the Nansei Islands) are considered established by aerial immigration, probably from Taiwan and the mainland.

Kiritani (1984) examined the records of insects collected in the East China Sea by a weather observatory ship during 1967–77. Over 300 putative species were recorded, and 127 of those could be identified to species (Table 3.4). Lepidoptera, Hemiptera, and Diptera comprised 91 percent of the total species. It should be remarked that the three species of Hemiptera are the brown planthopper, *Nilaparvata lugens*, the whitebacked planthop-

TABLE 3.4. Airborne insects collected in the East China Sea by a weather observatory ship during the period 1967–77 (Kiritani 1984).

	Species indigenous to Japan	Species occur only in the Nansei Islands	Species not established in Japan
Lepidoptera	56	5	4
Hemiptera	21	2	3
Diptera	23	2	0
Odonata	1	4	0
Neuroptera	2	0	0
Coleoptera	2*	0	0
Hymenoptera	1	0	0
Orthoptera	1	0	0

*Coccinella septempunctata and Necrobia violacea

per, *Sogatella furcifera*, and an egg predator of these planthoppers, *Tytthus chinensis*. All of them are long-distance migrants from mainland China. The two species of planthoppers normally account for 85 percent of the airborne insect population and invade Japan every year in large numbers, but they do not become established in Japan; they lack the ability to diapause and pass the winter. Absence of rice plants during the winter is also detrimental to such a monophagous insect as the brown planthopper.

Yoshimatsu (1991) examined 110 specimens of lepidopterous insects collected on the East China Sea during 1981–87 and found that of the 43 species recorded, 4 species (6 individuals) were not established in Japan.

Since no cases of exotic insects established in Japan have been reported as carried by ocean currents, almost all of the exotic species are introduced to Japan either deliberately or accidentally (by various means of transportation), except for some species belonging to Lepidoptera, Hemiptera, and Diptera. Indeed, the mark-recapture experiment conducted by releasing marked sterile male melon flies on Kikaijima Island revealed that the insect was able to fly as far as 265 km across the sea (Fig. 3.2). The melon fly, *Bactrocera cucurbitae*, was first reported in the Yaeyama Islands in 1919. It expanded its distribution range northward in a diffusion manner along the Nansei Islands and reached the northernmost island of Tanegashima in 1979. It is considered that spontaneous flight of the fruit fly was mainly responsible for its expansion over 1,000 km during sixty years.

It is significant that the Coleoptera is the largest order in Insecta, but few species, if any, have the potential as airborne immigrants over a long distance.

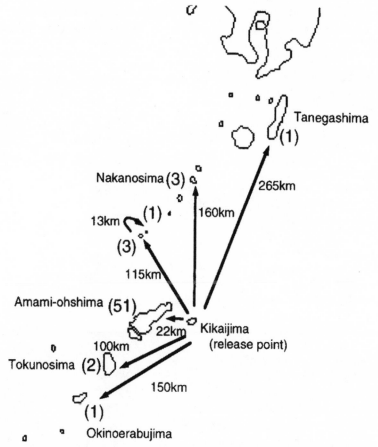

FIGURE 3.2. Recapture records of marked male adults of the melon fly release on Kikaijima Island. Numeral in parentheses shows the number of recaptured flies on each site (Kiritani 1983).

ACCIDENTAL INVASIONS

Accidental versus Spontaneous Invasions

It is interesting to compare migrant butterflies and invasive lepidopterous insects in terms of the place where they first became established (Table 3.5). Establishment of migrant butterflies refers to the cases where any species has been observed successively for ten years on the same island. Migrant butterflies from Taiwan usually establish in Japan by way of the Yaeyama Islands. On the other hand, invasive lepidopterous insects became established first in Okinawa Island, which is the center of trade and

TABLE 3.5. Comparison between migrant and invasive lepidopterous insects in terms of the place where they first became established (Y = Yaeyama Islands; O = Okinawa Island).

Migrant species (Watanabe 1998)		
Pachliopta aristolochiae interpositus (Papilionidae)	1968	Y
Artogeia canidia canidia (Pieridae)	1987	Y
Appias melania minato (Pieridae)	established for a fairly long time	Y
Appias lyncida formosana (Pieridae)	1966	Y
Catopsilia pomona pomona (Pieridae)	1974–75	O
Euploea mulciber barsine (Danaidae)	1986	Y
Cupha erymanthis erymanthis (Nymphalidae)	1973	Y
Athyma perius (Nymphalidae)	1967	Y
Megisba malaya sikkima (Lycaenidae)	1940s	Y
Hasora badra badra (Hesperiidae)	1974	Y
Suastus gremius gremius (Hesperiidae)	1973	Y
Invasive species (Kohama 1997)		
Pieris rapae curcivora (Pieridae)	1958	O
Erionota torus (Hesperiidae)	1971	O
Leucinodes orbonalis (Pyralidae)	1961–63	O
Omphisa anastomosalis (Pyralidae)	1941	Y
Crocidolomia binotalis (Pyralidae)	1987	Y
Chlumetia brevisigna (Noctuidae)	1992	Y
Parallelia palumba (Noctuidae)	1995	O
Parasa lepida (Limacodidae)	1982	O
Phthorimaea operculella (Gelechiidae)	1968	O
Statherotis discana (Tortricidae)	1990	O
Conopomorpha litchiella (Gracillariidae)	1989	O

transportation in the Nansei Islands, before extending their range outside the island in both northerly and southerly directions.

Typical examples of range expansion are illustrated by the comparison between a migrant butterfly, *Appias melania minato*, and the oriental chinch bug, *Cavelerius saccharivorus* (Fig. 3.3). *Appias melania minato*, which originated in Taiwan, had been distributed on the Yaeyama Islands for a fairly long time before expanding its range to the Miyako Islands in 1971. It expanded northward in stepwise fashion, reaching the Amami Islands in 1982.

The oriental chinch bug, a pest of sugarcane, was found in a sugarcane field on Okinawa Island in 1914. It is believed that the insect had invaded

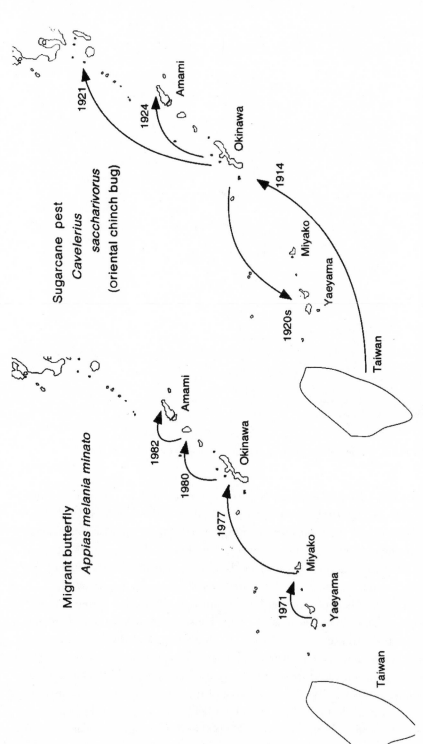

FIGURE 3.3. Range expansion differences between the migrant butterfly *Appias melania minato* and the oriental chinch bug, *Cavelerius saccharivorus*.

FIGURE 3.4. The number of exotic species becoming established in Japan.

Okinawa Island in association with sugarcane seedlings, imported from Taiwan during 1911. The insect spread all over the island by 1922. This also increased the probability of range expansion of the insect associated with the movement of sugarcane seedlings. In fact, this insect was reported on the Yaeyama and Amami Islands and Tanegashima Island during the 1920s, when seedlings were exported from Okinawa to these islands.

The number of unintentionally introduced species differs greatly among the areas and islands of Japan. Those areas with no records of the earliest occurrence of exotic insects (i.e., Shikoku Island, the Amami and the Miyako Islands) lack international air- and seaports. In other words, the exotic insects invaded these areas after they became established elsewhere in Japan (Fig. 3.4). The total land area of the southern islands accounts for 4,673 km² while that of the mainland, which includes Kyushu, Honshu, Shikoku, and Hokkaido, occupies 373,000 km². The southern islands share only 1.2 percent of the total Japanese land, but 43 percent of the exotic insects that invaded Japan after establishing in the southern islands.

TABLE 3.6. Number of exotic insect species established in Japan in relation to the time of introduction.

Taxonomic group	pre-1945	1945	1966	1986	Total
Termites	0	0	2	0	2
Cockroaches	1	1	2	0	4
Thrips	3	0	5	3	11
Aphids	3	1	5	4	13
Whiteflies, psyllids, etc.	0	1	2	5	8
Scales and mealybugs	8	1	3	7	19
Stinkbugs	2	0	1	0	3
Stored-product Coleoptera	4	15	0	2	21
Weevils	5	4	11	3	23
Ladybird beetles	1	0	0	8	9
Longhorn beetles	1	2	0	1	4
Leaf beetles	2	0	2	2	6
Stored-product Lepidoptera	0	1	0	2	3
Other lepidopteran species	2	5	8	6	21
Leafminer flies	0	0	1	3	4
Other dipteran species	4	4	4	1	13
Ants	2	0	5	0	7
Other hymenopteran species	3	0	0	0	3
Parasitoids	6	1	4	0	11
TOTAL	47	36	55	47	185

Chronological Changes in the Invasive Insects

When the species were arranged in chronological order from the earliest known years of occurrence, it was evident that the proportional representation of the orders and families has changed markedly during the last hundred years (Table 3.6).

Japan opened to the world in 1868 after a long national isolation policy. Since then, of the 185 species, 47 (approximately 25%) of them were introduced before World War II [1945]. Reflecting a rush to import nursery stock of fruit trees after the opening of Japan, scale insects and mealybugs took advantage and invaded in this first period. Their sessile habit increased the opportunity of invasion when their host plants were imported from overseas. If the host plants are raised under conditions favorable for their growth, this also favors the insects.

The second flux of invasive insects was characterized by the massive invasion of stored-product insects, particularly coleopterous insects, during the period 1945 to 1965 in association with the import of a great quantity of grain to cover food shortages. The homogenization of stored-product pests worldwide has resulted in exclusion of as many as 24 species from the checklist of plant quarantine inspection as of April 1, 1998. The third period, 1966 to 1985, was represented by the invasion of various kinds of weevils that infest upland crops, turf, vegetables, and ornamental trees. The fourth period (1986–99) is represented by the invasion of such greenhouse pests as thrips, aphids, whiteflies, scales, mealybugs, and leafminer flies, although influx of these groups began in the preceding period.

Homogenization of Pest Fauna

Kiritani et al. (1963) demonstrated that human-made protected environments such as flour mills and feed factories handling imported cereals act as reservoirs for new pests of stored products. More than 25 exotic species are considered to have invaded by this pathway.

The protected cultivation of various vegetables and flowers in green-

TABLE 3.7. Comparison of greenhouse pest fauna among Japan, Europe, and the United States (Kiritani 1999) (O = present; X = absent; # = indicates invasive pests; ? = remains to be known).

Species	Japan	Europe	USA
THRIPS			
Frankliniella occidentalis	O#	O#	O
F. intonsa	O	O	O
Thrips palmi	O#	O#	O#
T. simplex	O#	O#	O#
WHITEFLY			
Trialeurodes vaporariorum	O#	O#	O
Bemisia argentifolii	O#	O#	O
LEAFMINER			
Liriomyza trifolii	O#	O#	O
L. bryoniae	O#?	O	X
L. huidobrensis	X	O#	O
APHID			
Aphis gossypii	O	O	O
Myzus persicae	O	O	O
MITE			
Tetranychus urticae	O	O	O
Aculops lycopersici	O#	O#	O#
CURCULIONID			
Otiorhynchus sulcatus	O#	O	O#

houses provides greenhouse pests with a vacant niche or an ecological island, which was vulnerable to invasion by exotic insects. The combination of short day length and higher temperature created in greenhouses during the winter favors nondiapausing insects, most of which originate from subtropical or tropical countries (Kiritani 1999). The major greenhouse arthropod pest fauna among Japan, the United States, and Europe are compared in Table 3.7. Obviously the worldwide homogenization of the greenhouse pest fauna has taken place everywhere in the world through biological invasion.

With the increase of human-made habitats, insects originating from the tropics or the ones without diapause in their life cycle might become predominant in these habitats. In fact, most of the stored-product and greenhouse pests are tropical in origin and have no diapause.

Pathways of Invasive Insects

Major Pathways of Quarantine Pests

Since 1914 when the quarantine law was launched in Japan, it has been possible to identify the probable invasion pathway for 98 species of invasive insects. Plants or planting materials provide the pathway for more than 50 percent of invasive insects (Table 3.8). It should also be noted that invasive insects are often accidentally introduced through military activity; for example, the transportation of military commodities increases the opportunity for exotic insects attached to those commodities to hitchhike.

TABLE 3.8. Probable pathways of quarantine insects that accidentally invaded and became established in Japan during the period 1917–99.

Probable pathways	Number of species
Flight (air current, typhoon, etc.)	4
Plant materials (nursery stock, bulbs, etc.)	52
Military transportation	10
Hay (dried fodder)	9
Fruits (including potato)	8
Grain	8
Hitchhike	3
Sawn wood and packing materials	2
Cut flowers	1
Seeds	1
TOTAL	98

TABLE 3.9. Insect pests associated with different commodities at ports of entry (Yoshizawa 1996).

Commodities	Insect pests	Vectors
Nursery stock, grafts, bulbs, tubers, and seedlings	Aphids, thrips, whiteflies and phytophagous mites	Cargo
Cut flowers	Aphids, thrips, and phytophagous mites	Airfreight
Fruits and vegetables	Fruit flies, weevils, and leaf beetles	Airfreight and container ship
Grain, feed, and oil seeds	Bean weevils and khapra beetles	Cargo
Wood	Bark beetles, long-horn beetles, platypodids, etc.	Cargo

Also, the increasing importation of baled and rolled hay for livestock has facilitated colonization by curculionid weevils (Table 3.8).

U.S. military bases located in Japan were established by the bilateral agreement between Japan and the United States. The bases in Okinawa account for about 75 percent of the total area of the U.S. military bases in Japan and occupy almost half of the total area of Okinawa Island. Aircraft, warships, and military cargo were exempted from the regulation of plant quarantine before 1972, at which time the Okinawa Islands were returned to Japan. Consequently, of the 12 invasive species discovered on Okinawa Island from 1945 to 1973, half were first observed in or around the military bases (Teruya et al. 1973).

There is a more or less close relationship between the kind of commodity and the insect groups represented. The type of transportation also is related to the kind of commodity (Table 3.9)

Changes in the amount of consignments inspected during the period 1970–98 in Japan are shown in Table 3.10. Among the agricultural imports, cut flowers showed the most remarkable increase in amount (more than a 150-fold increase over the last twenty years), followed by planting materials and vegetables (Table 3.10). Rapid increases of these imported items are associated with an increasing number of exotic species, including thrips, aphids, whiteflies, scales, mealybugs, and leafminer flies that became established in Japan, particularly in the greenhouses (Table 3.7).

The two species of fruit flies, the oriental fruit fly and the melon fly,

TABLE 3.10. Changes in the amount of imported consignments* inspected during the period 1970–98 in Japan (MAFF 1999).

		1970	1975	1980	1985	1990	1994	1998
PLANT MATERIALS								
Seedlings	10^6 pieces	10	12	7	23	67	166	221
Bulbs	10^6 pieces	13	42	78	79	191	395	632
Seeds	10^3 tons	12	11	21	28	31	28	25
Cut flowers	10^6 pieces	—	9	84	122	358	808	1,376
Fruit	10^3 tons	997	1,268	1,254	1,323	1,487	1,756	1,528
Vegetables	10^6 tons	35	72	256	278	470	975	1,203
Grain	10^6 tons	16	20	26	28	28	31	28
Legumes	10^6 tons	4	4	5	5	5	5	5
Oil seeds & spices	10^6 tons	2	2	3	5	8	8	9
Wood	10^7 m³	4	4	4	3	3	2	2

*Total of ship cargoes, airfreight, containers, luggage, and postal packages.

TABLE 3.11. Number of interceptions of fruit fly–infested consignments at the ports of Japan during the period 1993–98. The consignments inspected consisted of hand luggage (93.3%), postal packages (6.0%), and cargoes (0.7%). All consignments contained fresh fruits or vegetables that are prohibited to import.

Species	1993	1994	1995	1996	1997	1998	Total
Bactrocera correcta	0	0	27	29	64	52	172
B. cucurbitae	19	14	10	11	27	12	93
B. dorsalis species complex	218	161	189	269	277	167	1,281
B. latifrons	0	11	76	146	155	115	503
Ceratitis capitata	6	4	7	8	5	5	35
Other fruit fly species	7	5	9	29	18	23	91
TOTAL	250	195	318	492	546	374	2,175

were eradicated in Japan in 1986 and 1993, respectively, at a cost of U.S.$250 million. Therefore, importation of fresh fruits and vegetables grown in the fruit fly–infested countries is prohibited in Japan, and interception of fruit flies at the ports of arrival is of paramount importance among the quarantine pests (Table 3.3). Import of produce from the designated countries on a commercial basis rarely occurs unless it is treated according to the protocol agreed to by Japan and the exporting countries. However, the hand luggage of travelers entering Japan frequently contains prohibited fruit from other countries. Surveys conducted over the last six years, 1993 to 1998, demonstrated as many as 2,100 cases in which fruit was infested by various kinds of fruit flies (Table 3.11). Of these cases, 93.3 percent were hand luggage, and the remaining 6.0 and 0.7 percent were postal packages and cargoes, respectively. This figure suggests that passengers carrying consignments infested by fruit flies arrive almost every day in Japan.

Other Pathways

In order to prevent the dissemination into Japan of plant pests, all means of conveyance, such as stores, luggage, mail, plants, plant products, soil, and garbage, are subject to inspection at the port of first arrival.

Lately, the rapidly developing international trade in used tires has occasioned widespread establishment of the Asian tiger mosquito, *Aedes alvopictus*, in several countries including the United States, Albania, Italy, Nigeria, Australia, New Zealand, and South Africa. The pathway for this mosquito is wet used tires exported for recycling, primarily from Japan. In New Zealand, all imported used tires became subject to inspection in 1994 (Laird et al. 1994).

The concentration of insects in crevices or corner spaces of containers makes them difficult to detect by inspection. It is well known that matured lepidopterous larvae wander about in search of a hidden pupation site, and that some species belonging to Coleoptera and Hemiptera aggregate as adults at a particular site for hibernation or aestivation in diapause state.

Parasitoids and predators are not subject to inspection, although importers are requested to obtain permission from the plant quarantine authority. Likewise, such sanitary insects as mosquitoes are free from regulation. Illegal introduction of some kinds of beetles and butterflies for breeding is blamed on some insect collectors. For example, a papilionid butterfly, *Sericinus montela*, in Honshu is believed to have been released by someone around the 1970s.

ESTABLISHMENT AND LATENCY PERIOD

Establishment Rate

The rate of establishment for exotic insects is difficult to determine, because outdoor experiments to measure establishment success are not allowed for phytophagous insect species. Concerning introduced arthropod natural enemies, Hall and Ehler (1979) and Ehler and Hall (1982) conducted a comprehensive analysis using 2,295 cases of biological control. An overall mean rate of establishment for all the examples was 34 percent,

TABLE 3.12. Number of cerambycid species discovered by inspection and those recorded and becoming established in Japan (Makihara 1986).*

No. of species discovered by inspection (a)	No. of species reported occurrence within Japan (b)	No. of species become established in Japan (c)
250 more	Ca. 14 b/a = 0.056	7 c/a = 0.028 c/b = 0.500

*The total number of cerambycid species in Japan is about 720.

TABLE 3.13. Establishing rate of migratory butterflies in the Yaeyama Islands (modified from Watanabe 1998).*

NUMBER OF BUTTERFLY SPECIES			
Total number of migrant species (a)	Migrants from south (b)	Migrants from north (c)	Migrants that established (d)
71 a = b + c + d	55	6	10 d/a = 0.14

*Establishment refers to the case where any migrant species has been observed successively for ten years on the same island. Total number of native species in the Yaeyama Islands is 66, including 2 species whose status remains unknown.

which varied from 43 percent for natural enemies of Homoptera to 13 percent of predacious mites (Hall and Ehler 1979).

In the case of cerambycid beetles, the figure was as low as 3 percent, and for the migrant butterflies it was 14 percent, both of which are significantly low compared with the natural enemies (Tables 3.12 and 3.13). Global warming may have facilitated the recent establishment of an increasing number of migrant butterflies in the Nansei Islands.

Latency Period

The period between the time of introduction or initial colonization and the initial detection may be quite long. For example, the medfly, *Ceratitis capitata*, was introduced into California more than fifty years before being detected in 1975 (anonymous 1993 cited in Carey 1996). This is because all incipient insect invasions consist of a small number of tiny organisms with a restricted distribution. This inability to detect invasions at an early stage makes it difficult frequently to determine the time of initial invasion.

A classic example for the latency period is the case of the gypsy moth, *Lymantria dispar.* The gypsy moth was imported into Medford, Massachusetts, by a French scientist to develop a resistant strain to pebrine disease, which was destroying the French silk industry. In 1869, some of these insects accidentally escaped from his laboratory. The gypsy moth remained practically hidden for twenty years, during which time it was breeding inconspicuously but efficiently in the lands around Medford until 1889, when an outbreak of caterpillars suddenly occurred in the town (Dethier 1976).

We have examined the latency periods of various invasive insects by reviewing the reports published in a monthly journal written in Japanese, *Shyokubutsu-Boeki (Plant Protection)*. All of the descriptions of the time of invasion were based on the circumstantial evidence, often stating that invasions occurred "several years ago." Sometimes, the latency period was measured as the interval between the first and the second occurrences, when the pests were first detected and believed to be completely eradicated by the authorities. When the suggested latency periods were more than ten years, it was difficult to preclude the possibility of multiple invasions during that time period.

The latency period ranged from 0.5 to 80.0 years, giving a mean of 11.8 years. Out of 35 cases, 25 had a latency period lower than 10 years, showing a mean of 3.8±2.4 years (Fig. 3.5). Kiritani (1998) calculated regression lines between the cumulative number of infected prefectures as a logarithm function (log Y) of time in years (X) for the first 5 years of initial detection. The mean number of prefectures occupied was calculated separately for greenhouse and outdoor pests (Fig. 3.6). It was found that the

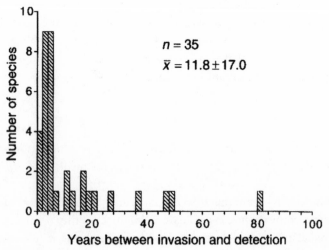

FIGURE 3.5. The frequency of the number of exotic species in relation to the period (years) between the time of invasion and the initial detection.

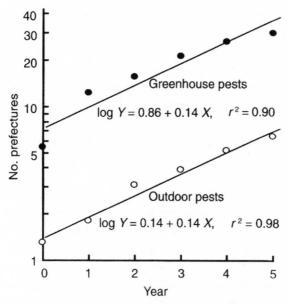

FIGURE 3.6. The mean rate of increase in the number of prefectures affected by invasive insect pests during the first five years after detection.

extension (spread) rate was the same for both greenhouse and outdoor pests, taking a value of 0.14 for slope. The difference was that a newly invaded greenhouse pest is detected after it has spread over 7.2 prefectures on average, but only 1.4 prefectures in the case of outdoor pests.

The number of years that elapsed before an invasive insect extends its range to a neighbor prefecture or island was also examined. Some species ranges are still confined within the same prefecture or island at the time of literature survey in 1999. The figures therefore give somewhat conservative estimates (Table 3.14). The difference in the duration between the prefecture in the mainland (Honshu) and the isolated island is due to geographical situation. Obviously, the sea is a great physical barrier in preventing the invasive insect pests from range expansion. It takes twice as

TABLE 3.14. The number of years elapsed before extending range out of the prefecture or the island where invasive insects were first detected.

	No. of cases	Mean	SE
Prefecture	8	4.25	2.25
Island	25	9.52	9.71

long as that required in a diffusion-type expansion on large islands. This result is consistent with the mathematical model of Shigesada et al. (1986), predicting that the velocity of the traveling wave is smaller in a heterogeneous environment including unfavorable habitats.

Our analyses suggested that invasive insects would have a latency period of at least four to ten years after initial colonization before they were detected.

PREVENTION OF INVASION

The possibility of accidental invasion of insect pests is increasing with the increasing amounts of imported commodities. International trade further accelerated after the establishment of the World Trade Organization (WTO) that superseded the General Agreement on Tariffs and Trade (GATT) in 1995. The transparency of phytosanitary requirements is one of the principal agendas under the WTO. For example, the validity of quarantine procedure of apples exported from the United States to Japan is under negotiation. To find an optimal solution to balance the prevention of biological invasion and the acceleration of free trade, we should determine a common index to evaluate the risk of invasion.

One possible index is the Expected Time required for Invasion (abbreviated by ETI). The successful invasion by a pest species is related to many factors, such as the climatic conditions at the destination, the existence of natural enemies, the availability of food resources, and the possibility of pest arrival. We can divide the probability of invasion into two components: (1) the probability that a reproductive pest passes the port, and (2) the probability that a reproductive pest establishes its population after passing the port. Let us use the following notation:

$k =$ number of consignments imported during a year;

$p =$ probability that a reproductive female establishes her filial population in the importing country when she passed the port;

$H_i =$ number of reproductive females passing the port in the ith consignment ($i = 1, 2, ..., k$);

$r_i =$ probability that establishment occurs via the ith consignment;

$R =$ probability that establishment occurs during a year; and

$T =$ years elapsed before the pest becomes established.

The probability that one or more females establish their populations when females passed the port is given by $1 - (1 - p)^h$. Hence, the probability that one or more females establish their populations via the ith consignment is given by:

$$r_i = \sum_{h=0}^{\infty} (1 - (1-p)^h) \Pr(H_i = h). \quad (1)$$

If p is sufficiently small, the above quantity is approximately given by:

$$r_i \approx \sum_{h=0}^{\infty} ph \Pr(H_i = h) = pE(H_i), \quad (2)$$

where E indicates the expectation. R is given by:

$$R = 1 - \prod_{i=1}^{k} (1 - r_i). \quad (3)$$

If r_i is sufficiently small, we obtain:

$$R \approx \sum_{i=1}^{k} r_i. \quad (4)$$

If the quantity of H_i is sufficiently small, we obtain the approximation:

$$E(H_i) \approx \Pr(H_i \geq 1). \quad (5)$$

From Equations (2), (4), and (5), we obtain:

$$R \approx p \sum_{i=1}^{k} \Pr(H_i \geq 1). \quad (6)$$

The probability distribution of T is given by a geometric distribution:

$$\Pr(T = t) = R(1 - R)^{t-1}. \quad (7)$$

Hence, ETI is given by:

$$\text{ETI} = E(T) = 1 / R. \quad (8)$$

From Equations (6) and (8), we obtain the following approximation if $\Pr(H_i \geq 1)$ is the same for all consignments:

$$\text{ETI} \approx \frac{1}{pk\Pr(H_i \geq 1)}. \quad (9)$$

Yamamura and Katsumata (1999) provided formulae to evaluate the probability of passing the port, $\Pr(H_i \geq 1)$, considering the effect of two prevention practices: one is disinfestation treatment before export and the other is the inspection of consignments by sampling after the disinfestation treatment. Figure 3.7 shows the example of the estimated effect of

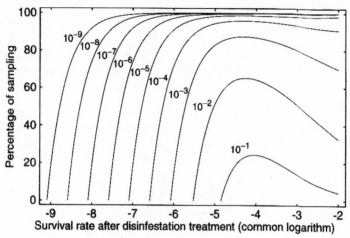

FIGURE 3.7. Contour of the probability of invasion per consignment, $\Pr(H_i \geq 1)$, for the Mexican fruit fly expressed as the function of the intensity of sampling inspection and disinfestation treatment (Yamamura and Katsumata 1999).

these practices for the Mexican fruit fly *Anastrepha ludens* (Loew) on citrus fruits. From Equation (9), we obtain the permissible limit of $\Pr(H_i \geq 1)$, which is denoted by $\Pr^*(H_i \geq 1)$, for a given set of p, k, and ETI:

$$\Pr^*(H_i \geq 1) \approx \frac{1}{pk\text{ETI}}. \qquad (10)$$

If we assume that $p = 10^{-5}$, $k = 10^4$, and ETI = 1,000, for example, we obtain $\Pr^*(H_i \geq 1) \approx 10^{-2}$. Then, we obtain the combination of the disinfestation treatment and sampling inspection to achieve the ETI by using the contour of 10^{-2} in Figure 3.7. If we want to achieve the probability without performing sampling inspection, for example, a disinfestation treatment with a survival rate of about $10^{-5.5}$ is required. If we inspect 40 percent of fruits after the disinfestation treatment, we can achieve ETI = 1,000 by using the disinfestation treatment with a survival rate of about $10^{-2.5}$. It will be important to note that some of the curves in Figure 3.7 are convex shaped, indicating the existence of an antagonistic interaction between the disinfestation treatment and sampling inspection; the efficiency of sampling inspection decreases with increasing intensity of disinfestation treatment.

CONCLUSION

Reflecting the large number of species involved in the class Insecta, the pathways that can be taken by insects for invasion are quite diverse. Ocean currents as a vector of invasive insects can be discounted, unless we consider the issue in a larger biogeographical scale. Invasion by long-distance flight (e.g., transoceanic dispersal) rarely takes place in Coleoptera. Most species belonging to Hymenoptera are introduced for biological control of phytophagous pests and for use as pollinators. Invasions by insects, therefore, are mostly represented by accidental invasions and a few species belonging to Lepidoptera, Hemiptera, and Diptera, which invade Japan by flight.

The kind of commodities imported affects greatly the fauna of invasive insects. Recent increases in the amount of imported plant materials, vegetables, fruits, and cut flowers, in association with the expansion of structured cultivation of various plants, have resulted in the establishment of many kinds of greenhouse pests in Japan (e.g., thrips, aphids, whiteflies, scales, mealybugs, and leafminer flies). They are generally small in size, feed by sucking, lack diapause, often reproduce parthenogenetically, and are reported to be highly tolerant of pesticides. This latter trait may be selected for during the latency period in the greenhouses. The increase of human-made habitats, such as flour mills, feed factories, and greenhouses, provided new habitats for insects originated in the tropics or insects without diapause in their life cycle. In fact, most of the stored-product and greenhouse pests are of tropical origin and have no diapause. In view of the increasing invasions of species of greenhouse pests, invasions by the leafminer fly *Liriomyza huidobrensis* and the spiraling whitefly *Aleurodicus dispersus*, which originate in tropical and subtropical America, are to be carefully guarded against.

Human activities play an important role when exotic insects invade accidentally into new areas across international borders. In contrast, biological characteristics of insects, such as the rate of population increase and flying ability, play an important role when they expand their distribution range within a country. Once an exotic insect species has invaded a country, it is very difficult to control its rate of expansion. Therefore, interception at the port of entry is most important to prevent the damage to ecosystems caused by invasive species.

There are two major pathways for accidental insect invasion: hand luggage and cargo. The total quantity of hand luggage is increasing year by year. Actually, at least one consignment infested by fruit flies arrives every day in Japan in passenger hand luggage. Since plant materials are subject to quarantine inspection by self-reporting, it is urgently necessary that passengers be well informed about the importance of the plant quarantine regulation. The possibility of invasion via cargoes is also increasing with the increasing pressure of free trade. In order to find a compromise between prevention of biological invasion and acceleration of free trade, we need to find an optimal solution in terms of risk assessment. International standard(s) are required to determine the optimal solution. ETI may be one of the possible candidates of such standards.

ACKNOWLEDGMENTS

We acknowledge Dr. D. A. Andow, University of Minnesota, for his comments on the manuscript. We are also grateful to Drs. N. Morimoto, M. Masumoto, E. Yano, S. Moriya, and K. Tanaka for their assistance in preparing the manuscript.

REFERENCES

Azuma, S. 1986. Colonizing insects in Okinawa—Invaders from the south. In *Insects in Japan—Ecology of invasion and disturbance*, K. Kiritani, ed., pp. 115–121. Tokyo: Tokaidaigaku Publishing Company. (In Japanese)

Carey, J. R. 1996. The incipient Mediterranean fruit fly population in California: Implication for invasion biology. *Ecology* 77: 1690–1697.

Dethier, V. G. 1976. *Man's plague?—Insects and agriculture*. Princeton, N.J.: The Darwin Press, Inc.

Ehler, L. E. and R. W. Hall. 1982. Evidence for competitive exclusion of introduced natural enemies in biological control. *Environmental Entomology* 11: 1–3.

FAO. 1996. *Glossary of phytosanitary terms*. Secretariat of the international plant protection convention, FAO, ISPN no. 5.

Hall, R. W. and L. E. Ehler. 1979. Rate of establishment of natural enemies in classical biological control. *Bulletin of the Ecological Society of America* 26: 280–282.

Kiritani, K. 1983. Colonizing insects—Range extension by pioneers. *Insectarium* 20: 284–291. (In Japanese)

Kiritani, K. 1984. Colonizing insects—What prevented the brown planthopper from establishing itself in Japan? *Insectarium* 21: 136–143. (In Japanese)

Kiritani, K. 1998. Exotic insects in Japan. *Entomological Science* 1: 291–298.

Kiritani, K. 1999. Formation of exotic insect fauna in Japan. In *Biological invasions of ecosystem by pests and beneficial organisms*, E. Yano, M. Matsuo, M. Shiyomi, and D. A. Andow, eds., pp. 49–65. Tukuba: National Institute of Agro-Environmental Sciences.

Kiritani, K., ed. 2002. Alien insects in Japan. In *Handbook of alien species in Japan*, Ecological Society of Japan, pp. 124–125. Tokyo: Chijinshokan Publishing Company.

Kiritani, K. and N. Morimoto. 1993. Exotic insects in Japan. *Insectarium* 3: 120–129. (In Japanese)

Kiritani, K., T. Muramatsu, and S. Yoshihara. 1963. Characteristics of mills in faunal composition of stored product pests: Their role as a reservoir of new imported pests. *Japanese Journal of Applied Entomology and Zoology* 7: 49–58.

Kohama, T. 1997. Colonized insects. In *Naturalized animals in Okinawa,* K. Takehara, M. Toyama, T. Kohama, Y. Kochi, M. Chinen, and Y. Higa, eds., pp. 136–185. Okinawa Publishing Company. (In Japanese)

Laird, M., L. Calder, R. C. Thornton, R. Syme, P. W. Hoder, and M. Mogi. 1994. Japanese *Aedes albopictus* among four mosquito species reaching New Zealand in used tires. *Journal of the American Mosquito Control Association* 10: 14–23.

Lindroth, C. H. 1957. *The faunal connections between Europe and North America.* Stockholm: Almqvist & Wiksell.

MAFF. 1999. *Plant quarantine statistics.* Japan: Plant Protection Station, MAFF. No. 65, p. 380.

Makihara, H. 1986. Cerambycid beetle—range expansion and geographical variation. In *Insects in Japan—Ecology of invasion and disturbance,* K. Kiritani, ed., pp. 96–106. Tokyo: Tokaidaigaku Publishing Company.

Mooney, H. A. 1996. *The SCOPE initiatives: The background and plans for a global strategy on invasive species.* Conclusions and recommendations from the UN/Norway Conference on Alien Species, Trondheim, Norway, 1–5 July 1996.

Morimoto, N. and K. Kiritani. 1995. Fauna of exotic insects in Japan. *Bulletin of National Institute of Agro-Environmental Sciences* 12: 87–120.

Sailer, R. I. 1983. In *Exotic plant pests and North American agriculture,* C. L. Wilson and C. L. Graham, eds., pp. 15–38. New York: Academic Press.

Shigesada, N., K. Kawasaki, and E. Teramoto. 1986. Traveling periodic waves in heterogeneous environments. *Theoretical Population Biology* 30: 143–160.

Teruya, T., Y. Araki, and M. Osada. 1973. *Erionota torus,* a new insect pest of banana plant. *Shokubutu Boeki* 27: 17–19.

Watanabe, K. 1998. Fauna of butterfly in the Yaeyama Islands. *Kontyuto Shizen* 33(12): 8–14. (In Japanese)

Waterhouse, D. F. 1991. Biological control: Mutual advantages of interaction between Australia and the Oceanic Pacific. *Micronesica Supplemental* 3: 83–92.

Yamamura, K. and H. Katsumata. 1999. Estimation of the probability of insect pest introduction through imported commodities. *Researches on Population Ecology* 41: 275–282.

Yoshimatsu, S. 1991. Lepidopterous insects captured on East China Sea from 1981 to 1987. *Japanese Journal of Entomology* 59: 811–820. (In Japanese with English summary)

Yoshizawa, O. 1996. Plant quarantine in Japan. In *Current topics on newly occurred agricultural pests in Japan,* Zenkoku Nôson Kyoiku Kyokai, ed., pp. 191–210. Takeda Plant Protection Ser. 9. (In Japanese)

Chapter 4

Invasion Pathways for Terrestrial Vertebrates

Fred Kraus

Terrestrial vertebrates have been introduced to areas outside their native ranges since at least the Stone Age (Lever 1985), beginning the process of faunal intermixing that exploded in the twentieth century. These vertebrates have provided many of the best-known examples of ecological loss caused by alien species introduction (Greenway 1967, Honegger 1981, Morgan and Woods 1986, Ebenhard 1988, Case and Bolger 1991, Henderson 1992), both because the organisms themselves are often large and conspicuous and because their ecological effects are frequently obvious, even to the unecolate (cf Hardin 1993: 15–16).

Despite this long history of vertebrate introduction and a relatively extensive record of ecological damage compared to other organisms, quantitative analyses of the primary pathways by which these organisms are being moved appear lacking. This is due, in part, to the familiarity of many of these animals, giving the impression that the underlying dynamics of their invasions, and the human role in promoting them, are well understood. Although this may be true for many of those familiar with the biology of these particular organisms, it is not so for a broad spectrum of nonspecialists (e.g., policymakers, managers, other scientists, and the public). Hence, quantification of the patterns and underlying causes of vertebrate introductions serves to inform the general public and other interested parties, and provides both the knowledge and the catalyst for effective management and policy actions.

In this chapter, I provide an overview of pathways and trends in terrestrial vertebrate introductions, focusing especially on amphibians and reptiles. By "pathways" I mean those human cultural practices and attendant institutions that transport vertebrates outside of their native ranges, whether intentionally or not. This is the first detailed analysis of invasion patterns for amphibians and reptiles. In addition, I provide a more limited pathway analysis of previously compiled databases of bird (Long 1981, Lever 1987) and mammal (Lever 1985) introductions, allowing some comparisons across these vertebrate groups.

MATERIALS AND METHODS

Reptiles and Amphibians

Data were taken from the literature to analyze global patterns of introductions of amphibians, reptiles, birds, and mammals. Data for amphibians and reptiles were combined for analysis and are here referred to as the "herp" (short for *herptile*, a term referring to amphibians and reptiles combined) data. These data are scattered throughout a large, and frequently obscure, literature and have not previously been summarized. Hence, to analyze patterns for these taxa, I constructed a database for as many taxa as I could obtain. This database is the subject of continual updating and addition with the view toward eventual publication. The database included 577 records, where a single record (hereafter referred to as an "introduction") consisted of an instance of transportation of a particular species to a particular location, typically a political entity (e.g., country or state). Introduction, as used here, does not imply that the transportation event resulted in successful establishment of the species in the introduced location. I subsequently distinguished such introduction events as "successful" versus "unsuccessful" introductions.

Data collected on introductions included species identity, locality to which introduced, whether the introduction has led to a current established population, date of first introduction (sometimes known precisely; more often approximated by date of first report when it clearly followed a recent introduction event), reason for introduction, number of independent introductions for the same species to the same locality, and references. Not all of these data are available for many species, but some are available for most. Consequently, sample sizes for the analyses presented below vary depending on the category of information utilized for each analysis. Although the reason for introduction is frequently obvious for a particular entry, I left that information cell blank unless the source (author) of the introduction record explicitly mentioned it. For example, the blind snake,

Ramphotyphlops braminus, was introduced around the world in nursery materials, and many authors have pointed to this introduction pathway, and to no other, in explaining its recent arrival in their particular geographic area. Nonetheless, for several areas in the introduced range of the species, the introduction pathway was not explicitly identified; thus, I did not ascribe a pathway for those areas. Once a successful introduction has been made to a political entity, I have not tracked information on its subsequent spread to new localities therein. Introductions presumed to have occurred prior to 1800 were not included in the herp database, except for seven species introduced prehistorically to Hawaii. This exception was made to provide a complete record for that locality when contrasting patterns and trends among different regions of the United States.

Birds and Mammals

The same categories of information discussed above were taken for birds from the excellent summary by Long (1981). For mammals, data on species identity, locality of introduction, date of earliest successful introduction, and reason for introduction were taken from the sample of taxa provided in Table 26 of Lever (1985). Data for both birds and mammals are not as current as those for herps; and in the case of the mammals, certain important categories of information (information on unsuccessful introductions and detailed information for many successfully introduced species) were not available in Lever (1985). Consequently, analyses for birds and mammals are not as detailed as those for herps, although they provide a useful, broad outline of pathways and trends of introduction for these taxa.

Analyses

I created cumulative growth curves by plotting the cumulative number of introductions over time, using only the earliest instance of an introduction for those localities for which multiple introductions are known. This is because, in most cases, exact dates of introductions are unknown or because reports referred to an ongoing problem (e.g., the decades-long introduction of snakes and frogs in shipments of bananas). For the herp data, the large majority of multiple introductions (48 of 56 total) occurred within the United States (e.g., King and Krakauer 1966, Bury and Luckenbach 1976); such instances were more widely distributed in the bird dataset (Long 1981). Use of only the earliest introduction date in the analyses will tend to underestimate total numbers of introductions and depress estimates for introduction rates in later time periods.

For all taxa, instances in which an alien species first appeared in a political entity due to population growth and range expansion from a neighboring political entity to which introduced were not counted as introductions. Reestablishments of native species to portions of their former ranges for conservation purposes were also not included in the databases.

For the purposes of geographic analysis of pathway importance and trends, Hawaii was included in the geographic category "United States" instead of the category "Pacific," as was southern Florida in lieu of placing it in the category "Caribbean." The category "Caribbean" included the adjacent coastal areas of South and Central America as well as the Caribbean islands themselves.

Evaluation of Methods

Before discussing the results of the analyses, it is important to bear in mind several caveats that pertain to the data from which they are derived. First, it should be clear that only a fraction of all recorded herp introductions have been sampled in compiling the herp database. The most obvious bias in this sample is geographic: far more records have been taken from the United States, Caribbean region, and Pacific islands than for the remainder of the world. This is a reflection of library bias in Hawaii, my primary source of data, and makes pathway analyses for other regions unreliable at this point.

Another caveat pertains to the pathways themselves. By their very nature, introductions from cargo shipping (including the nursery trade) and from the pet trade (and other "aesthetic" pathways) are not likely to come to the attention of professional herpetologists or to be deemed scientifically important by them unless resulting in establishment. Hence, these events are liable to be grossly underreported and, therefore, grossly undersampled in my analyses. As an illustration of this problem, the numbers of literature reports for Hawaiian introductions for the past 1,600 years (42 introductions/42 species) are considerably less than introductions documented by the Hawaii Department of Agriculture for the period 1991–98 (278 introductions/100 species). These data clearly demonstrate that literature reports of Hawaiian herp introductions represent only a fraction of actual activity, and similar patterns no doubt exist elsewhere. Consequently, conclusions indicating the degree to which the cargo and pet pathways contribute to global introduction of herps will be significant underestimates. This bias should not apply to deliberate pathways of introduction (e.g., biocontrol, human food use) to the same extent, because of the greater involvement of professionals with a vested interest in documenting introduction outcomes.

The primary caveat applying to the bird data is that these data are twenty years out of date. Hence, suggestive rate changes noted for the end of the sampled period need to be verified by reference to data published since 1980.

The quality of the mammal data is far more limited than that for herps or birds. Due to time constraints in producing this paper, my sample of mammal data was limited to a summary table of some of the better-documented established introductions presented in Lever (1985). As noted above, this table lacked information on introductions that did not lead to establishment, making assessment of relative pathway success impossible. It also lacked records for a huge number of established introductions. To cite just one consequence, it is impossible to perform an adequate analysis (such as one can for birds) of geographic bias in introduction numbers or rate, because the vast number of oceanic islands to which rats, mice, cats, and other mammals have been introduced is not presented by Lever (1985). One final caveat is that mammal introductions extend so far back into antiquity that many of them, and their pathways, cannot be known with a fair degree of certainty. All of these caveats make conclusions from the mammal data set the most tentative, and a more rigorous and detailed pathway analysis for this vertebrate class would be desirable.

Results

Herps

Introductions of reptiles and amphibians have risen approximately at an exponential rate since at least 1860, doubling in numbers every forty years, and appear to exhibit a rate decrease in the past decade (Fig. 4.1). Most introductions have involved lizards and frogs (71% of 577 total introductions), snake and turtle introductions have been common (26%), and salamander and crocodilian introductions have been rare (3%). The numbers of established invasions among taxa are similar to numbers of introductions, with frogs and lizards having the greatest success at establishment, followed by turtles, snakes, salamanders, and crocodilians (Table 4.1)

Ten pathway categories were identified as contributing to the flow of alien herps (Table 4.2). Herp introductions are dominated in importance by two pathways, accidental transport in cargo shipments and intentional

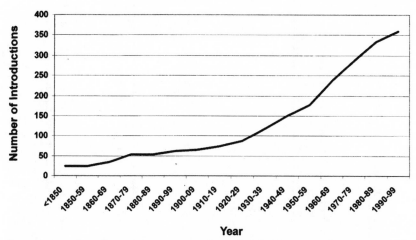

FIGURE 4.1. Cumulative increase in numbers of global herp introductions since 1850.

movement via the pet trade, which together account for 63 percent of all introduction events for which a pathway could be identified. Four other categories also contribute significantly to the introduction of herp species: intentional introductions for biocontrol (8%), human food consumption (9%), or aesthetic purposes (7%), and accidental introductions associated with the nursery trade (7%). The remaining categories associated with bait use, duck food, scientific research, and zoo releases are relatively minor, contributing only 6 percent of introductions in aggregate.

Several pathway categories are separately tracked manifestations of the same underlying problem. The nursery trade is just one component of cargo shipping, but is of sufficient importance in its own right to warrant

TABLE 4.1. Taxonomic differences in introduction numbers and success rate among reptile and amphibian orders for the database under discussion.

Taxon	Number of Introductions	Number Established	Percentage of Introductions Established
Frogs	177	134	76
Salamanders	9	3	33
Lizards	231	153	66
Snakes	81	36	44
Turtles	70	39	56
Crocodilians	9	3	33

TABLE 4.2. Pathways of amphibian and reptile introductions, with relative contributions of each herp order to the pathway.

Pathway	All Introductions		Established Introductions Only		Percent Established	Percent Contribution by Taxon to Each Pathway					
	Number of events	Number of species	Number of events	Number of species		Frogs	Salamanders	Lizards	Snakes	Turtles	Crocodilians
Bait	3	2	2	1	67	0	100	0	0	0	0
Biocontrol	24	12	18	10	75	92	0	8	0	0	0
Cargo	85	57	46	31	54	12	0	51	37	0	0
Duck food	2	2	1	1	50	100	0	0	0	0	0
Food	27	15	21	10	78	65	0	0	0	35	0
Nursery	20	11	17	10	85	45	0	35	20	0	0
Pet	102	72	48	36	47	22	2.5	36.5	8.5	24.5	6
Research	6	5	3	2	50	67	0	33	0	0	0
Aesthetic	20	19	13	12	65	53	0	47	0	0	0
Zoo	7	7	4	4	57	14	0	86	0	0	0

Note: Category definitions: Bait = intentional release of animals used for angling; Biocontrol = intentional release of animals to provide population control of insects or other animals; Cargo = accidental shipment of animals hiding in cargo or cargo containers; Duck food = intentional release of animals to provide food source for domestic ducks; Food = intentional release or accidental escape of animals farmed for human consumption; Nursery = accidental shipment of animals hiding in horticultural materials; Pet = intentional release or accidental escape of animals kept as pets or being transshipped in the pet trade; Research = accidental escape or deliberate release of animals used for biomedical research, or deliberate release of animals for experimental ecology experiments; Aesthetic = deliberate release of animals with a view to establishing local populations, but not necessarily connected to the pet trade per se; Zoo = intentional release of animals by zoo personnel or accidental escape from a zoo facility.

separate consideration. And the "pet," "aesthetic," and "zoo" categories are all separate reflections of an underlying aesthetic motivation leading to the rampant release and/or escape of alien herps into new environments. Clearly, combining these separately tracked categories only reinforces the importance that cargo shipping and the aesthetic nexus have in leading to herp introductions. Further discussion of pathway patterns and trends focuses on only the six categories providing greatest input to the problem. The four minor categories of "bait," "duck food," "research," and "zoo" are not considered further.

The relative importance of the major pathways of herp introduction has varied temporally (Fig. 4.2), with cargo being the most important pathway until the 1960s, except for the decade of the 1930s, when the large number of biocontrol releases made that pathway of comparable magnitude. Beginning in the 1930s, the pet trade pathway has shown dramatic growth in numbers of introductions, far surpassing the role of cargo transport by the 1960s. That growth appears to have leveled off in the past decade, so that the combination of cargo and nursery materials transport now accounts for an equivalent number of introductions to that of the pet trade alone (Fig. 4.2). The same pattern effectively holds if only introductions leading to successful establishment of populations are considered, except that introductions via the pet trade do not outstrip the importance of the cargo pathway to nearly the same extent.

The relative importance of the cargo and pet trade pathways in leading to established populations of alien herps is largely a result of the vast numbers of introductions generated from those sources; the relative rate

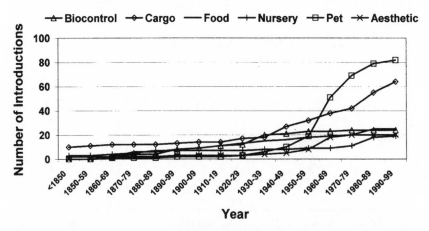

FIGURE 4.2. Cumulative numbers of herp introductions since 1850 distinguished by pathway.

of successful establishment for those pathways is the lowest of the pathways examined (Table 4.2). When tracked temporally for each pathway, the proportion of introductions leading to establishment varied dramatically since 1850, largely as a result of small sample sizes prior to the 1930s. Two notable patterns are apparent, however. The proportion of successfully established herps introduced via the pet trade rose beginning in the 1910s, peaked in the 1960s and 1970s, and has declined dramatically since (Kraus, unpublished data). Conversely, the proportion of successfully established herps introduced via cargo transport has risen steeply since the 1970s (Kraus, unpublished data). This appears to result from a sharp increase in rates of introduction via cargo in the past two decades as opposed to increased colonization efficiency (Table 4.3). In contrast, the decline in establishments via the pet trade does not result from a decrease in introduction rate and appears to reflect a real decrease in colonization efficiency (Table 4.3).

The number of herp introductions shows considerable geographic variation, with most entries in the database being from the United States (n = 250 introductions/121 established), Caribbean region (n = 123/104), and Pacific islands (n = 117/77). For the remainder of the globe, information on only 84 introductions and 66 establishments has been available. For each of these geographic regions, the trend through much of the twentieth century was toward increasing rates of herp introductions, although recently these rates may have declined for most regions (Fig. 4.3).

The importance of each pathway in generating herp introductions varies geographically (Table 4.4). For the United States, herp introductions have been dominated by the pet trade (53.8%), with a lesser but considerable contribution from cargo shipping (23.4%). Conversely, introductions

TABLE 4.3. Recent changes in global herp pathway importance.

Pathway	Percent Contribution to Global Total		Introductions/yr.	
	Since 1850	Since 1980	Since 1850	Since 1980
Biocontrol	10.8	0	0.17	0
Cargo	23.3	45.8	0.36	1.10
Food	9.9	10.4	0.15	0.25
Nursery	6.9	16.7	0.11	0.40
Pet	40.5	27.1	0.63	0.65
Aesthetic	8.6	0	0.13	0

Note: Table compares overall importance and rate of introductions by pathway for all introductions since 1850 and contrasts that overall average against the pattern seen for introductions since 1980.

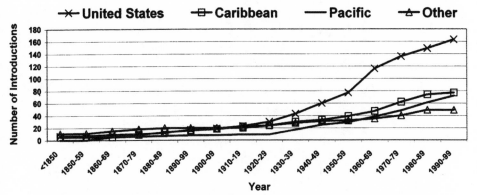

FIGURE 4.3. Cumulative numbers of herp introductions since 1850 distinguished by recipient region.

of herps in the Caribbean region have primarily been due to the nursery trade (27.5%) and to cargo shipping (30.0%), with lesser contributions from other pathways. In the Pacific, most herp introductions have resulted from cargo shipping (45.6%), but introductions for biocontrol purposes (16.2%) and through the pet trade (20.6%) have also been important. Sufficient records for the remainder of the globe are not available to provide a compelling portrait of pathway importance at this time.

Even within a major geographic region, pathway importance and trend can vary on a finer scale. For example, if one examines those states within the United States having the greatest numbers of herp introductions, striking differences in pathway patterns and importance emerge. Most herp introductions in the United States have been to Hawaii (n = 42 introductions/27 established), Florida (n = 67/39), and California (n = 43/11). For the remaining forty-seven states there is a total of 98 introductions and 44 establishments. The rate of herp introductions to California appears to have stabilized in the past few decades, but introductions elsewhere in the United States continue to grow, especially in Florida (Fig. 4.4).

Herp introductions to Hawaii have largely been due to cargo shipping, but in recent decades deliberate release of pets has become the primary route of new species additions (Table 4.5). In contrast, Florida has long been dominated by herp introductions via the pet trade, although cargo shipping has also accounted for numerous introductions, especially before 1940. The pet trade has also dominated introductions to California with secondary importance given to introductions for human food consumption, mostly made during the late 1800s and early 1900s. Introductions to

TABLE 4.4. Geographic variation in numbers of herp introductions and relative importance of pathways for three major regions, and the remainder, of the globe. Numbers represent percent contributions of a particular pathway to each region's introduction load.

Percent Contribution to Each Region of World

Pathway	United States		Caribbean		Pacific		Other	
	All Introductions	Established Introductions Only	All Introductions	Established Introductions Only	All Introductions	Established Introductions Only	All Introductions	Established Introductions Only
Biocontrol	5.5	5.3	7.5	9.1	16.2	27.0	8.0	5.6
Cargo	23.4	26.7	30.0	27.3	45.6	27.0	32.0	33.3
Food	5.5	9.3	7.5	6.1	8.8	13.5	36.0	44.4
Nursery	4.1	8.0	27.5	33.3	1.5	0	8.0	5.6
Pet	53.8	41.3	20.0	15.2	20.6	27.0	12.0	5.6
Aesthetic	7.6	9.3	7.5	9.1	7.4	5.4	4.0	5.6

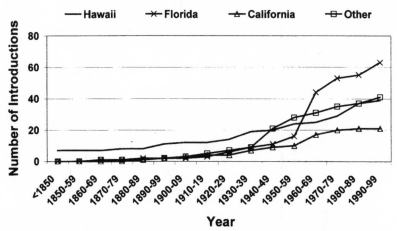

FIGURE 4.4. Cumulative numbers of herp introductions since 1850 distinguished by recipient state of United States.

the forty-seven remaining states combined result from a more even mix of pathways, although the pet trade still holds a preeminent role.

Differences between continental and island locations are also apparent in herp introduction numbers, success, and pathway importance. In the database, more introductions (n = 316) and more establishments (n = 226) occurred on islands than on continents (n = 261 and 156, respectively). Thus, 72 percent of introductions to islands led to successful establishment compared to 60 percent on continents. Introductions to islands occurred primarily via cargo shipping, while those to continents primarily involved the pet trade (Table 4.6). Biocontrol and nursery trade introductions were also significant sources of herp introductions on islands, but were relatively unimportant in continental locations.

Birds

Bird introductions occurred at a relatively low rate until the 1860s, at which point the rate of introduction boomed for a century, finally showing indications of slowing in the 1960s and 1970s (Fig. 4.5). Data are currently not available to determine if this apparent decline holds since the late 1970s.

Seven pathway categories were identified as contributing to the flow of alien birds (Table 4.7). Bird introductions are dominated in importance by two pathways: intentional release as game animals and intentional movement via the pet trade. These pathways have dominated since the begin-

TABLE 4.5. Geographic variation in numbers of herp introductions and relative importance of pathways for the three states of the United States with the greatest number of herp introductions, and for the remainder of the country. Numbers represent percent contributions of a particular pathway to each region's introduction load.

	Percent Contribution to Each Region of the United States							
	Hawaii		Florida		California		Other	
Pathway	All Introductions	Established Introductions Only	All Introductions	Established Introductions Only	All Introductions	Established Introductions Only	All Introductions	Established Introductions Only
Biocontrol	17.5	13.8	1.8	0	0	0	0	0
Cargo	37.5	34.5	16.4	32.1	7.7	0	29.6	8.3
Food	7.5	10.3	0	0	15.4	33.3	11.1	25.0
Nursery	10.0	13.8	1.8	3.6	0	0	3.7	8.3
Pet	27.5	28.6	65.5	50.0	73.1	55.6	48.2	41.7
Aesthetic	0	0	14.5	14.3	3.9	11.1	7.4	16.7

TABLE 4.6. Relative importance of pathways in herp introductions to islands versus continents, measured as percentage of total contribution to each geographic area.

Pathway	Islands		Continents	
	All Introductions	Established Introductions Only	All Introductions	Established Introductions Only
Biocontrol	14.9	17.5	1.6	1.5
Cargo	42.6	33.0	17.1	20.0
Food	9.5	11.3	8.5	15.4
Nursery	9.5	13.4	3.9	7.7
Pet	16.9	19.6	59.7	43.1
Aesthetic	5.4	5.2	9.3	12.3

ning of the introduction boom in the 1860s (Fig. 4.6). Of the remaining categories, only intentional releases for "aesthetic" purposes, and to serve as biocontrol agents, have contributed significant numbers of introductions. Of all pathways, release for "domestic" purposes (pigeon rearing) has the highest rate of successful establishment, while the high-volume pathways of game and pet introductions are relatively low (Table 4.7). Recently, the pet trade has taken on a larger relative share of responsibility for bird introductions compared to the average since 1700 (Table 4.7).

There are striking geographic differences in numbers of bird introductions; most introductions were made to island environments, and the majority of these to the Pacific region (Table 4.8). Among continents, most

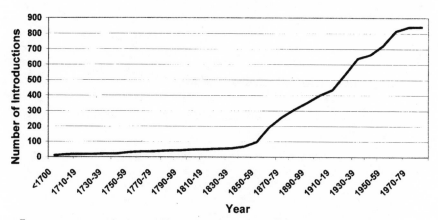

FIGURE 4.5. Cumulative increase in numbers of global bird introductions since 1700.

TABLE 4.7. Relative importance of pathways for global bird introductions, with recent trends.

Pathway	Number of Introductions	Percent Success in Establishment	Percent Contribution to Global Total Since 1700	Percent Contribution to Global Total Since 1960
Aesthetic	50	32.0	9.6	0
Biocontrol	37	70.3	1.3	2.0
Cargo	7	57.0	7.1	2.0
Domestic	8	100.0	1.3	0
Food	18	61.0	3.5	0
Game	167	34.7	35.4	32.7
Pet	218	42.7	41.7	63.3

Note: Category definitions are taken from Long (1981), with some modification: Aesthetic = deliberate release of animals with a view to establishing local populations, but not necessarily connected to the pet trade per se; Biocontrol = intentional release of animals to provide population control of insects or other animals; Cargo = accidental shipment of animals hiding in cargo or cargo containers; Domestic = accidental or intentional release of animals kept as domestic stock; Food = intentional release or accidental escape of animals farmed for human consumption; Game = intentional release of animals for sporting purposes; Pet = intentional release or accidental escape of animals kept as pets or being transshipped in the pet trade.

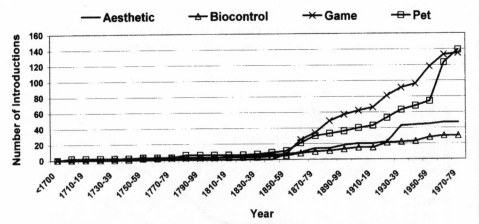

FIGURE 4.6. Cumulative numbers of bird introductions since 1700 distinguished by pathway.

TABLE 4.8. Geographic variation in numbers of bird introductions and recent trends in relative importance for each.

Pathway	Number of Introductions	Percent Contribution to Global Total	
		Since 1700	Since 1960
Europe	69	5.8	8.4
North America	119	9.9	23.4
South America	24	2.0	1.9
Australia	96	8.0	4.7
Southeast Asia	15	1.3	0
Africa–Arabia	31	2.6	2.8
Hawaii	162	13.5	37.4
New Zealand	133	11.1	1.9
Other Pacific islands	208	17.4	8.4
Caribbean	107	8.9	3.7
Other Atlantic islands	78	6.5	1.9
Indian Ocean islands	155	12.9	5.6
CONTINENTAL TOTAL	354		
ISLAND TOTAL	843		

bird introductions were made to North America and Australia (Table 4.8). The recent trend (since 1968) for most geographic regions is a decrease in numbers of bird introductions (Table 4.8), but this does not hold everywhere. Introductions to Hawaii, North America, and, to a lesser extent, Europe have increased in global importance in the past few decades relative to their long-term averages (Table 4.8).

Mammals

Mammal introductions occurred at a relatively low rate from the Stone Age until the mid-1700s, when the rate of introduction increased tremendously, finally showing indications of declining in the late twentieth century (Fig. 4.7). Data are currently not available to determine if this apparent decline holds since the early 1970s.

Ten pathways have contributed to the global flow of alien mammals (Table 4.9). Introductions for food and game purposes have historically been the most important pathways, but large numbers of mammals have also been introduced for aesthetic reasons, for their skins, for use as draught or conveyance animals, and accidentally in cargo. Introductions to provide human food have dominated in importance since at least the

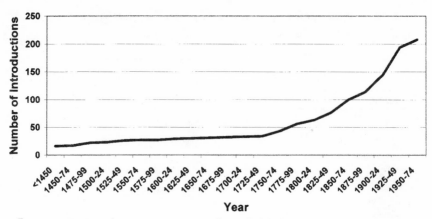

FIGURE 4.7. Cumulative increase in numbers of global mammal introductions since 1450.

middle of the Renaissance, and probably earlier; however, introductions for game purposes increased dramatically in the early 1800s and, by the late 1800s, surpassed in importance introductions for food (Fig. 4.8). Similarly, introductions for aesthetic purposes largely began in the mid-1800s and continued to be an important pathway for a century, after which it seems to have declined dramatically. Introductions for the provision of skins were of considerable importance in the twentieth century, being second only to game introductions during that time (Fig. 4.8). Introductions for game, skins, and aesthetic reasons had greater relative importance during the twentieth century than before (Table 4.9),

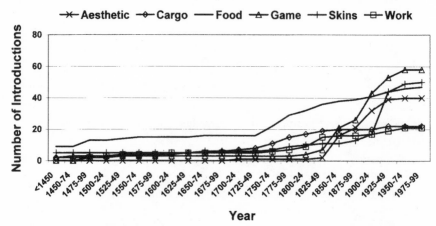

FIGURE 4.8. Cumulative numbers of mammal introductions since 1850 distinguished by pathway.

TABLE 4.9. Relative importance of pathways for global mammal introductions, with recent trends.

Pathway	Number of Introductions	Percent Contribution to Global Total	
		Since Stone Age	Since 1900
Aesthetic	52	13.4	19.0
Biocontrol	17	4.4	3.4
Cargo	40	10.3	1.7
Domestic	10	2.6	0
Food	74	19.0	6.9
Game	98	25.2	28.4
Research	3	0.8	0.9
Skins	58	14.9	31.9
Tourism	5	1.3	3.4
Work	32	8.2	4.3

Note: Category definitions are taken, with some modification, from Lever (1985): Aesthetic = deliberate release of animals with a view to establishing local populations, but not necessarily connected to the pet trade per se; Biocontrol = intentional release of animals to provide population control of insects or other animals; Cargo = accidental shipment of animals hiding in cargo or cargo containers; Domestic = accidental or intentional release of animals kept as domestic stock; Food = intentional release or accidental escape of animals farmed for human consumption; Game = intentional release of animals for sporting purposes; Research = accidental escape or deliberate release of animals used for biomedical research; Skins = accidental or intentional release of animals raised for their skins, whether furs or hides; Tourism = intentional release of animals to provide tourist attractions; Work = accidental or intentional release of animals used in providing work, either draught or personal conveyance.

although aesthetics may have been a relatively short-lived phenomenon (Fig. 4.8).

There are striking geographic differences in numbers of mammal introductions in the dataset; most represented introductions were made to continental environments, especially Europe (Table 4.10). Among continents, most mammal introductions were made to Europe and the Americas, with lesser numbers to Australia, Asia, and Africa (Table 4.10). Among islands, New Zealand and Hawaii have suffered the largest number of mammalian introductions. The recent trend for all island regions represented in the dataset is a decrease in numbers of mammal introductions (Table 4.10). In contradistinction, introductions of mammals to all continents except Australia were higher during the twentieth century than the long-term average, with those to South America and Asia most noticeably so (Table 4.10).

TABLE 4.1O. Geographic variation in numbers of mammal introductions and recent trends in relative importance for each.

| Pathway | Number of Introductions | Percent Contribution to Global Total | |
		Since Stone Age	Since 1900
Europe	115	32.5	34.3
North America	43	12.1	13.9
South America	43	12.1	18.5
Australia	25	7.1	0.9
Asia	18	5.1	8.3
Africa	19	5.4	8.3
Hawaii	20	5.6	3.7
New Zealand	31	8.8	5.6
Other Pacific islands	19	5.4	4.6
Caribbean	19	5.4	0.9
Other Atlantic islands	2	0.6	0.9
CONTINENTAL TOTAL	263		
ISLAND TOTAL	91		

DISCUSSION

For each of the vertebrate classes considered herein, several pathways for introduction exist, but two have been predominant in each case. Except for unintentional introductions of herps via cargo shipping, all other major pathways involved deliberate introductions and/or release. Of these, the pet industry was of predominant importance in the introduction of alien herps and birds. The other important pathway for birds, game introductions, was also the most important pathway for mammals. Most terrestrial vertebrate introductions were deliberate (whether release itself was intentional or not). This will be an unsurprising conclusion to those who research these animals, but stands in contrast to the perception that most alien invasions are an unintended side effect of accepted human activities, such as trade (e.g., Jenkins 1996); thus, such generalizations cannot reliably be made without explicitly noting the taxa for which they apply. Although unintentional pathways of introduction have received some attention in recent efforts to reduce the spread of alien invaders (Reeves 1997), intentional pathways involving the pet industry (and its aesthetic partner, the horticulture industry) and state game programs remain unaddressed. In rectifying this situation, it is relevant to remember that present estimates of pet and cargo pathways contributing to the global introduction of herps are significant underestimates of the actual magnitude, because of the reporting biases discussed earlier. The same is probably true of bird releases stemming from the pet trade.

The temporal patterns of terrestrial vertebrate introduction shed additional light on the cultural foundations creating and sustaining this problem. Mammal introductions have been occurring since the Stone Age but blossomed into a global problem with the advent of European colonialism and its transportation of entire agricultural and commensal communities to new continents (Crosby 1986). This transfer process increased dramatically beginning in the mid-1700s (Fig. 4.7), presumably reflecting the addition of non-Eurasian mammals to the repertoire of already-utilized Eurasian species and the introduction of many of these new species back to Europe. These introductions continue today, although no longer driven by the food and work needs of European colonists so much as by providing amenities such as furs and sport (Fig. 4.8). A current manifestation of this pattern is the growth of "game ranching" in many parts of the United States, accounting for the introduction of many ungulates into wild or semiwild conditions. The extent to which this practice will lead to additional established alien mammal species remains to be seen.

Bird introductions were of low frequency until the advent of acclimatization societies in the 1850s, at which point introductions for pets, for related aesthetic reasons, and for game hunting skyrocketed (Figs. 4.5, 4.7). Although the acclimatization societies may not have been directly responsible for each of these early introductions, they widely promulgated and validated the underlying ethic that created this phenomenon (Lever 1992). To perhaps an even greater extent than mammal introductions, bird introductions result from European, particularly English, cultural practices. This is illustrated by the fact that at least 77 percent of all bird introductions were to countries and islands colonized by the English or by their cultural offspring (another 10% of the remaining introductions were to areas colonized by the French). Although some small number of these introductions was made prior to colonial rule, they do show the strong influence British culture and its colonial derivatives have contributed to alien bird introductions.

Herp introductions were a later phenomenon, beginning to increase significantly in the 1930s as a result of biocontrol releases, and booming since the 1950s with the tremendous growth of the pet trade and global movement of cargo (Figs. 4.1, 4.2). Again, both pathways largely reflect Western cultural desires, although cargo movement has a more global ambit and serves to negate any simple geographic relationship between number of introductions and colonial history, as is seen for birds. Indeed, what is clear from the herp data is that there is considerable geographic variation in pathway importance, even within single, albeit large, countries such as the United States.

Recent changes in importance of pathways, in terms of establishments, are of great interest. Of particular importance in the United States are indications that the pet industry, which accounted for the large majority of herp establishments in the 1960s and 1970s, has been replaced in the 1980s and 1990s by an upsurge in establishments resulting from cargo movement. As noted earlier, the increase in relative importance of the cargo pathway for establishment of alien herps results not from an increase in establishment efficiency (Table 4.3), but instead from an increase in rates of introduction, reflecting increasing trade volume. The same is true for the nursery trade (Table 4.3). In contrast, introduction rates via the pet trade in the last two decades have remained approximately the same as the long-term average (Table 4.3). The dramatic increase in the 1950s and 1960s of establishments via the pet trade is almost certainly due to the then-common practice among animal dealers of releasing large numbers of weakened or unsaleable animals. In such instances, numbers of released animals could be in the hundreds per species (King and Krakauer 1966), increasing probabilities of population establishment. Development of better herp husbandry techniques since the 1970s has led to increased survivorship of retail stocks and, consequently, less incentive for dealers to release large numbers of animals. The continued high release rates of pets (at low density) by owners is less likely to result in establishment, except for species (such as for the red-eared slider, *Trachemys scripta*) that many owners keep and release, thereby increasing the probability that independently released animals will encounter each other and mate.

One pattern apparent for each vertebrate dataset is a decline in introduction rates in recent decades (Figs. 4.1, 4.5, 4.7). It is not clear whether this pattern reflects an actual decline in vertebrate introductions or is an artifact of analytical methods used and the lag time involved in discovering and reporting recent introductions in the scientific literature. Lending support to the latter hypothesis is the gross disparity between recent published and unpublished Hawaiian records noted above. Also, it should be remembered that in the herp and bird datasets there are many species that have been introduced multiple times to the same political entity, yet only the earliest introduction was used in constructing the cumulative plots. This conservative practice will underestimate the total number of herp and bird introductions as well as the introduction rates for more recent time periods. One test of the hypothesis that recent declines are an artifact would be to determine if the apparent decline in previously reported introduction rates of birds (during the 1960s and 1970s) is falsified by reports for that time period published since the last analysis (1981). Without passing such a test, recent declines in vertebrate introduction rate should be viewed skeptically. What should be clear is that there is no compelling evidence yet that overall terrestrial vertebrate introduction rates are declining. Furthermore, there is evidence that introductions by some pathways continue to rise sharply (e.g., cargo pathway for herps, pet trade pathway for birds), suggesting that vertebrate introductions and associated effects will continue for the foreseeable future.

It has been suggested that approximately 10 percent of alien species imported to an area appear in the wild, 10 percent of those appearing in the wild establish self-sustaining populations, and 10 percent of established species become pestiferous. This pattern, the so-called "tens" rule, was first noted for a variety of British alien species (Williamson and Brown 1986) and later argued to cover a wider geographic array of alien introductions (Williamson and Fitter 1996). The present data would appear to conflict with the second claim for the tens rule—that 10 percent of species released to the wild become established. Based on the present data, 30–76 percent of herp introductions have become established, depending on the taxon (Table 4.1). Similarly, 32–100 percent of bird introductions also appear to have established successfully (Table 4.7). The herp introductions include a mix of what Williamson and Fitter (1996) distinguish as importations (animals intercepted in cargo or other contained situation) and introductions (animals recorded outside control or captivity), meaning that the total proportion of introductions (in their sense) leading to successful establishment for the herp dataset is even larger than that indicated in Table 4.1. The bird data represent only introductions as defined by Williamson and Fitter (1996).

This discrepancy from the expectations of the tens rule could result for a couple of reasons. First, the tens rule could be inapplicable, at least to animals. And, indeed, most of the exceptions to the rule discussed by Williamson and Fitter (1996) involved animals. Some of their exceptions (e.g., Hawaiian birds) could be fit to the rule by focusing on establishment in native ecosystems only, but this is a difficult argument to use objectively, because most of those ecosystems are now invaded or disturbed to some extent by alien pests and categorization of Hawaiian habitats as "native" versus "nonnative" becomes an arbitrary partition of a continuum. This same problem will hold for many or most areas receiving herp introductions, perhaps calling into question the utility of a rule that may not be severable from such arbitrary distinctions in a thoroughly human-modified world.

A second possibility is that the apparently high rate of successful establishment for herps could be an artifact of my reliance on literature reports combined with the likelihood that most alien introductions will never be known or reported. However, most scientific assessments of importation and establishment rates, including those of Williamson and Brown (1986) and Williamson and Fitter (1996), should suffer from the same problem, and it is difficult to see how such limitations can be quantified or controlled for among studies. Further, as noted earlier, the discrepancy between the actual levels of herp introductions to the wild and those reported in the literature is likely to be in the range of two or three orders of magnitude, not the one order required for the herp data to better fit the tens rule. In any event, an apparently large and growing number of herp introductions are leading to establishment of feral populations and at rates higher than what would be predicted from the general literature on this topic.

Given the deliberate and predictable nature of most terrestrial vertebrate introductions, it is worth considering what preventive measures have been taken to curb this problem. The scope of this issue is far larger than can be treated fully here; nonetheless, consideration of measures adopted by the United States, and its two most affected states, provides some perspective on the inadequacy of response measures so far attempted.

- The United States Injurious Wildlife section of the Lacey Act (U.S. Code of Federal Regulations: Title 50 CFR Section 16.11–16.15) prohibits the introduction without permit of eleven genera and one species of mammal, four species of birds, and the brown tree snake (*Boiga irregularis*). It also prohibits release into the wild of any terrestrial vertebrate unless conducted or approved by a state wildlife conservation agency.

- Federal public health regulations (Title 42 CFR Section 71.52–53) prohibit the importation of primates for pet uses and prohibit the importation of turtles smaller than four inches, or turtle eggs, or bats, without a permit. However, imported primates have entered the commercial pet trade upon completion of the scientific or zoological purposes for which they were imported (G. Phocas, U.S. Fish and Wildlife Service Special Agent, pers. comm., Oct. 1999).

- The State of Florida prohibits release of any alien wildlife, the importation of sea snakes, and the ownership of many other animals as personal pets. However, one only needs to obtain a $25 license having minimal requirements to exhibit/sell wildlife in order to own such species (P. Moler, Florida Fish and Wildlife Conservation Commission, pers. comm., Oct. 1999).

- Hawaii prohibits the release of alien wildlife, prohibits transport and commercial export of fourteen taxa of "injurious wildlife" that are already established but of limited distribution in the state, and has a very restricted list of species allowed entry into the state.

- Hawaii also has a state-funded quarantine program to inspect arriving cargo, although this program is considerably underfunded. (Florida lacks such an inspection program.)

A notable feature of these restrictions is that they are all species-focused, generally prohibiting importation or ownership of a small list of taxa deemed harmful for one reason or another. Prohibitions against release of animals are generally broader but unenforceable in all but the most exceptional circumstances. The combination of these two limitations means that most terrestrial vertebrate introductions in the United States remain unregulated, and the underlying pathways driving them have not been addressed. Federal acquiescence in allowing state game departments to release whatever they wish further hinders recognition of the general problem by validating vertebrate releases and by ignoring the interstate effects of these releases. Consequently, it should not be surprising that there has been little progress to stem the flow of alien terrestrial vertebrates into the United States. A broader-based, not narrowly focused and species-specific, approach that regulates underlying pathways of introduction may be a sensible means of reducing the magnitude of introductions in the future and should be a primary consideration in directing control, monitoring, and research. Such an approach would necessarily involve a significant public-education element to modify the underlying attitudes that drive this particular component of the alien invasion crisis.

ACKNOWLEDGMENTS

I thank E. Campbell, D. Cravalho, P. Moler, and G. Phocas for providing useful information, references, and/or discussion for this study; E. Campbell, J. Carlton, G. Perry, G. Rodda, and G. Ruiz for providing useful comments on the manuscript; J. Lim for kindly entering the bird and mammal data into spreadsheets; and J. Carlton and G. Ruiz for inviting me to review this topic and present the results at their conference on pathways of alien species invasions.

REFERENCES

Bury, R. B. and R. A. Luckenbach. 1976. Introduced amphibians and reptiles in California. *Biological Conservation* 10: 1–14.

Case, T. J. and D. T. Bolger. 1991. The role of introduced species in shaping the distribution and abundance of island reptiles. *Evolutionary Ecology* 5: 272–290.

Crosby, A. W. 1986. *Ecological imperialism: the biological expansion of Europe, 900–1900.* Cambridge: Cambridge University Press.

Ebenhard, T. 1988. Introduced birds and mammals and their ecological effects. *Viltrevy* 13: 1–107.

Greenway, J. C. 1967. *Extinct and vanishing birds of the world.* New York: Dover Publishing.

Hardin, E. 1993. *Living within limits.* New York: Oxford University Press.

Henderson, R. W. 1992. Consequences of predator introductions and habitat destruction on amphibians and reptiles in the post-Columbus West Indies. *Caribbean Journal of Science* 28: 1–10.

Honegger, R. E. 1980–1981. List of amphibians and reptiles either known or thought to have become extinct since 1600. *Biological Conservation* 19: 141–158.

Jenkins, P. T. 1996. Free trade and exotic species introductions. *Conservation Biology* 10: 300–302.

King, W. and T. Krakauer. 1966. The exotic herpetofauna of southeast Florida. *Quarterly Journal of the Florida Academy of Science* 29: 144–154.

Lever, C. 1985. *Naturalized mammals of the world.* Essex: Longman Scientific and Technical.

Lever, C. 1987. *Naturalized birds of the world.* Essex: Longman Scientific and Technical.

Lever, C. 1992. *They dined on eland: the story of the acclimatisation societies.* London: Quiller Press.

Long, J. L. 1981. *Introduced birds of the world.* New York: Universe Books.

Morgan, G. S. and C. A. Woods. 1986. Extinction and the zoogeography of West Indian land mammals. *Biological Journal of the Linnean Society* 28: 167–203.

Reeves, M. E. 1997. Techniques for the protection of the Great Lakes from infection by exotic organisms in ballast water. *Zebra mussels and aquatic nuisance species,* F. M. D'Itri, ed., pp. 283–299. Chelsea: Ann Arbor Press, Inc.

Williamson, M. H. and K. C. Brown. 1986. The analysis and modeling of British invasions. *Philosophical Transactions of the Royal Society of London. Series B: Biological Sciences* 314: 505–522.

Williamson, M. H. and A. Fitter. 1996. The varying success of invaders. *Ecology* 77: 1661–1666.

Chapter 5

Pathways of Introduction of Nonindigenous Land and Freshwater Snails and Slugs

Robert H. Cowie and David G. Robinson

The world's biota is being homogenized as a result of the decline of native species and their replacement by a relatively small number of species that are deliberately and accidentally moved beyond their natural ranges by people (McKinney and Lockwood 1999). Land and freshwater snail (including slug) faunas are no exception (Solem 1964, Kay 1995, Cowie 1998a, 1998b, 2000a, 2002a, Robinson 1999). In some cases, these introduced species are actively replacing native species, by predation and perhaps by competition, but in other instances they may simply be occupying modified habitat from which the native species have already vanished (Cowie 1998a, 1998b). In addition to replacing the native snail faunas, many of these species cause other ecological, agricultural, and medical problems (Cowie 2000b, 2002b).

This chapter describes the pathways by which these snails and slugs are being introduced throughout the world, looks for trends in the relative significance of the various pathways, and briefly outlines actions needed to prevent the continuing increase of introductions that is likely to occur as global trade barriers are relaxed. We have provided many examples from the United States and the islands of the Pacific, in large part because these are among the best documented (Britton 1991, Robinson 1999, Cowie 2000b, 2002a) but also because they reflect our experience. However, the principles involved, with some country-specific modifications, are global, and examples are drawn from around the world.

KNOWN (OR PRESUMED) PATHWAYS

The term *pathway* is considered narrowly here as referring to the purpose or activity by which an alien species was introduced, either deliberately (e.g., for biological control) or inadvertently in association with an imported commodity (e.g., horticultural products). Pathways are distinguished from the vector of introduction, or the actual vehicle/mechanism used to transport the alien species (ship, airplane, etc.), as well as from the geographic route of introduction, both of which have been considered pathways elsewhere.

In general, deliberate introductions of snails have been better documented than inadvertent introductions, even though the latter are far more numerous. Thus, knowledge of the pathways by which snails are inadvertently introduced is far more speculative.

Deliberate Introductions

Species introduced deliberately (sometimes legally) may have a better chance of survival and subsequent invasion than species introduced inadvertently. The reasons for this are that (1) at least some deliberate introductions involve many individuals (though this may also be the case in some inadvertent introductions), and (2) these individuals usually receive active care to promote their growth and reproduction. The following list includes the major pathways of deliberate transfer or introduction of land and freshwater snails.

Aquarium Industry

Many species of freshwater snails are moved around the world to be sold in pet stores for use in domestic aquaria. Then they either escape or are released. These include various species of Ampullariidae (apple snails, mostly *Pomacea* spp. and *Marisa cornuarietis*), Viviparidae (e.g., *Cipangopaludina chinensis*), Planorbidae (ramshorn snails, e.g., *Helisoma* spp., *Planorbis* spp.), Physidae, and Lymnaeidae (e.g., *Radix auricularia*) (Britton 1991, Perera and Walls 1996, Mackie 1999a). These and other species may also be transported inadvertently in association with domestic aquarium plants and fish (see below).

Food (Includes Aquaculture)

Many species of European land snails (mostly Helicidae) have been deliberately introduced around the world as escargot (e.g., *Helix aspersa* to

New Caledonia—Gargominy et al. 1996; *Otala lactea* to Bermuda—Simmonds and Hughes 1963), becoming serious pests in many areas (e.g., California—Gammon 1943). They are often transported in personal baggage, or even sewn into clothing (Smith 1989), to be released into backyards or vacant lots for harvest once the populations have become large enough (Robinson 1999) or used for large-scale commercial enterprises (Britton 1991). The South American freshwater ampullariid *Pomacea canaliculata* was taken to Southeast Asia in about 1980, with the intention of developing aquaculture programs to supplement local food resources but also to develop a gourmet export industry (Cowie 2002b). It was also taken to Hawaii (probably from Southeast Asia) for the same reasons (Cowie 1995, 1997). This snail escaped, or was released, and is now a serious agricultural pest in both Southeast Asia and Hawaii (Cowie 2002b), has spread to natural ecosystems (Lach and Cowie 1999), and has been implicated in the decline of native snail species (Halwart 1994). The viviparids *Cipangopaludina chinensis* and *C. japonica* (which may be the same species; Jokinen 1982), also freshwater snails, were probably introduced to the United States from Asia as food resources (Hanna 1966, Chace 1987, Mackie 1999b); *C. chinensis* has been in Hawaii since at least 1900 (Cowie 1995, 1997). The giant African snail, *Achatina fulica*, has been introduced to many areas as a food resource (Godan 1983, Clarke et al. 1984), not only for humans but also for domestic ducks (Civeyrel and Simberloff 1996), as well as for other reasons (see below). Other achatinids such as *Archachatina marginata* may have been introduced for this purpose.

Medicinal Purposes

Some snail species have been used for a variety of medical purposes, mostly for extracting medicinally useful compounds (references in Mead 1979), but whether snails have been introduced widely for such purposes is unknown. For example, *Achatina fulica* was introduced to Hawaii for unspecified medicinal purposes, as well as for other reasons (van der Schalie 1969, 1970), and was much earlier introduced to Réunion to make snail soup as a remedy for a chest infection (Petit 1858).

Biological Control

Cowie (2001) has reviewed the use of snails as biological control (biocontrol) agents. The best publicized example is the introduction of predatory snails, notably *Euglandina rosea* and *Gonaxis* spp. (but also a large number of other species), in ill-conceived attempts to control the giant African snail, *Achatina fulica* (e.g., Griffiths et al. 1993, Civeyrel and Simberloff

1996). *Euglandina rosea* has been introduced to many islands of the Pacific and Indian Oceans and elsewhere (Griffiths et al. 1993, Civeyrel and Simberloff 1996). In Hawaii, fifteen predatory species were introduced in the 1950s and 1960s in these attempts to control *Achatina fulica*, although only three of them became established (Cowie 1998a). There is no good evidence that these predatory snails reduce populations of *A. fulica* (Christensen 1984) but ample evidence of their devastating effects on native snail faunas (Hadfield 1986, Murray et al. 1988, Cowie 1992, Hadfield et al. 1993, Cowie and Cook 2001). *Euglandina rosea* has also been used against other snail pest species, such as *Otala lactea* in Bermuda (Simmonds and Hughes 1963, Bieler and Slapcinsky 2000).

An additional example of the use of nonindigenous predatory snails in attempts to control other invasive snail species is the widespread use of the southern European species *Rumina decollata*. This facultative predator was introduced to California to "control" the European *Helix aspersa* (Fischer and Orth 1985, Sakovich 1996), although evidence that populations of *H. aspersa* were reduced as a direct result of predation by *Rumina decollata* is scanty (Cowie 2001). *Rumina decollata* has been introduced to other areas both deliberately as a biocontrol agent (D. G. Robinson, unpublished data) and inadvertently by the horticultural trade, discussed below (Britton 1991).

A number of freshwater species (e.g., *Marisa cornuarietis*) have been introduced, especially in the Caribbean region, to eradicate or control invasive aquatic weeds such as water lettuce (*Pistia stratiotes*) (Perera and Walls 1996, Simberloff and Stiling 1996). Various species of Ampullariidae and Thiaridae have also been introduced to control snail vectors of human schistosomes by either predation or competition (Pointier et al. 1988, 1991, 1994). Impacts on native vegetation and associated fauna are generally not discussed or simply ignored.

Such biological control agents have frequently been introduced simply because they were reported to feed on the target, without demonstrating (1) that they are specific to the target and will not attack native species, and (2) that they can indeed reduce populations of the target. Most official biocontrol programs now make some attempt to demonstrate specificity to the target, but the much more complex problem of demonstrating potential reduction of pest populations is rarely addressed. Furthermore, putative biocontrol agents are frequently introduced unofficially with no testing at all. In the United States, there is little regulation of the import or spread of biocontrol agents. Australia has enacted legislation explicitly governing biological control, but this act is limited in scope and in provision of standards (Miller and Aplet 1993).

Pets

A major infestation of *Achatina fulica* in Florida in the 1960s resulted from the deliberate importation of snails from Hawaii by a child (Mead 1979). Interception of *A. fulica* in personal luggage of tourists returning from Hawaii to the U.S. mainland is routine (Robinson 1999), and deliberate smuggling of *A. fulica* by the pet trade in the United States has occurred (OTA 1993, p. 84). *Helix pomatia,* intercepted coming into California, was being imported for snail races (Hanna 1966)! Probably there are other instances involving these and other species.

Aesthetics

Snails have been deliberately imported and released for aesthetic reasons. One of the original introductions of *Achatina fulica* into Hawaii was for such ornamental purposes (van der Schalie 1969, 1970). The European land snail *Otala lactea* was introduced to the United States on at least one occasion to remind immigrants of their homeland (Murray 1968), as well as for food (see above). Shell collectors and dealers may have introduced aesthetically attractive species (e.g., the European helicid *Cepaea nemoralis* to the United States; Britton 1991). Other large or brightly colored species, especially Caribbean species of *Liguus* and *Orthalicus,* have been released for similar reasons (e.g., species from Cuba and the Bahamas into Florida; Clapp 1919) and perhaps continue to be imported (Robinson 1999).

Biological Research

Snails imported for research purposes, perhaps even with appropriate permits, may have escaped or even been released into the wild. For instance, ten species of *Cerion* from the Bahamas, Puerto Rico, and Cuba were transplanted to the Florida Keys to investigate the relative importance of genetic and environmental control of shell morphology (Pilsbry 1946); one of them became established, hybridizing with the local endemic species (Woodruff 1978). Three of these species of *Cerion* were also released in Hawaii for the same purpose but did not become established (Cowie 1996).

Inadvertent Introductions

Many species of snails and slugs are transported in or on a huge range of products, vehicles, containers, and the like. Some species probably travel as eggs, or perhaps as small juveniles, which are much less readily detected than are adult snails. Reproductive ecology may facilitate colonization in many cases. Although pulmonate land (Stylommatophora) and freshwater (Basommatophora) snails and slugs are hermaphrodites, many species must cross-fertilize in order to reproduce (Geraerts and Joosse 1984, Tompa 1984). However, a significant number of species can self-fertilize (Duncan 1975, Foltz et al. 1982, 1984, Geraerts and Joosse 1984, Tompa 1984), such that introduction of a single individual could lead to successful invasion. European slugs introduced to North America are disproportionately represented by selfing species (Foltz et al. 1984). Also, sperm storage, which is well known in pulmonates (Duncan 1975, Tompa 1984), would permit even single individuals of cross-fertilizing species to invade successfully. Sperm can be stored for long periods—for instance, for more than a year in *Arianta arbustorum* (Baminger and Haase 1999). Some cross-fertilizing species are also facultative selfers (Tompa 1984). Furthermore, although most "prosobranchs" (Prosobranchia is now considered polyphyletic; Ponder and Lindberg 1997) have separate sexes and must cross-fertilize (Fretter 1984), some species (e.g., *Melanoides tuberculata*) appear to be predominantly parthenogenetic (Pace 1973), which would permit colonization by a single individual. Although single individuals could lead to invasions, there may be long-term genetic consequences of reduced genetic variation in these introduced populations that influence invasion success and dynamics.

Below, we highlight the major pathways of inadvertent introductions for snails and slugs. Certainly this is not a comprehensive list. Other miscellaneous pathways, not discussed in detail, include association with various wood products (e.g., pallets, crating), flowerpots and other earthenware, quarry products (e.g., ornamental rocks), machinery and heavy equipment, and possibly hiking and camping equipment (Robinson 1999).

Agricultural Products (Excluding Horticulture)

Agriculture (products for consumption as food) is here distinguished from horticulture (for the most part, decorative or propagative products), although the latter has elsewhere been considered a subdivision of agriculture. Many species of snails and slugs have been found associated with shipments of agricultural produce (Godan 1983). In the United States, for instance, numerous species are intercepted on a wide variety of fruit and vegetables, with no obvious patterns of association, except, notably, several species of Arionidae with European mushrooms (D. G. Robinson, unpublished data).

Horticultural Products

Cut flowers, live plants, seeds, turf, leaves used for mulch, and similar products provide a ready pathway for introduction, and many species of snails and slugs are frequently intercepted in shipments of these products

(Godan 1983). The scale of the international horticultural trade is huge (Devine 1998, Ewel et al. 1999). Many nurseries are infested with snails and slugs (Hara and Hata 1999), especially small species. Often, new records of alien species in an area are associated with nurseries, garden stores, botanical gardens, or recent landscaping activities (e.g., *Parmarion martensi* and *Polygyra cereolus* in Hawaii—Cowie 1998c; *Bulimulus guadalupensis* in Florida—Thompson 1976).

Many of the species associated with this pathway may have been imported to a nursery from elsewhere and then exported to one or more additional regions. For example, *Liardetia doliolum* (alien in Hawaii) has been exported to the U.S. mainland from a nursery in Hawaii (Cowie 1999), and the neotropical *Guppya gundlachi* is regularly intercepted in horticultural shipments to the United States from Thailand (Robinson 1999). Leaves used as mulch have been suggested as the pathway by which litter-dwelling snails could be introduced (Roth and Pearce 1984), and hay used as mulch was the cause of the spread of *Theba pisana* in California (Hanna 1966).

Commercial and Domestic Shipments

Any number of commodities, usually but not always shipped in containers, may have snails and slugs associated with them (Robinson 1999). Snails and slugs may be inadvertently transported from the source with the product itself, or may attach themselves to the product, its packaging, or the shipping container at any point en route. Particularly important may be the time the shipment or container spends on a dockside or cargo holding area (Britton 1991). Especially notable in the United States is shipment of domestic tiles from the Mediterranean region that are frequently infested with European snails (Robinson 1999).

Snails and slugs can be transported in association with domestic shipments of household goods in the same way as they can with commercial shipments.

Military Shipments

Snails and slugs can also be transported with military supplies and other goods associated with military campaigns, as well as during routine peacetime transportation of military equipment and supplies. Recent U.S. involvement overseas has resulted in many interceptions of alien snails—for example, 231 snail interceptions on military equipment returning from Europe to the United States in the immediate aftermath of the 1999 involvement in the Balkans (D. G. Robinson, unpublished data).

Vehicles/Rail/Airplanes

Snails may attach themselves to vehicles (e.g., parked cars) and then be transported to wherever the vehicle is driven (R. H. Cowie unpublished observations of the helicid *Theba pisana*; also Cowie 1987). Many snails are intercepted on private cars and commercial trucks crossing the U.S.–Mexico border (D. G. Robinson, unpublished data). Similar transport by rail is likely. Snails may also attach themselves to airplanes; however, because of the extreme cold at high altitude and unlikely survival on landing in the inhospitable environment of large, open runways, inadvertent introduction of snails attached directly to airplanes is not a major pathway. Nevertheless, prior to reaching the extreme cold of high altitude, attached snails may be blown off as a result of the speed of the airplane, resulting in expansion of their range by relatively short-distance dispersal.

Soil

Snails and slugs, and especially their eggs, are readily transported in soil. This could result from (1) soil deliberately transported with agricultural or horticultural products (e.g., potted plants), (2) soil to be used for landscaping or top-dressing, (3) soil accidentally transported with agricultural and horticultural products, or (4) soil transplanted with vehicles, shoes, and the like. The United States prohibits import of any shipment with soil for these reasons.

Aquarium Industry

Small freshwater snails are easily transported inadvertently attached to aquarium plants (Smith 1989). These include species of Physidae, Thiaridae (e.g., *Melanoides tuberculata, Tarebia granifera*), Planorbidae, and Lymnaeidae (Frandsen and Madsen 1979, Britton 1991, Perera and Walls 1996, Mackie 1999a, Pointier 1999). In addition, snails may be associated with aquaculture of aquarium and other alien fish (Britton 1991) and have been transported with them (e.g., *Biomphalaria straminea* from Hong Kong to Australia—Dudgeon and Yipp 1983). This pathway was considered of major importance by Pointier (1999).

Aquaculture

Aquatic snails can be introduced accidentally (or even intentionally) along with a different species (not necessarily snails) specifically introduced for aquaculture (Carlton 1992). This has been suggested as the mechanism of introduction for the freshwater snail *Potamopyrgus antipodarum* to the Snake River, from elsewhere within the United States (Britton 1991, OTA 1993, p. 85). In addition, aquaculture of aquarium fish may be important in this respect, as mentioned above.

Ships/Boats

Although ballast water is more often implicated in the introduction of marine species (Smith et al. 1999), if a ship takes on ballast in a freshwater harbor and then discharges it at the end of its voyage in another freshwater harbor, it is possible that freshwater snails (or their larvae) could be introduced (e.g., *Potamopyrgus antipodarum, Bithynia tentaculata, Radix auricularia* introduced to North America; Mackie 1999c, Ricciardi and MacIsaac 2000). Dry ballast is no longer used and thus no longer constitutes a pathway for introduction of terrestrial species (OTA 1993, p. 100, MacIsaac 1999). Hull fouling may also constitute a mechanism of transport, possibly if snails or their eggs (e.g., Physidae, Lymnaeidae) were attached to hulls or associated algae of small boats or barges being moved from one body of freshwater to another, although no instances appear to have been reported in the literature.

Canals and Other Modified Waterways

When canals are built and two or more formerly unconnected bodies of water become joined, they become "biotic corridors" (Carlton 1992), permitting the formerly distinct faunas to mix. For instance, two pleurocerid species (*Elimia livescens, Pleurocera acuta*) and a valvatid (*Valvata piscinalis*) have been introduced to the Hudson River Basin in the northeastern United States via this route (Mills et al. 1997). In addition, canals can facilitate the spread of alien species from a focus of introduction (e.g., *Potamopyrgus antipodarum* in Britain; Kerney 1966). Dams and other modifications of existing rivers may alter the aquatic habitat in ways that facilitate the spread of alien snails.

Roads

By acting as corridors for the introduction of alien plants (Andrews 1990, Trombulak and Frissell 2000), roads may create habitat more suitable for invasion by alien snails, thereby facilitating the spread of a species from a focus of introduction. However, there are no instances in which this has been demonstrated to have taken place. In addition, roadside ditches may act as corridors along which freshwater snails may be able to invade.

SIZE AND STRENGTH OF PATHWAYS

Clearly, all the pathways mentioned above are not equally important in facilitating the introduction and establishment of alien snails and slugs. To address the question of which pathways are more or less important requires data on not only the occurrence of the introduced species but also the pathways via which they were introduced. There are good summaries of the occurrences of alien snails and slugs for some regions of the world—for example, New Zealand (Barker 1999), New Caledonia (Gargominy et al. 1996), the Samoan Islands (Cowie 1998d), the Hawaiian Islands (Cowie 1997), and Pacific islands in general (Cowie 2000b, 2002a). Occurrences in most other regions for which there are no explicit listings would have to be gleaned from the primary literature, and to compile this information on a global basis would be a daunting task. All such information is dependent on recording of the species' presence in the first place. Traditionally, most mollusk specialists have tended to ignore the introduced species, concentrating their research efforts on what they considered to be the more interesting native species. Thus, records of introduced species are woefully inadequate for many parts of the world (Smith 1989), and even the lists mentioned above must be used with this caveat in mind.

While the information about the global occurrence of introduced snails and slugs is inadequate, information about the pathways of introduction is almost nonexistent. Unlike vertebrates, for which the pathways are frequently well documented because these animals (except for mice and rats) are usually introduced deliberately, documentation of pathways of introduction for invertebrates is poor because, with notable exceptions (e.g., biological control agents—Howarth 1991, Cowie 2001), most invertebrate introductions are inadvertent.

The following example illustrates this lack of knowledge. The occurrences of nonindigenous snails and slugs on the islands of the Pacific have been documented by Cowie (2000b, 2002a), but the pathways via which they arrived in these islands remain extremely poorly understood (Table 5.1), being simply educated guesses based on experience of the species involved and the likely reasons for their introduction. For most species, it is not even possible to guess. Further, in most cases the time and provenance of the introduction cannot be determined.

Most of the data (below) that address specific pathways are interception data from various quarantine and inspection agencies around the world and hence not associated with known instances of introduction (establishment). These data are highly biased, because they are derived primarily from efforts to prevent the introduction of agricultural pests (and to a lesser degree environmental pests and human health pests), targeting only particular pathways and ignoring others. Also, these data are the numbers of interceptions (each species intercepted on one occasion counts as one interception) and not the numbers of successful invasions that result from failure of interception; the latter numbers are the data that are most

TABLE 5.1. Pathways of introduction of non-native land snails and slugs established on Pacific islands (68 species from 26 islands/archipelagos; island distribution data from Cowie 2002a; pathway data from R.H. Cowie, unpublished).

Pathway	Introductions
ACCIDENTAL	
Horticulture/agriculture	12
Unknown/multiple	169
DELIBERATE	
Food	1
Biological control	29
Unknown/multiple	24
TOTAL SPECIES × ISLAND RECORDS	235

needed (cf., Britton 1991). Nevertheless, we can use interception data to show minimum levels of transfer associated with pathways.

The only major dataset published recently for snail and slug interceptions is that of Robinson (1999), documenting interceptions of snails and slugs coming into the United States from other countries between 1993 and 1998. These data (over 4,900 interceptions; Table 5.2) show that association with horticultural products (cut flowers and plants for propagation) constitutes the biggest single pathway. Second largest is the import of tiles (roofing tiles, household tiles); the majority of tiles came from southern Europe, an area with a high diversity of snail species, many of which can withstand long periods exposed to harsh conditions. The United States imports almost all of its tiles; the packaging and containerization of these shipments provide ideal opportunities for snails and slugs to be transported with them (Robinson 1999). Attachment to containers in general constitutes the third largest pathway, and association with agricultural (excluding horticultural) products the fourth. These introductions are all inadvertent. Deliberate introductions constitute a very small fraction of interceptions. However, simply because they are deliberate, and great effort is made to introduce the snails and keep them alive, any one deliberate introduction may have a relatively greater chance of establishment than many inadvertent introductions. Earlier U.S. data (1984 to mid-1991; Britton 1991), based on almost 3,000 interceptions, show a similar pattern, although the data were categorized slightly differently (Table 5.3).

Interception data from Canada, taken from Godan (1983), are summarized in Table 5.4, and from New Zealand, taken from Barker (1979), in

TABLE 5.2. The commodity or article with which snails and slugs intercepted by the U.S. Department of Agriculture were associated (data from 1993 to 1998; over 4,900 interceptions; from Robinson 1999).

Pathway/association	Percent of interceptions
Horticultural products (plants and cut flowers)	29
Household tiles	23
Containers	16
Agricultural (nonhorticultural) products	7
Aquarium plants	4
Baggage (deliberate smuggling)	4
Military cargo	1
Mail, other similar services (e.g., FedEx)	<1
Import for consumption (permit)	<1
Other	>14

TABLE 5.3. The commodity or article with which snails and slugs intercepted by the U.S. Department of Agriculture were associated (data from 1984 to mid-1991; 2,889 interceptions; from Britton 1991).

Pathway/association	Percent of interceptions
Plants or plant products	46
Containers	16
Crating	11
Other commerce	16
Other	11

TABLE 5.4. The commodity or article with which snails and slugs intercepted by the Canadian Department of Agriculture were associated (data from 1963 to 1971; compiled from Godan 1983).

Pathway/association	Number of interceptions
Horticultural products (plants and cut flowers)	50
Agricultural (nonhorticultural) products	7
Packaging	3
Baggage	2
TOTAL	62

TABLE 5.5. The commodity or article with which slugs intercepted entering New Zealand were associated (data from 1955 to 1978; compiled from Barker 1979).

Pathway/association	Number of interceptions
Horticultural products /plants	11
Agricultural (nonhorticultural) products	10
Packaging	1
TOTAL	22

TABLE 5.6. The commodity or article with which snails and slugs intercepted by the State of Hawaii Plant Quarantine Branch were associated (data from 14 December 1994 to 6 August 1999; raw data provided by N. Reimer, pers. comm.).

Pathway/association	Number of interceptions
Agricultural products	164
Horticultural (nonagricultural) products	67
Other (container, pallet, vehicle, unknown)	6
TOTAL	237

Table 5.5. These data, although not as recent as the U.S. data, again emphasize the horticultural trade.

Interception data from the Hawaii Department of Agriculture are given in Table 5.6. These data refer only to domestic interceptions, that is, interceptions of shipments from the mainland United States to Hawaii (international shipments to Hawaii are dealt with by the federal U.S. Department of Agriculture; Holt 1996). Again the horticultural trade is emphasized, though it is second to the shipment of agricultural products. Undoubtedly, however, these data are biased inasmuch as inspectors focus their attentions in these two areas, although this reflects past experience of levels of interceptions.

Thus not surprisingly, these few datasets heavily implicate the horticultural trade in the worldwide spread of alien snails and slugs. The trade in agricultural products is also important; and there is a general association with containers. Just as the import of tiles is a major pathway into the United States, there are probably many other country-specific pathways associated with the specific import patterns of any given country.

TEMPORAL AND SPATIAL CHANGES IN PATHWAYS

Given the paucity of data available to identify the relative importance of different pathways of introduction of snails and slugs in general, it is even more difficult to ascertain with any certainty whether there have been any changes in these relative proportions over time, changes that might permit us to predict the future significance of the pathways. We do know, however, that rates of introduction of aquatic snails increased in the 1970s and 1980s because of expansion of the aquarium trade (Britton 1991, OTA 1993, p. 96, Pointier 1999).

Clearly, modern transport permits more rapid and wider movement of alien species than in the past (Carlton 1996). Long ocean voyages, perhaps involving extended periods at high latitudes (e.g., to round Cape Horn), may not have been conducive to the survival of any but the most hardy snails and slugs. In contrast, modern shipping, facilitated by the Panama and Suez Canals, is rapid and may not involve long periods in harsh environments. The burgeoning of air travel and air cargo no doubt has had a major impact on the ease and speed with which these species can be introduced. Although in general these statements must be true, it is not possible to provide a more detailed analysis on the basis of the very limited data available.

One example, however, illustrates these general changes. Introduced snails and slugs in Hawaii have come from many parts of the world. These geographic origins have changed over time (Table 5.7), reflecting the changing relative importance of different pathways. The first period (pre-1778) is prior to the arrival of Westerners in the islands; introductions by Pacific islanders during this period include a Pacific island species and two species of unknown origin. The second period (1778–1909) is dominated by taxonomic description of introduced species, often without realizing they were not native; Asian and Australasian (i.e., Pacific Rim) species dominate. During the third period (1910–99), many species were deliberately introduced or at least were immediately acknowledged as nonindigenous; introductions are heavily dominated by New World species, undoubtedly in large part because of rapidly increased commerce with the United States following its annexation of Hawaii, but introductions during this period also include a large number of African species introduced in the 1950s and 1960s as putative biological control agents against the giant African snail, *Achatina fulica*. For additional discussion of these changing geographic origins of the Hawaiian nonindigenous snail and slug faunas, see Cowie (1998a).

TABLE 5.7. Region of origin of snails and slugs introduced to the Hawaiian Islands, summarized by the period during which they arrived (data from Cowie 1997, 1998a, 1999, 2000b).

Region of origin	pre-1778	1778–1909	1910–1999	Total
Pacific	1	—	1	2
Asia/Australasia	—	9	5	14
New World	—	6	23	29
Europe	—	3	9	12
Africa	—	2	10	12
Unknown	2	8	6	16
Total	3	28	54	85

PREDICTED CHANGES

Introductions do not appear to be leveling off. For instance, in Hawaii the rate of introductions of nonmarine snails and slugs does not appear to be slowing (Fig. 5.1). Up until 1989, 76 species had been introduced (Cowie 1997, 1998a), and an additional 9 species were recorded during the decade 1990–99 (Cowie 1998a, 1999, 2000a). It is possible, however, that this is to some extent an artifact of an increased scientific interest in and attempt to document these alien species during the 1990s. Whether the rate of intro-

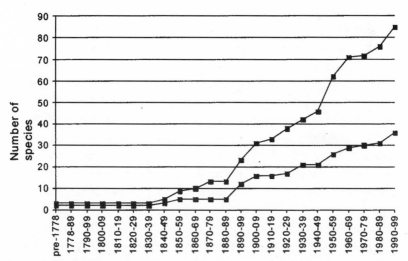

FIGURE 5.1. Cumulative numbers of land and freshwater species of snails and slugs recorded as introduced to the wild (upper line) and established in the wild (lower line) in the Hawaiian Islands, by decade. (Data from R. H. Cowie 1997, 1998a, 1999, 2000b.)

ductions is changing cannot be assessed. At least since the mid–nineteenth century, the rate of introduction of snails and slugs to Hawaii has fluctuated widely (Cowie 1998a) just as it has more generally for all alien species in the United States (OTA 1993, p. 95).

Britton (1991) suggested that the rate of introduction of terrestrial mollusks to the United States had remained fairly constant from 1955 to 1990, perhaps declining slightly from 1980 to 1990, but that the rate for freshwater species had increased. The number of recorded interceptions of nonindigenous mollusks entering the United States between 1984 and mid-1991 was 2,889 (Britton 1991). More recently, but in the slightly shorter period of 1993 to 1998, more than 4,900 interceptions were made (Robinson 1999). However, Britton's (1991) data were primarily for those species that were considered of quarantine significance, whereas Robinson's (1999) data were for all species. Only about 45 percent of Robinson's data (about 2,200 interceptions) were for species of quarantine significance. There is, therefore, no obvious trend of increasing interceptions; but there is considerable temporal stochasticity in the rate of introduction (Britton 1991), and a general increase in the rate is to be expected.

It is possible to make only general predictions regarding future changes in the relative importance of different pathways. Increased facilitation of global trade by international trade agreements, in particular under the auspices of the World Trade Organization, will lead to a continuing increase in the number of pathways and hence the movement of alien species (McNeely 1996). If trade routes become even more global, this will lead to the increasing global homogenization of snail and slug faunas. Similarly, increased facility for international travel will expand pathways and facilitate biotic homogenization.

With most quarantine efforts focused on plant (especially agricultural) pests (see below), it may be that agricultural exporters are becoming more concerned about contamination (and hence rejection) of their products. For example, introduced European snails (notably *Theba pisana, Cernuella virgata, Cochlicella acuta*) that are pests of Australian cereals have caused great concern because contamination by snails can lead to rejection of shipments of grain by countries to which Australia exports (Baker 1986, 1996). Exporters may increase their efforts in the country of origin to ensure their shipments are not contaminated, thereby reducing the importance of the pathway associated with the export/import of agricultural products. Such an effort has yet to be seen in the horticultural trade, which generally seems to be very poorly regulated in regard to contamination by alien organisms. This appears especially problematic as horticultural activities are predicted to increase (OTA 1993, p. 291).

Whether the aquarium trade and associated introductions of aquatic snails will continue to expand rapidly, as observed during the 1970s (Britton 1991, OTA 1993, p. 96), is less certain.

PREVENTION MEASURES

A comprehensive summary of what is needed to prevent the continued loss of biodiversity resulting from the introduction of alien species has been presented by the IUCN—the World Conservation Union (IUCN 1999). In a short but eloquent and compelling article, Simberloff (1998) summarized these needs. The following comments are presented with these general needs as a backdrop. Most of the ideas presented are not specific to snails and slugs; they are general in nature but also appropriate to addressing the continuing introduction of many species of nonmarine mollusks. However, the emphasis placed on different aspects of the following measures should be tailored to those pathways outlined above that appear to be important in the introduction of snails and slugs.

Quarantine

Preventing entry of a harmful species is always preferable to attempting to eradicate or control it after it has been introduced (Simberloff 1998), although complete prevention of the entry of all alien species is an unachievable goal, even with much increased resources (Devine 1998, Steinke and Walton 1999). Most countries lack (or have minimal) quarantine restrictions or quarantine enforcement agencies; where inspection/enforcement agencies exist, they may be ineffective and/or underfunded.

Generally, quarantine efforts are focused on potential plant pests (OTA 1993, p. 173, Devine 1998). Indeed, some quarantine and inspection agencies may be authorized to routinely inspect only agricultural and horticultural products, with aliens arriving by other pathways being intercepted only incidentally (e.g., the Hawaii Department of Agriculture). In the United States, the quarantine agency (Plant Protection and Quarantine, U.S. Department of Agriculture) can currently prevent entry of an intercepted organism only if it is a proven plant pest or there is at least strong evidence that it might be such (Devine 1998, Robinson 1999), thus restricting its authority where other nonindigenous species are concerned. Introduction of biological control agents is barely regulated at all (Miller and Aplet 1993). Because of this focus on terrestrial plant pests, aquatic mollusks are far less restricted than terrestrial species (Britton 1991); and predatory snails such as species of *Euglandina*, which are known to have devastated native snail faunas elsewhere (Murray et al. 1988, Cowie 1992, Hadfield et al. 1993, Civeyrel and Simberloff 1996), are poorly regulated (Miller and Aplet 1993). International agreements (e.g., the International Plant Protection Convention) often define quarantine pests only in terms of economic importance (OTA 1993, p. 296, Campbell 1998, Steinke and Walton 1999), although the United States now recognizes the potential threat of invasive pest species, whether they are of agricultural concern or not. Daehler (1998) has shown that only 25 percent of plant species that are important invaders in natural areas were also serious agricultural weeds, indicating that quarantine regulations focused only on preventing agricultural pest introductions are unlikely to prevent the introduction of most natural area invaders.

Many countries base their quarantine on so-called blacklists, lists of species that are known to cause problems; authorities may prevent the entry only of proven potential pests. Simberloff and Stiling (1996) and Simberloff (1998) have argued that instead what is needed are "whitelists," such that only species that have been proven benign may be introduced, as has recently been implemented in Australia for plants (Steinke and Walton 1999). However, most countries are unlikely to implement whitelists, for practical reasons and because of international agreements. Proving that a species is or is not potentially damaging is difficult.

Models have been developed for plants that allow authorities to evaluate the likely invasiveness of an alien species (Tucker and Richardson 1995, Reichard and Hamilton 1997, Pheloung et al. 1999). Many factors, changing over time, mean that prediction of invasion based on such models is unlikely ever to be entirely accurate (Carlton 1996). Although no such formal model is available for snails and slugs, Robinson (1999) has used the term *traveling species*, first introduced by Smith (1989), for those species that seem particularly susceptible to being distributed in association with human activities; and determination of a species as a traveling species depended on application of a suite of basic criteria (a simple model), modified from Harry (1964, 1966) and Smith (1989). As used by Robinson (1999), the term traveling is not exactly synonymous with invasive. Most of the species identified by Smith (1989) and Robinson (1999) are both traveling species and invasive species. However, traveling is confined to those species transported by humans to new areas, whereas invasive also includes species that may be increasing their range without humans providing the means of transport. In some instances, traveling species may not be invasive. It may eventually become possible to predict which snail and slug species will or will not be traveling/invasive species by development of models expanding on these criteria.

Only a small proportion of what enters a country can be inspected by quarantine officials, even in relatively wealthy countries (e.g., the United States; see Devine 1998), because of limited resources and time constraints (Dahlsten 1986). Even when a major pathway has been identified (e.g., tiles imported into the United States) and inspections are targeted at this pathway, inspections may not be comprehensive because of the sheer scale of commerce compared to available resources (Robinson 1999). Throughout the world, many alien snails and slugs undoubtedly get through quarantine inspections (if such exist); it is difficult to see that current levels of application of quarantine measures, in a limited number of countries, in fact do much more than scratch the surface of the global tide of invasions.

Nevertheless, some success can be mentioned. The giant African snail, *Achatina fulica*, has been successfully prevented from becoming established in the continental United States, Australia, and Fiji; this has largely been accomplished through the vigilance of quarantine officials, both in preventing its initial entry and in rapidly eradicating it once it had been detected in the wild (Mead 1979, Colman 1977, 1978, Ikin 1983). Also, some species that were often intercepted by quarantine agencies in the past are now much less frequently intercepted, suggesting perhaps that certain quarantine regulations are effective in preventing entry of similar species (e.g., the United States now permits no soil to be carried with horticultural plants for propagation).

Many shippers and exporting countries have improved their export procedures because of the frequency of interceptions in the past. However, this could also reflect changes in pathways. For instance, certain plants are no longer imported because they are now readily available locally (Robinson 1999), so the snails and slugs associated with them would be less likely to be introduced. Larger species may be more readily intercepted and more easily eradicated than smaller species, so the above-mentioned successes may in part be a reflection of this.

The success of quarantine measures cannot logically be assessed in terms of interception data. Success of quarantine measures in preventing establishment of alien species can be measured only in terms of the numbers of species getting through quarantine and becoming established. If this number is going down, then that might represent correlative (but not causative; there may be other causes) evidence that quarantine measures are working. Of course, such data currently are almost impossible to generate. Thus, evaluation of the efficacy of quarantine measures is fraught with difficulty; in fact, given the level of inspection that is usually possible, current quarantine measures may in general have a small influence on the worldwide spread of alien species, especially inadvertently introduced invertebrates.

Current levels of quarantine activities may have only minor effects on the continuing spread of alien snails and slugs into most countries, but clear steps to significantly reduce the flow of species include creation of quarantine agencies in countries currently lacking them, enhanced levels of inspection, stricter provision for treatment or rejection of infested shipments, and wider authority to inspect, treat, and reject shipments. Infractions should be strongly discouraged by whatever means possible. However, faced with increasing global demands—especially from Western/Northern governments and corporate businesses—to facilitate trade (short-term benefits), regardless of the negative issues (long-term costs) associated with it (Jenkins 1996a, 1996b, 1998), there may be little hope of imposing regulations to deal with the introduction of alien species, which would inevitably be seen as restrictive in terms of global trade (but see Yu 1994, 1996).

Eradication

It is becoming an automatic refrain when addressing problems associated with alien species that eradication is difficult or more often impossible once a species has become established (Crooks and Soulé 1996, Ewel et al. 1999). This is especially the case where invertebrates are concerned. Usually, an invasion of invertebrates such as snails and slugs is noticed only when it has become widespread and population densities are already high. At this stage, eradication is likely to be impossible. Even if a new invasion is detected early and could be eradicated, authorities can seldom justify allocation of resources for eradication, often because of lack of information about potential problems. They will therefore wait until the invasion becomes a serious problem, by which time it is too late. Also, legal challenges to eradication programs (often but not always related to pesticide applications) can become extremely complex and may delay the program to a point when it will no longer have a chance of success (Getz 1989), although recent legislative changes (e.g., in California) may allow such legal challenges to be overridden (C. Daehler, pers. comm.).

Nevertheless, eradication may occasionally be possible. *Achatina fulica* was eradicated from Florida after a major effort extending over six years (Mead 1979); and *Theba pisana* was probably eradicated from California, but only after many years of intensive effort (Gammon 1943, Armitage 1949). *Theba pisana* subsequently reappeared in California, and on the basis of genetic evidence this may have resulted from a new introduction (Cowie 1987, Roth et al. 1987).

Eradication can therefore be achieved under certain circumstances, especially if attempted in the early stages of invasion. Authorities should continue to be receptive to the idea of eradication before a new invasion is demonstrably an agricultural, environmental, or human health problem.

Environmental Responsibility and Education

The original shipper in the country of origin, and the importer in the receiving country, should be encouraged to make sure that the product is free of infestation as it is shipped (Courtenay and Williams 1992, in relation to aquaculture; Dudgeon and Yipp 1983, in relation to the aquarium trade; see also Devine 1998). This approach, which may require education and outreach in some cases, is far better than having to deal with alien species once they are detected in the new locality or having to rely on inspection agencies to detect them as they are imported. Noncompliance should be discouraged by whatever means possible. In the case of deliberate introductions (e.g., aquarium snails), shippers would have to agree, prior to shipment, to pay the costs associated with an invasion resulting from their shipment (Bean 1996); though identification of the source of a particular invasion would be difficult if not impossible. An additional difficulty is that, in some countries, species that are already widely utilized cannot be legally regulated (e.g., most aquatic species in the aquarium industry in the United States).

Where deliberate and legal import of snails is concerned (e.g., the aquarium trade), the person or organization importing the animals should be encouraged (or required) to act responsibly by importing only species that are not considered invasive, ensuring that their animals do not escape, and informing their customers of the need to act equally responsibly. This requires major efforts to educate all those involved in the import/export of alien species, including the public whose demand drives the trade. Lack of public knowledge of the potential impacts of alien snails and slugs is a core problem that can be addressed only by much greater involvement in all aspects of the process on the part of the scientists (malacologists) who possess that knowledge.

More broadly, problems associated with the continued spread of alien species will never be resolved until the issue has the support of the politicians and business people who have the power to do something about them on a global scale. Many of these people will care about the issue only if they see their actions (or inactions) leading to loss of votes and/or money. Thus, the general populace, who can pressure these politicians and business people, must be educated and become enthused about the issue to a sufficient extent that they will influence the people with power. This call for public education is often repeated, but few organizations involved in dealing with the problems of alien species expend much effort in this direction (OTA 1993). Furthermore, the brochures and leaflets that are produced often do not reach an appropriate and broad audience. Education is a huge challenge, which goes beyond issues of alien species and addresses the entire biodiversity crisis. It probably requires a major turn-around in the way people, especially Westerners/Northerners, view the world.

CONCLUSION

Alien snails and slugs are being introduced to most parts of the world via numerous pathways. Rigorous data to quantify the relative and perhaps changing importance of these pathways are scarce. Nevertheless, some generalities appear. The majority of species probably travel accidentally in association with commerce. The horticultural trade (specifically the trade in cut flowers and propagative plants) seems particularly responsible for the inadvertent introduction of many species of terrestrial snails and slugs, while the domestic aquarium trade seems responsible for many of the aquatic species. Some pathways are specific to certain countries; for example, import of tiles to the United States is a major pathway for introduction of terrestrial snails and slugs from Europe. Deliberate introductions are few by comparison, but may sometimes be more successful and include some of the worst invaders in both the terrestrial and freshwater realms. Deliberate introductions may be legal, illegal, or simply unregulated. Attempted biological control of the giant African snail (*Achatina fulica*) using numerous predatory snails (most notably *Euglandina rosea*), and the introduction of ampullariids (most notably *Pomacea* sp.) as food resources to be developed in aquaculture, have probably been the worst cases in terrestrial and freshwater systems, respectively.

From a worldwide perspective the prognosis is not good. There is no evidence that introductions are declining; in fact, the opposite appears to be true. The lack of adequate quarantine and authority to deal with alien species outside the agricultural/economic realm, even in the best-regulated countries, means that most introductions probably can only be delayed. Increasing emphasis on free trade globally means that the necessary, more restrictive measures to slow the spread of alien species will probably not be implemented by many countries, even though the economic costs (not to mention the ecological, aesthetic, and public health costs) associated with alien species are enormous (Pimentel et al. 2000, 2001). As for the world's biota as a whole, the land and freshwater snail and slug fauna is becoming and will continue to become increasingly uniform. However, some success in preventing or at least slowing this trend may be possible. We must continue to address the issue of alien species and advocate for solutions, however this conflicts with the current climate of economic globalization.

ACKNOWLEDGMENTS

We thank Greg Ruiz and Jim Carlton for inviting RHC to participate in the Global Invasive Species Programme workshop on invasion pathways, November 1999, from which this paper is derived. Geoff Baker, Joe Britton, Domingo Cravalho, Curt Daehler, Lu Eldredge, and Neil Reimer provided much valuable information and/or made useful suggestions. We also thank Dave Britton, Carl Christensen, Lori Lach, and Rebecca Rundell for discussion and/or for reviewing the manuscript.

REFERENCES

Andrews, A. 1990. Fragmentation of habitat by roads and utility corridors: a review. *Australian Zoologist* 26: 130–141.

Armitage, H. M. 1949. Bureau of entomology. Thirtieth annual report. *California Department of Agriculture Bulletin* 38: 157–216.

Baker, G. H. 1986. The biology and control of white snails Mollusca: Helicidae, introduced pests in Australia. *CSIRO Division of Entomology Technical Paper* 25: 1–31.

Baker, G. H. 1996. Population dynamics of the Mediterranean snail, *Cernuella virgata*, in a pasture-cereal rotation in South Australia. *British Crop Protection Council Symposium Proceedings* 66: 117–124.

Baminger, H. and M. Haase. 1999. Variation in spermathecal morphology and amount of sperm stored in populations of the simultaneously hermaphroditic land snail *Arianta arbustorum. Journal of Zoology* 249: 165–171.

Barker, G. M. 1979. The introduced slugs of New Zealand Gastropoda: Pulmonata. *New Zealand Journal of Zoology* 6: 411–437.

Barker, G. M. 1999. *Naturalised terrestrial Stylommatophora (Mollusca: Gastropoda). Fauna of New Zealand no. 38.* Canterbury: Manaaki Whenua Press.

Bean, M. J. 1996. Legal authorities for controlling alien species: a survey of tools and their effectiveness. In *Proceedings of the Norway/UN conference on alien species*, O. T. Sandlund, P. J. Schei, and A. Viken, eds., pp. 204–209. Trondheim:

Directorate for Nature Management and Norwegian Institute for Nature Research.

Bieler, R. and J. Slapcinsky. 2000. A case study for the development of an island fauna: recent terrestrial mollusks of Bermuda. *Nemouria, Occasional Papers of the Delaware Museum of Natural History* 44: 1–100.

Britton, J. C. 1991. *Pathways and consequences of the introduction of non-indigenous freshwater, terrestrial, and estuarine mollusks in the United States.* Unpublished report to the Office of Technology Assessment, Congress of the United States, contract number H3-5750.0.

Campbell, F. T. 1998. Fatal flaws. *World Conservation; quarterly bulletin of IUCN—The World Conservation Union* 4/97–1/98: 6–7.

Carlton, J. T. 1992. Dispersal of living organisms into aquatic ecosystems as mediated by aquaculture and fisheries activities. In *Dispersal of living organisms into aquatic ecosystems*, A. Rosenfield and R. Mann, eds., pp. 13–46. College Park: Maryland Sea Grant College, University of Maryland.

Carlton, J. T. 1996. Pattern, process, and prediction in marine invasion ecology. *Biological Conservation* 78: 97–106.

Chace, P. G. 1987. *Viviparus*, the Chinese field snail, a historic archaeological enigma. *Pacific Coast Archaeological Society Quarterly* 23(2): 69–79.

Christensen, C. C. 1984. Are *Euglandina* and *Gonaxis* effective agents for biological control of the Giant African Snail in Hawaii? *American Malacological Bulletin* 2: 98–99.

Civeyrel, L. and D. Simberloff. 1996. A tale of two snails: is the cure worse than the disease? *Biodiversity and Conservation* 5: 1231–1252.

Clapp, G. H. 1919. Cuban mollusks colonized in Florida. *The Nautilus* 32: 104–105.

Clarke, B., J. Murray, and M. S. Johnson. 1984. The extinction of endemic species by a program of biological control. *Pacific Science* 38: 97–104.

Colman, P. H. 1977. An introduction of *Achatina fulica* to Australia. *Malacological Review* 10: 77–78.

Colman, P. H. 1978. An invading giant. *Wildlife in Australia* 15: 46–47.

Courtenay, W. R. Jr., and J. D. Williams. 1992. Dispersal of exotic species from aquaculture sources, with emphasis on freshwater fishes. In *Dispersal of living organisms into aquatic ecosystems*, A. Rosenfield and R. Mann, eds., pp. 49–81. College Park: Maryland Sea Grant College, University of Maryland.

Cowie, R. H. 1987. Rediscovery of *Theba pisana* at Manorbier, South Wales. *Journal of Conchology* 32: 384–385.

Cowie, R. H. 1992. Evolution and extinction of Partulidae, endemic Pacific island land snails. *Philosophical Transactions of the Royal Society of London* B 335: 167–191.

Cowie, R. H. 1995. Identity, distribution and impacts of introduced Ampullariidae and Viviparidae in the Hawaiian Islands. *Journal of Medical and Applied Malacology* 5 [1993]: 61–67.

Cowie, R. H. 1996. New records of land and freshwater snails in the Hawaiian Islands. *Bishop Museum Occasional Papers* 46: 25–27.

Cowie, R. H. 1997. Catalog and bibliography of the nonindigenous nonmarine snails and slugs of the Hawaiian Islands. *Bishop Museum Occasional Papers* 50: 1–66.

Cowie, R. H. 1998a. Patterns of introduction of non-indigenous non-marine snails and slugs in the Hawaiian Islands. *Biodiversity and Conservation* 7: 349–368.

Cowie, R. H. 1998b. Homogenization of Pacific island snails. *World Conservation; quarterly bulletin of IUCN—The World Conservation Union* 4/97–1/98: 18.

Cowie, R. H. 1998c. New records of nonindigenous land snails and slugs in the Hawaiian Islands. *Bishop Museum Occasional Papers* 56: 60.

Cowie, R. H. 1998d. *Catalog of the nonmarine snails and slugs of the Samoan Islands. Bishop Museum Bulletin in Zoology 3.* Honolulu: Bishop Museum Press.

Cowie, R. H. 1999. New records of alien nonmarine mollusks in the Hawaiian Islands. *Bishop Museum Occasional Papers* 59: 48–50.

Cowie, R. H. 2000a. New records of alien land snails and slugs in the Hawaiian Islands. *Bishop Museum Occasional Papers* 64: 51–53.

Cowie, R. H. 2000b. Non-indigenous land and freshwater molluscs in the islands of the Pacific: conservation impacts and threats. In *Invasive species in the Pacific: a technical review and regional strategy*, G. Sherley, ed., pp. 143–172. Apia: South Pacific Regional Environment Programme.

Cowie, R. H. 2001. Can snails ever be effective and safe biocontrol agents? *International Journal of Pest Management* 47: 23–40.

Cowie, R. H. 2002a. Invertebrate invasions on Pacific islands and the replacement of unique native faunas: a synthesis of the land and freshwater snails. *Biological Invasions* 3(3)[2001]: 119–136.

Cowie, R. H. 2002b. Apple snails (Ampullariidae) as agricultural pests: their biology, impacts and management. In *Molluscs as crop pests*, G. M. Barker, ed., pp. 145–192. Wallingford: CABI Publishing.

Cowie, R. H. and R. P. Cook. 2001. Extinction or survival: partulid tree snails in American Samoa. *Biodiversity and Conservation* 10: 143–159.

Crooks, J. and M. E. Soulé. 1996. Lag times in population explosions of invasive species: causes and implications. In *Proceedings of the Norway/UN conference on alien species*, O. T. Sandlund, P. J. Schei, and A. Viken, eds., pp. 39–46. Trondheim: Directorate for Nature Management and Norwegian Institute for Nature Research.

Daehler, C. C. 1998. The taxonomic distribution of invasive angiosperm plants: ecological insights and comparison to agricultural weeds. *Biological Conservation* 84: 167–180.

Dahlsten, D. L. 1986. Control of invaders. In *Ecology of biological invasions of North America and Hawaii*, H. A. Mooney and J. A. Drake, eds., pp. 275–302. New York: Springer.

Devine, R. S. 1998. *Alien invasion. America's battle with non-native animals and plants.* Washington: National Geographic Society.

Dudgeon, D. and M. W. Yipp. 1983. A report on the gastropod fauna of aquarium fish farms in Hong Kong, with special reference to an introduced human schistosome host species, *Biomphalaria straminea* (Pulmonata: Planorbidae). *Malacological Review* 16: 93–94.

Duncan, C. J. 1975. Reproduction. In *Pulmonates. Volume 1. Functional anatomy and physiology*, V. Fretter and J. Peake, eds., pp. 309–365. London: Academic Press.

Ewel, J. J., D. J. O'Dowd, J. Bergelson, C. C. Daehler, C. M. D'Antonio, L. D. Gómez, D. R. Gordon, R. J. Hobbs, A. Holt, K. R. Hopper, C. E. Hughes, M. LaHart, R. R. B. Leakey, W. G. Lee, L. L. Loope, D. H. Lorence, S. M. Louda, A. E. Lugo, P. B. McEvoy, D. M. Richardson, and P. M. Vitousek. 1999. Deliberate introductions of species: research needs. *BioScience* 49: 619–630.

Fischer, T. W. and R. E. Orth. 1985. Biological control of snails. *Occasional Papers, Department of Entomology, University of California, Riverside* 1: i–viii, 1–111.

Foltz, D. W., H. Ochman, J. S. Jones, S. M. Evangelisti, and R. K. Selander. 1982. Genetic population structure and breeding systems in arionid slugs (Mollusca: Pulmonata). *Biological Journal of the Linnaean Society* 17: 225–241.

Foltz, D. W., H. Ochman, and R. K. Selander. 1984. Genetic diversity and breeding systems in terrestrial slugs of the families Limacidae and Arionidae. *Malacologia* 25: 593–605.

Frandsen, F. and H. Madsen. 1979. A review of *Helisoma duryi* in biological control. *Acta Tropica* 36: 67–84.

Fretter, V. 1984. Prosobranchs. In *The Mollusca. Volume 7. Reproduction*, A. S. Tompa, N. H. Verdonk, and J. A. M. van den Biggelaar, eds., pp. 1–45. London: Academic Press.

Gammon, E. T. 1943. Helicid snails in California. *Bulletin of the State of California Department of Agriculture* 32: 173–187.

Gargominy, O., P. Bouchet, M. Pascal, T. Jaffré, and J. C. Tourneur. 1996. Conséquences des introductions d'espèces animales et végétales sur la biodiversité en Nouvelle-Calédonie. *Revue d'Ecologie (la Terre et la Vie)* 51: 375–402.

Geraerts, W. P. M. and J. Joosse. 1984. Freshwater snails (Basommatophora). In *The Mollusca. Volume 7. Reproduction*, A. S. Tompa, N. H. Verdonk, and J. A. M. van den Biggelaar, eds., pp. 141–207. London: Academic Press.

Getz, C. W. 1989. Legal implications of eradication programs. In *Eradication of exotic species*, D. L. Dahlsten and R. Garcia, eds., pp. 66–73. New Haven: Yale University Press.

Godan, D. 1983. *Pest slugs and snails. Biology and control.* New York: Springer-Verlag.

Griffiths, O., A. Cook, and S. M. Wells. 1993. The diet of the carnivorous snail *Euglandina rosea* in Mauritius and its implications for threatened island gastropod faunas. *Journal of Zoology* 229: 79–89.

Hadfield, M. G. 1986. Extinction in Hawaiian achatinelline snails. *Malacologia* 27: 67–81.

Hadfield, M. G., S. E. Miller, and A. H. Carwile. 1993. The decimation of endemic Hawai'ian [sic] tree snails by alien predators. *American Zoologist* 33: 610–622.

Halwart, M. 1994. The golden apple snail *Pomacea canaliculata* in Asian rice farming systems: present impact and future threat. *International Journal of Pest Management* 40: 199–206.

Hanna, G. D. 1966. Introduced mollusks of western North America. *Occasional Papers of the California Academy of Sciences* 48: 1–108.

Hara, A. and T. Hata. 1999. Insects, mites, and other pests. In *Growing dendrobium orchids in Hawaii*, K. Leonhardt and K. Sewake, eds., pp. 29–45. Honolulu: College of Tropical Agriculture and Human Resources, University of Hawaii.

Harry, H. W. 1964. The foreign freshwater snails now established in Puerto Rico. *American Malacological Union Annual Reports* 1964: 4–5.

Harry, H. W. 1966. Land snails of Ulithi Atoll, Caroline Islands: a study of snails accidentally distributed by man. *Pacific Science* 20: 212–223.

Holt, A. 1996. An alliance of biodiversity, agriculture, health, and business interests for improved alien species management in Hawaii. In *Proceedings of the Norway/UN conference on alien species*, O. T. Sandlund, P. J. Schei, and A. Viken, eds., pp. 155–160. Trondheim: Directorate for Nature Management and Norwegian Institute for Nature Research.

Howarth, F. G. 1991. Environmental impacts of classical biological control. *Annual Reviews of Entomology* 36: 485–509.

Ikin, R. 1983. Giant African snail in Fiji. *South Pacific Commission Information Circular* 91: 1.

International Union for the Conservation of Nature and Natural Resources— the World Conservation Union. 1999. IUCN guidelines for the prevention of biodiversity loss due to biological invasion. *Species* 31–32: 28–42.

Jenkins, P. T. 1996a. Free trade and exotic species introductions. *Conservation Biology* 10(1): 300–302.

Jenkins, P. T. 1996b. Free trade and exotic species introductions. In *Proceedings of the Norway/UN conference on alien species*, O. T. Sandlund, P. J. Schei, and A. Viken, eds., pp. 145–147. Trondheim: Directorate for Nature Management and Norwegian Institute for Nature Research.

Jenkins, P. T. 1998. Re-joining the continents. *World Conservation; quarterly bulletin of IUCN—The World Conservation Union* 4/97–1/98: 5–6.

Jokinen, E. H. 1982. *Cipangopualudina chinensis* Gastropoda: Viviparidae in North America, review and update. *The Nautilus* 96: 89–95.

Kay, E. A. 1995. Which molluscs for extinction? In *The conservation biology of mollusks*, E. A. Kay, ed., pp. 1–7. Gland: IUCN.

Kerney, M. P. 1966. Snails and man in Britain. *Journal of Conchology* 26: 3–14.

Lach, L. and R. H. Cowie. 1999. The spread of the introduced freshwater apple snail *Pomacea canaliculata* Lamarck Gastropoda: Ampullariidae on Oʻahu, Hawaiʻi. *Bishop Museum Occasional Papers* 58: 66–71.

MacIsaac, H. J. 1999. Biological invasions in Lake Erie: past, present and future. In *The state of Lake Erie. Past, present and future*, M. Munawar, T. Edsall, and I. F. Munawar, eds., pp. 305–322. Leiden: Backhuys Publishers.

Mackie, G. L. 1999a. Mollusc introductions through aquarium trade. In *Nonindigenous freshwater organisms. Vectors, biology, and impacts*, R. Claudi and J. H. Leach, eds., pp. 135–147. Boca Raton: Lewis Publishers.

Mackie, G. L. 1999b. Introduction of molluscs through the import for live food. In *Nonindigenous freshwater organisms. Vectors, biology, and impacts*, R. Claudi and J. H. Leach, eds., pp. 305–313. Boca Raton: Lewis Publishers.

Mackie, G. L. 1999c. Ballast water introductions of Mollusca. In *Nonindigenous freshwater organisms. Vectors, biology, and impacts*, R. Claudi and J. H. Leach, eds., pp. 219–254. Boca Raton: Lewis Publishers.

McKinney, M. L. and J. L. Lockwood. 1999. Biotic homogenization: a few winners replacing many losers in the next mass extinction. *Trends in Ecology and Evolution* 14: 450–453.

McNeely, J. A. 1996. The great reshuffling: how alien species help feed the global economy. In *Proceedings of the Norway/UN conference on alien species*, O. T. Sandlund, P. J. Schei, and A. Viken, eds., pp. 53–59. Trondheim: Directorate for Nature Management and Norwegian Institute for Nature Research.

Mead, A. R. 1979. *Pulmonates. Volume 2B. Economic malacology with particular reference to* Achatina fulica. London: Academic Press.

Miller, M. and G. Aplet. 1993. Biological control: a little knowledge is a dangerous thing. *Rutgers Law Review* 45: 285–334.

Mills, E. L., M. D. Scheuerell, J. T. Carlton, and D. L. Strayer. 1997. Biological invasions in the Hudson River Basin: an inventory and historical analysis. *New York State Museum Circular* 57, viii + 51 p.

Murray, H. D. 1968. *Otala lactea* in San Antonio, Texas. *The Nautilus* 81: 141–143.

Murray, J., E. Murray, M. S. Johnson, and B. Clarke. 1988. The extinction of *Partula* on Moorea. *Pacific Science* 42: 150–153.

Office of Technology Assessment, U.S. Congress. 1993. *Harmful non-indigenous species in the United States*. Washington: U.S. Government Printing Office.

Pace, G. L. 1973. The freshwater snails of Taiwan (Formosa). *Malacological Review Supplement* 1: 1–118.

Perera, G. and J. G. Walls. 1996. *Apple snails in the aquarium*. Neptune City: T. F. H. Publications.

Petit, S. 1858. Nota. *Journal de Conchyliologie* 7: 268.

Pheloung, P. C., P. A. Williams, and S. R. Halloy. 1999. A weed risk assessment model for use as a biosecurity tool evaluating plant introductions. *Journal of Environmental Management* 57: 239–251.

Pilsbry, H. A. 1946. Land Mollusca of North America (north of Mexico). Volume 2, part 1. *Academy of Natural Sciences of Philadelphia Monographs Number 3*. Philadelphia: Academy of Natural Sciences.

Pimentel, D., L. Lach, R. Zuniga, and D. Morrison. 2000. Environmental and economic costs of nonindigenous species in the United States. *BioScience* 50: 53–65.

Pimentel, D., S. McNair, J. Janecka, J. Wightman, C. Simmonds, C. O'Connell, E. Wong, L. Russel, J. Zern, T. Aquino, and T. Tsomondo. 2001. Economic and environmental threats of alien plant, animal, and microbe invasions. *Agriculture, Ecosystems and Environment* 84: 1–20.

Pointier, J. P. 1999. Invading freshwater gastropods: some conflicting aspects for public health. *Malacologia* 41: 403–411.

Pointier, J. P., R. N. Incani, C. Balzan, P. Chrosciechowski, and S. Prypchan. 1994. Invasion of the rivers of the littoral central region of Venezuela by *Thiara granifera* and *Melanoides tuberculata* Mollusca: Prosobranchia: Thiaridae and the absence of *Biomphalaria glabrata*, snail host of *Schistosoma mansoni*. *The Nautilus* 107: 124–128.

Pointier, J. P., A. Théron, and D. Imbert-Establet. 1988. Decline of a sylvatic focus of *Schistosoma mansoni* in Guadeloupe (French West Indies) following competitive displacement of the snail host *Biomphalaria glabrata* by *Ampullaria glauca*. *Oecologia* 75: 38–43.

Pointier, J. P., A. Théron, D. Imbert-Establet, and G. Borel. 1991. Eradication of a sylvatic focus of *Schistosoma mansoni* using biological control by competitor snails. *Biological Control* 1: 244–247.

Ponder, W. F. and D. R. Lindberg. 1997. Towards a phylogeny of gastropod molluscs: an analysis using morphological characters. *Zoological Journal of the Linnaean Society* 119: 83–265.

Reichard, S. H. and C. W. Hamilton. 1997. Predicting invasions of woody plants introduced into North America. *Conservation Biology* 11: 193–203.

Ricciardi, A. and H. J. MacIsaac. 2000. Recent mass invasion of the North American Great Lakes by Ponto-Caspian species. *Trends in Ecology and Evolution* 15: 62–65.

Robinson, D. G. 1999. Alien invasions: the effects of the global economy on nonmarine gastropod introductions into the United States. *Malacologia* 41: 413–438.

Roth, B., C. M. Hertz, and R. Cerutti. 1987. White snails (Helicidae) in San Diego County, California. *The Festivus* 19: 84–88.

Roth, B. and T. A. Pearce. 1984. *Vitrea contracta* (Westerlund) and other introduced land mollusks in Lynnwood, Washington. *The Veliger* 27: 90–92.

Sakovich, N. J. 1996. An integrated pest management (IPM) approach to the control of the brown garden snail (*Helix aspersa*) in California citrus orchards. *British Crop Protection Council Symposium Proceedings* 66: 283–287.

Simberloff, D. 1998. Facing the future. *World Conservation; quarterly bulletin of IUCN—The World Conservation Union* 4/97–1/98: 21–23.

Simberloff, D. and P. Stiling. 1996. Risks of species introduced for biological control. *Biological Conservation* 78: 185–192.

Simmonds, F. J. and I. W. Hughes. 1963. Biological control of snails exerted by *Euglandina rosea* (Ferussac) in Bermuda. *Entomophaga* 8: 219–222.

Smith, B. J. 1989. Travelling snails. *Journal of Medical and Applied Malacology* 1: 195–204.

Smith, L. D., M. J. Wonham, L. D. McCann, G. M. Ruiz, A. H. Hines, and J. T. Carlton. 1999. Invasion pressure to a ballast-flooded estuary and an assessment of inoculant survival. *Biological Invasions* 1: 67–87.

Solem, A. 1964. New records of New Caledonian nonmarine mollusks and an analysis of the introduced mollusks. *Pacific Science* 18: 130–137.

Steinke, E. and C. Walton. 1999. Weed risk assessment of plant imports to Australia: policy and process. *Australian Journal of Environmental Management* 6: 157–163.

Thompson, F. G. 1976. The occurrence in Florida of the West Indian land snail *Bulimulus guadalupensis*. *The Nautilus* 90: 10.

Tompa, A. S. 1984. Land snails Stylommatophora. In *The Mollusca. Volume 7.*

Reproduction, A. S. Tompa, N. H. Verdonk, and J. A. M. van den Biggelaar, eds., pp. 47–140. London: Academic Press.

Trombulak, S. C. and C. A. Frissell. 2000. Review of ecological effects of roads on terrestrial and aquatic communities. *Conservation Biology* 14: 18–30.

Tucker, K. C. and D. M. Richardson. 1995. An expert system for screening potentially invasive alien plants in South African fynbos. *Journal of Environmental Management* 44: 309–338.

van der Schalie, H. 1969. Man meddles with nature—Hawaiian style. *The Biologist* 51(4): 136–146.

van der Schalie, H. 1970. Snail control problems in Hawaii. *Annual Reports of the American Malacological Union 1969:* 55–56.

Woodruff, D. S. 1978. Evolution and adaptive radiation of *Cerion*: a remarkably diverse group of West Indian land snails. *Malacologia* 17: 223–239.

Yu, D. W. 1994. Free trade is green, protectionism is not. *Conservation Biology* 8: 989–996.

Yu, D. W. 1996. New factor in free trade: reply to Jenkins. *Conservation Biology* 10: 303–304.

Chapter 6

Freshwater Aquatic Vertebrate Introductions in the United States: Patterns and Pathways

Pam L. Fuller

Freshwater fish, amphibians, reptiles, birds, and mammals have been moved—intentionally and unintentionally—within and between continents for millennia. These translocations can have societal benefits and often unforeseen impacts in native ecosystems. Today, there are very few freshwater ecosystems that have not been altered by introductions. In this chapter, I examine patterns in time and space of freshwater vertebrate introductions (excluding birds) in the continental United States and the Hawaiian Islands. Furthermore, I discuss whether the likelihood of a successful invasion can be correlated to pathways, making some recommendations about pathways of freshwater introductions and ways they can be managed to reduce numbers of aquatic introductions.

An introduction is defined as the occurrence of a species outside its native range, regardless of whether the species is of foreign or domestic origin (i.e., native species within a country that are transplanted outside their native range). Here, introductions include not only species that have become established but also introductions of species that did not survive to become established populations. Species that are not known to be established are included because (1) they give insight into invasion pathways and potential, and (2) the population status is often unknown, especially as the scale of analysis (spatial or temporal) gets smaller.

This chapter primarily focuses on fishes. The vertebrate groups

123

TABLE 6.1. Origin of introduced species of aquatic animals reported nationwide in the United States. (Note: These numbers include failed introductions.)

	Total Number Introduced	Native to United States (%)	From Another Country (%)
Fishes[a]	565	60	36
Amphibians[b]	40	55	45
Reptiles[b]	51	65	35
Mammals[c]	5	60	40

[a] Remaining 4 percent of fish species are of hybrid origin.
[b] Reptile and amphibian numbers are probably much higher than reported, as probably only a small percentage of introductions are documented.
[c] Nutria (*Mycocaster coypu*), capybara (*Hydrochoerus hydrochaeris*), muskrat (*Ondatra zibethicus*), beaver (*Castor canadensis*), and California sea lion (*Zalophus californicus*).

included in the U.S. Geological Survey (USGS) database are fishes, amphibians, reptiles, and mammals (Table 6.1; see description below). However, there are relatively few introduced aquatic mammals, and relatively few amphibians and reptiles that are established; most amphibian and reptile introductions are single individuals that are released pets (see also Kraus, this volume, for analysis of these groups).

METHODS

The information presented is largely based on the USGS Nonindigenous Aquatic Species database http://nas.er.usgs.gov. This database has compiled occurrence records from published scientific papers, biologists' field collections, reports from state and federal agency biologists, and museum specimens (collections of species outside their native ranges that have been deposited in natural history museums). An animal is defined as being aquatic if it depends on water for at least some stage of its life cycle, or if it spends a significant portion of time in aquatic habitats. For example, many amphibians lead a terrestrial life but have an obligatory aquatic larval stage. Some reptiles, such as water snakes, do not live in the water as fish do, but are closely associated with aquatic habitats.

Species occurrences are often recorded in reference to political boundaries such as state and county. Since organisms do not observe political boundaries, this creates some difficulties in interpreting distributions, especially for aquatic habitats. Rivers often form the political boundaries between counties or states; hence, the river is associated with two states.

Conversely, a state or county can have multiple drainages, which are discrete biogeographic units.

The USGS program takes a unique approach in this problem by georeferencing all records within a drainage. The system used was devised by the USGS and is referred to as hydrologic unit codes (HUCs). Hydrologic units are a hierarchical series of drainages set up by the USGS in 1972. Boundaries are determined by the topography of the area, which, in turn, defines the direction of water flow. Each unit is assigned a number, a hydrologic unit code (HUC), to identify the hydrologic area. The largest scale has two digits; each successive level is a smaller section of the 2-digit HUC until an 8-digit HUC is reached.

The HUC system divides the United States into 21 regions (2-digit), 222 subregions (4-digit), 352 accounting units (6-digit), and 2,262 cataloging units (8-digit). Each hydrologic unit is assigned an 8-digit attribute code that uniquely identifies each of the four levels of classification within four 2-digit fields. A complete listing of all the hydrologic numbers and names can be found in the publication "Hydrologic Unit Maps" (USGS 1982).

Identifying the drainage where a species occurs can help predict regions susceptible to its spread and can be invaluable in protecting them or in knowing where to set up monitoring programs for detection of first arrivals. Unless there is a dramatic habitat change or barrier, downstream areas are at high risk of acquiring the introduced species. Upstream areas may also be at risk if the organism is either mobile or likely to be transported upstream by some other means (boat; carried on fish; bait release). Adjacent drainages are less at risk because most aquatic organisms cannot move over land (by themselves) to invade the next water body, except for some of the more amphibious creatures (e.g., the Asian swamp eel, the walking catfish, and some crayfish).

RESULTS

Spatial Patterns

On a regional basis, the Southeast Atlantic–Gulf is, by far, the region with the highest number of introduced fish species (Fig. 6.1). This region is followed by the California region and, more distantly, the Mid-Atlantic region. The South Atlantic–Gulf is also the region with the most introduced aquatic reptile and amphibian species, followed by New England and California.

When examining the country on the basis of drainage systems (Fig.

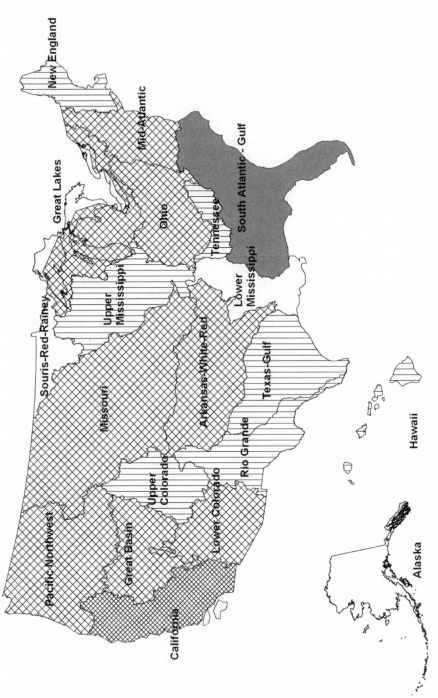

Figure 6.1. Number of fish taxa introduced by region (2-digit hydrologic unit).

6.2), the East and West Coasts in general have had many species introduced, whereas the central portion of the country has had fewer species
introduced. Certain drainages stand out as having high numbers of introductions. In the case of introduced fishes, the drainages with the most
species introduced include the upper Tennessee, Kanawha, Oahu, South
Florida, Rio Grande headwaters, Sacramento, Chowan–Roanoke, East
Texas coastal, Lower Colorado–Lake Meade, Edisto, South Platte, Southern
California, Humboldt, Salton Sea, Susquehanna, and Potomac. Many of
these drainages occur in areas with large human population centers
nearby.

Sometimes the number of introduced species in a state reflects state
management practices, or it may be a consequence of some other state-
based factor such as the size of the state, the number of drainages, an active
stocking program, or the presence of aquaculture facilities. On a national
basis, the states with the most fish species introduced are California,
Florida, Colorado, Texas, and Nevada (Fig. 6.3).

California has had a long history of stocking fishes, dating back to the
1870s when East Coast species were first brought to the West Coast and
introduced into the Sacramento River in California. In the early 1950s, the
state also attempted to stock many marine species in the Salton Sea. Stocking is not the only factor contributing to fish introductions. California's
warm climate allows some privately released tropical aquarium species to
survive. In addition, the extensive canal system delivering water to urban
areas has dispersed fishes into nonnative areas. Species have also been
introduced unintentionally by commercial ships to San Francisco Bay and
Long Beach.

The majority of introductions in Florida are due to the warm climate
and the presence of tropical fish farms. Exotic fish either escape these
farms or are released from someone's aquarium.

Colorado and Texas have traditionally had active stocking programs.
Some southern portions of Texas allow introduced tropical species to survive. The presence of a tropical fish farm in the headwaters of the Rio
Grande drainage in Colorado contributed numerous species to that state's
list. Nevada has numerous fish species introduced into Lake Meade and
has had tropical species introduced into thermal springs surrounding that
reservoir.

Smaller states, states in cold or remote areas, and states with few
human inhabitants have lower numbers of introduced species. The states
of Rhode Island, Vermont, New Hampshire, North Dakota, and Alaska are
included in these categories. Fewer human inhabitants mean fewer opportunities and less reason for introductions. For example, if the overall num-

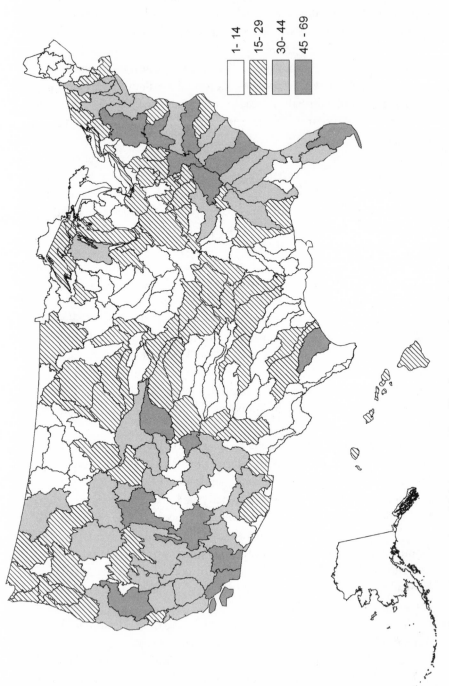

FIGURE 6.2. Number of fish taxa introduced by drainage (4-digit hydrologic unit).

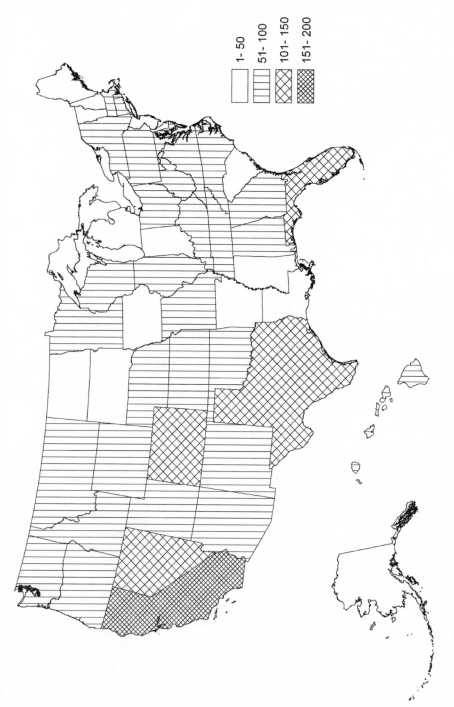

FIGURE 6.3. Number of fish taxa introduced into each state.

ber of aquarium hobbyists is smaller than in other states, chances are there will be fewer aquarium releases. Also, many of the intentional fish introductions have resulted from stocking (for angling) and were focused on lakes and reservoirs near metropolitan areas, to bring good fishing to close proximity of large population centers. For the above states that lack large metropolitan areas (typified by Atlanta, Los Angeles, Dallas), there is less reason to stock fish; this is coupled with the fact that fish stocks are relatively accessible already in less populated regions.

Temporal Patterns

The transportation and introduction of species has become a growing problem in recent years with an increasingly mobile society, with access to more expedient transportation and better shipping techniques. The number of fish species introduced in fifty-year time blocks has skyrocketed in recent years (Fig. 6.4). This increase has been not only in transplantation of native species (i.e., those native to some portion of the United States) but also in introduction of foreign species, many of them tropical ornamentals in the aquarium trade. By 1984, forty-one foreign fish species were known to be established in the forty-eight continental states (Courtenay et al. 1984). Just fifteen years later, that number had risen to seventy-five established species in fifty states (Fuller et al. 1999).

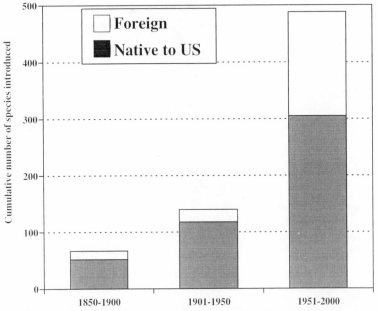

FIGURE 6.4. Number of fish taxa introduced over time (1850–2000).

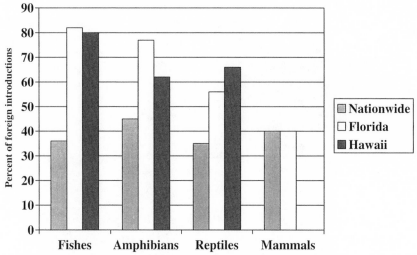

FIGURE 6.5. Percentages of foreign introductions of vertebrate groups.

Origins of Introduced Species in the United States

In order to identify the introduction pathway and to determine where best to intercept it, we need to identify the origins of introduced species. Most introduced species are native to the United States but are moved outside of their native range. Importantly, this type of introduction can be as detrimental as introducing an animal from another country.

The majority of introductions in all freshwater vertebrate groups are species that are native to the United States but are introduced to a nonnative area of the country (Table 6.1). Florida and Hawaii, however, have a much higher percentage of foreign introductions (Fig. 6.5).

Origins of Species Exotic to the United States

Together, Asia and South America contribute more than 50 percent of the foreign fish species reported (Fig. 6.6). Fishes from Asia include the Chinese carps (silver [*Hypophthalmichtys molitrix*], bighead [*H. nobilis*], grass [*Ctenopharyngodon idella*], common [*Cyprinus carpio*], and black [*Mylopharyngodon piceus*]) and many species in the aquarium trade (such as barbs [*Puntius* spp.], anabantids [air breathers], belontiids [gouramis], and channids [snakeheads]). Loaches (Cobitidae), Clariid catfish, and gobies also originate in Asia. Two of the most recent arrivals from the Asian continent are the swamp eel (*Monopterus albus*) and the bull's-eye snakehead (*Channa maurulius*).

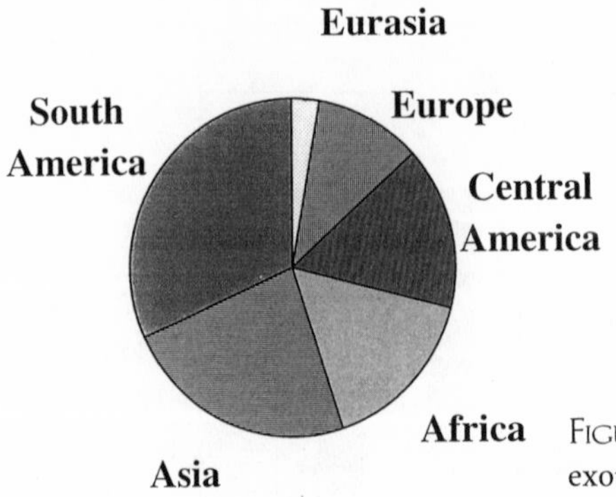

Eurasia

FIGURE 6.6. Origins of introduced exotic fishes.

Fishes from South America include species belonging to the families Cichlidae (genera *Cichla, Cichlasoma, Geophagus, Astronotus*), Characidae, Rivulidae, and several families of catfishes in the aquarium trade (Callichthyidae, Loricariidae, Doradidae, Pimelodidae, and Auchenipteridae).

Cichlids of the genera *Oreochromis, Tilapia, Labeotropheus, Melanochromis,* and *Sarotherodon* are the main group of fishes from Africa. A few species of Nile perch (*Lates* spp.) (Centopomidae) have been imported and stocked as potential sport fish.

Livebearers (such as the platys [*Xiphophorus* spp.], swordtails [*Xiphorphorus* spp.], and mollies [*Poecilia* spp.]) and some cichlids (*Cichlasoma* spp.) are the main groups from Central America.

A smaller proportion of introductions come from Europe, including the brown trout (*Salmo trutta*), ruffe (*Gymnocephalus cernuus*), rudd (*Scardinius erythrophthalmus*), and tench (*Tinca tinca*).

Vertebrate Groups Other Than Fish

Two species of exotic aquatic mammals, the nutria (*Mycocaster coypu*) and the capybara (*Hydrochoerus hydrochaeris*), are from South America. Nutria were originally imported for fur, whereas the capybaras were imported as zoo animals and as pets.

The majority of introduced reptiles come from South America and Asia (Fig. 6.7). The introduced amphibians are mainly from Asia, South America, and the Caribbean. Species from the Caribbean often get to the United States as stowaways in shipments of products, such as plants. These species may continue to expand their range as stowaways in products such as plants and cypress mulch shipped from Florida to other parts of the United States. See Kraus, this volume, for more details on amphibian and reptile introductions.

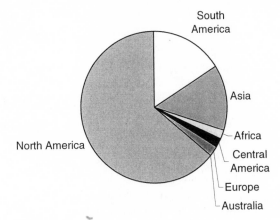

FIGURE 6.7. Origins of introduced reptiles.

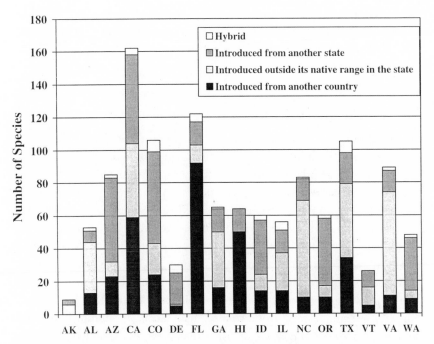

FIGURE 6.8. Origins of introduced fishes in selected states.

Origins of Introduced Species by States

Because of their relatively low diversity of native fish species, and the early westward transplantation of fishes, the western states have relatively high percentages of species from other states (Fig. 6.8). Many of the introduced species in the western states originate from the eastern United States and were moved by humans to the West Coast.

The states of Florida and Hawaii have high percentages of foreign introduced species. This is probably because their warmer climates allow more of these species (mostly tropical in origin) to survive in the wild. Florida and Hawaii are also where many of the tropical exotic animals are bred and grown.

The states of North Carolina and Virginia have high percentages of species that are actually native to these states but were moved outside of their historical native range. Factors contributing to such intrastate introductions are high species diversity within states and the presence of a major drainage divide (the Appalachian Mountains) that separates the Atlantic Slope from the Ohio basin, creating unique fauna by region within the state. Fish species are generally native to only one side of the divide, and many have been transplanted to the other side of the divide. Similarly, although Texas doesn't have a mountain range as a drainage divide, it does contain three major drainages (the Rio Grande, Texas coastal, and the Arkansas-Red-White), which have experienced similar interdrainage translocations.

Pathways

Introductions, intentional or unintentional, occur through a variety of pathways. As Shelton and Smitherman (1984) noted, importation for whatever purpose presents the opportunity for naturalization.

The prime pathway for fish introductions is stocking for food, sport, or forage (44% of all fish species introduced). The second is aquarium release or escape from tropical fish farms (26%), followed by bait release (16%), miscellaneous (escape from aquaculture, release of lab animals, and canal connections) (3%), stocking for conservation (3%), ballast release (2%), and stocking for biocontrol (2%). There are a few species (4%) for which the method of introduction is uncertain (Fig. 6.9). The pathways of aquarium release and escape from tropical fish farms are lumped together because it is often not possible to distinguish between these two pathways, especially where farms are present (e.g., South Florida).

The prime pathway for reptile introductions is pet escapes or releases (Fig. 6.10), whereas amphibians lack a dominant pathway (Fig. 6.11).

Stocking

Animals are stocked for a variety of reasons. Fishes are the group most commonly stocked. Reasons for stocking fish include sportfishing, forage, food source, biocontrol, stock contamination or misidentification, and con-

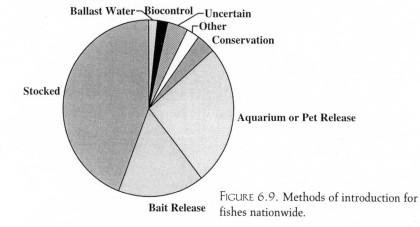

FIGURE 6.9. Methods of introduction for fishes nationwide.

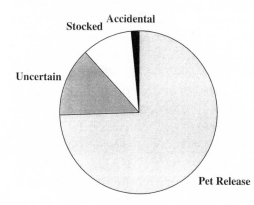

FIGURE 6.10. Methods of introduction for aquatic reptiles nationwide.

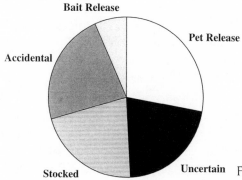

FIGURE 6.11. Methods of introduction for amphibians nationwide.

servation purposes. These stockings may be legal (authorized) or illegal (unauthorized), intentional or unintentional.

SPORT (INTENTIONAL; AUTHORIZED AND UNAUTHORIZED)

Stocking for sportfishing is the method responsible for the most fish transplantation. Sportfishing is a multimillion-dollar industry enjoyed by millions of anglers every year. However, many of the popular sport species are stocked outside their native range. Large, predaceous fish are commonly stocked because they are often targeted by anglers. Unfortunately, these are also among the species that can most disrupt an ecosystem or place native species at risk. Species commonly stocked for sport include basses (*Micropterus* spp. and *Morone* spp.), sunfish (*Lepomis* spp.), catfish (*Ictalurus* spp.), pikes (*Esox* spp.), trout (*Salvelinus* spp.), and salmon (*Oncorhynchus* spp. and *Salmo* spp.).

Most of the fish species stocked for sport are native to the United States but are transplanted outside their native range (bass, sunfish, bullhead, trout). Relatively few are foreign species. Examples of foreign species include brown trout (*Salmo trutta*), Ohrid trout (*S. letnica*), Amur pike (*Esox reicherti*), tench (*Tinca tinca*), two species of peacock cichlids (*Cichla ocellaris* and *C. temensis*), and Nile perch (*Lates niloticus*).

Flathead catfish (*Pylodictis olivaris*), a species native to the central portion of the country, have been stocked both legally and illegally outside of their native range as a sport fish. This large, predaceous catfish has been implicated in extirpating native bullheads (*Ameiurus* spp.) and sunfish. The State of Georgia now has a program to try to reduce the number of flatheads in some of its Atlantic Slope rivers. They have also begun a campaign to educate anglers not to transplant this species into any more rivers.

FOOD (INTENTIONAL; AUTHORIZED AND UNAUTHORIZED)

Some of the earliest fish transfers were food species taken from the East Coast to the West Coast of the United States in the early 1870s, to provide new settlers with a source of food. These included American shad (*Alosa sapidissima*), striped bass (*Morone saxatilis*), American eel (*Anguilla rostrata*), walleye (*Stizostedion vitreum*), largemouth bass (*Micropterus salmoides*), yellow perch (*Perca flavescens*), and bullheads (*Ameiurus*) (Baird 1874, 1876). All but the eels are now well established in the West.

The common carp (*Cyprinus carpio*) is another species that became widely distributed for use as a food fish. Early on, the species was prized for its food value. However, later it was looked down upon, not only as a food fish but also for the disturbances it caused in waters where it was introduced.

More recent food fish introductions include the unauthorized release of Asian snakeheads. The two species recently introduced via this pathway include the bull's-eye snakehead (*Channa maurulius*), now established in South Florida, and the northern snakehead (*C. argus*), eradicated from a pond in Maryland.

Fishes are not the only animals stocked as a food source. Two turtle species, the Chinese softshell (*Pelodiscus sinensis*) and the wattleneck softshell (*Palea steindachneri*), were stocked in California and Hawaii during World War II for food (McKeown 1996). Bullfrogs were also taken to Hawaii for the restaurant industry (McKeown 1996). The frogs provided the added benefit of pest control.

Forage (intentional; authorized and unauthorized)

Often after predatory fish are stocked for sportfishing, a forage base is added to sustain them. Shiners (*Notropis* spp., *Notemigonus chrysoleucas*), minnows (*Pimephales* spp.), shad (*Alosa* spp. and *Dorosoma* spp.), alewives (*Alosa pseudoharengus*), and rainbow smelt (*Osmerus mordax*) are usually stocked for this purpose.

Stock Contamination or Misidentification (unintentional)

Occasionally, unintended species are stocked because of either stock contamination or misidentification. This can happen when the intended species and the contaminant look similar to one another, or if other species were picked up when collecting the intended stock. Some of the most common examples of this include green sunfish (*Lepomis cyanellus*) in shipments of bluegill (*L. macrochirus*), madtoms (*Noturus* spp.) in shipments of small bullheads (*Ameiurus* spp.), suckers (*Catostomus* spp.) in shipments of trout (*Oncorhynchus* spp.), and flathead catfish in shipments of channel catfish (*Ictalurus punctatus*). Populations of African clawed frog (*Xenopus laevis*) in California are also the result of stock contamination, in one case with a shipment of goldfish (*Carassius auratus*) (St. Amant et al. 1973). These unintentional introductions can be prevented by careful examination of the stock prior to release.

It is noteworthy that parasites and diseases are also easily and unintentionally introduced along with their aquatic vertebrate hosts (Stewart 1991). There are many examples of ailments introduced with the introduction of their host. The African tapeworm (*Cephalochlamys namaquensis*) was introduced along with the African clawed frog (*Xenopus laevis*) (Lafferty and Page 1997); the swimbladder fluke (*Anguillicola crassus*) was introduced worldwide with eels (*Anguilla* spp.) (Barse and Secor 1999); whirling disease (*Myxobolis cerebralis*) is spread by stocking infected trout, or even by transporting mud from infected waterways (Modin 1998, Nehring 1999).

BIOCONTROL (INTENTIONAL; AUTHORIZED)

Biocontrol, the use of one species to control another, is used to control a pest. Often the control agent is a nonnative. Some of the classic examples of nonnative biocontrol are grass carp to control aquatic vegetation; silver carp and bighead carp to control algal blooms; mosquitofish (*Gambusia affinis, Gambusia holbrooki*) to control mosquitos; blue tilapia (*Oreochromis aureus*), Mozambique tilapia (*O. mossambicus*), and redbelly tilapia (*Tilapia zillii*) to control vegetation; and marine toad (*Bufo marinus*) to control sugarcane pests. Peacock cichlids (*Cichla ocellaris* and *C. temensis*) were stocked in Florida not only as a sport fish but also with the hope that they would prey on the other species of introduced cichlids, primarily the spotted tilapia (*Tilapia mariae*) in the canals (Courtenay and Robins 1989, Shafland 1995).

Biocontrol offers an alternative to the use of chemicals and in some cases is quite effective. However, in other cases it does not have the intended control level or has other deleterious effects that were not predicted. Mosquitofish were distributed all over the world, yet in many areas the native fishes are more effective at controlling the mosquito population (Danielsen 1968, Courtenay and Meffe 1989). Because mosquitofish are so aggressive, they often reduce populations of these native fish species (Myers 1967, Courtenay and Meffe 1989).

Inland silversides (*Menidia beryllina*) were stocked to control the Clear Lake gnat (*Chaoborus astictopus*) in Clear Lake, California. However, it had several unforeseen effects, including the extinction of an endemic fish, the Clear Lake splittail (*Pogonichthys ciscoides*) (Moyle et al. 1974, Moyle 1976).

In a bizarre attempt to control carp at a California reservoir, which supplied water to the nearby human population, California sea lions (*Zalophus californicus*) were stocked. Things seemed to work fine for a while, but when the carp became less plentiful the sea lions left and headed back to the ocean (Smith 1896, Dill and Cordone 1997).

Another aquatic mammal stocked as a biocontrol agent is the nutria (*Myocaster coypus*). It was stocked for vegetation control in Texas and parts of Florida (King 1968). However, nutria were a poor choice, because they frequently preferred native plants to the species they were intended to control (Griffo 1957, Conniff 1989, Jackson 1990, Haramis and Colona 1999). Nutria are now a major threat to an inland stand of the endangered rice species *Zizania texana* in Texas (Power 1996) and do serious damage to *Spartina* marshes along the Atlantic and Gulf coasts (Haramis and Colona 1999, Patuxent Wildlife Research Center 1999, Carter et al. 1999).

CONSERVATION (INTENTIONAL; AUTHORIZED)
Some species with very limited distributions are stocked outside of their native range in an effort to preserve the species, should something happen to their native habitat. This is especially true of some of the southwestern cyprinids such as the woundfin (*Plagopterus argentissimus*) and the tui chub (*Gila bicolor*); some pupfish in the Southwest including the Devil's Hole pupfish (*Cyprinodon diabolis*) and desert pupfish (*C. macularius*); and some rare trout species including the Gila trout (*Oncorhynchus gilae*), Apache trout (*O. apache*), and cutthroat trout (*O. clarki* spp.). Since this pathway is for species preservation, it is generally not considered to be negative; however, it could have negative consequences for native invertebrate prey.

Bait Release (intentional; unauthorized)

At the end of the fishing day, anglers often release unused live bait without considering that the baitfish may not be from that area. Many species of baitfish are raised in large farms in Arkansas and shipped around the country. Species that have been distributed in this way include rudd (*Scardinius erythrophthalmus*), red shiner (*Cyprinella lutrensis*), golden shiner (*Notemigonus chrysoleucas*), and fathead minnow (*Pimephales promelas*). Releasing these baitfish can have consequences beyond the expected direct competition with, or predation on, natives. Research has shown that rudd are capable of hybridizing with native golden shiners, thereby compromising the native species' genetic makeup (Burkhead and Williams 1991).

In another case, the Asian tapeworm (*Bothriocephalus acheilognathi*) was imported into this country in grass carp. Red shiners, reared in ponds in Arkansas along with grass carp, picked up the parasite and transported it to the southwestern desert where they were shipped for use as bait. After the shiners were released by an unknowing angler, the parasite colonized the woundfin (*Plagopterus argentissimus*), an endangered minnow in southern Nevada. The parasite is now one of the most serious threats to the survival of this endangered species (Deacon 1988).

In addition to fishes, salamanders are also used as bait. Releases of bait are responsible for tiger salamander (*Ambystoma tigrinum*) and blackbelly salamander (*Desmognathus quadramaculatus*) introductions into new areas. The author has been told of a fisherman from the Carolinas who collects coolers of salamanders and takes them to northwestern Montana to use as bait.

Aquarium Release/Tropical Fish Farm Escape (intentional/unintentional)

More and more species are being imported into the United States for use in home aquaria. They are much more likely to survive their journey now than in decades past because of the increased speed of travel and better packaging techniques (e.g., Styrofoam containers to maintain temperature and the addition of pure oxygen to the shipping bags).

In certain areas of Florida, where both tropical fish farms and feral aquarium fish are plentiful, it is often not possible to determine whether an ornamental species was released by its hobbyist owner or escaped a local fish farm. For this reason, these two pathways are combined here. In fact, in Florida, the number of introduced fish species, the presence of tropical fish farms, dense human populations, and presence of disturbed habitat all appear to be intimately related (Nico and Fuller 1999).

As of the early 1980s, approximately 1,200 to 1,500 species of ornamental fish were regularly imported into the United States for the aquarium trade (Radonski et al. 1984). It has been estimated that as many as 1,000 species may be in transit at one time and that up to 6,000 species are of potential interest to aquarists (Welcomme 1984). In 1968 alone, over 64.2 million tropical fish were imported (Lachner et al. 1970). Lachner et al. further stated that 1,000 species were available from wholesalers in the United States in 1970. These startling figures do not even include the myriad of marine species imported for the aquarium trade (Lachner et al. 1970). Whatever the number, it has undoubtedly grown in the past sixteen years as more new species continue to appear in stores. Market surveys, conducted in the late 1970s and early 1980s, found that 60–70 percent of the ornamental fish imported into the United States come from Southeast Asia, another 25 percent come from South America, and the remaining 5–15 percent are raised in the United States (Ford 1981 in Shotts and Gratzek 1984). The fishes raised in the United States, primarily in Florida, are not U.S. natives, but species that have already been imported from other countries.

Fuller et al. (1999) documented 126 freshwater exotic aquarium fish species collected from the wild. Most introduced tropical fish do not survive. Of the 126 species released, at least 42 have reproducing populations in the United States. Examples of established ornamentals include the Oriental weather loach (*Misgurnus anguillicaudatus*); numerous genera of cichlids (*Oreochromis, Tilapia, Heros, Cichla, Cichlasoma*, etc.); and several genera of South American catfish (*Ancistrus, Liposarcus* [=*Pterygoplicthys*], *Hypostomus, Hoplosternum*) (Fuller et al. 1999).

Most of these established populations are in the subtropical states of Florida and Hawaii. Although it is true that many of these species pose no threat to northern areas because they would not survive the winters, a few aquarium species are cold tolerant and have become established in northern regions. Goldfish (*Carassius auratus*) are widely established around the country; Oriental weatherfish are established in Michigan, Illinois, Idaho, Washington, and Oregon (Fuller et al. 1999). Both species are from Asia and are widely distributed there in similar latitudes. Others, such as platys, swordtails, and some cichlids, can survive in northern climates in geothermal waters and have established populations as far north as Montana, Idaho, and Wyoming (Fuller et al. 1999). Some species have thrived in northern climates by living in heated discharges from power plants. Though not an aquarium species, this was the case with a population of blue tilapia in the Susquehanna River of Pennsylvania. The population was eventually eradicated by shutting down the plant during a period of cold weather (Skinner 1984, 1986, Stauffer et al. 1988).

One of the earliest species to be transported was the goldfish (*Carassius auratus*). This species began its worldwide travels in the 1600s. It is still one of the most available, and probably the most released, species in the United States.

Although most hobbyists probably release a fish thinking they are being kind to it, they are not doing their pet a favor, since most don't survive. These species are not adapted to the climate, ecosystem, food, and predators where they are released. They are generally condemned to a slow death caused by starvation, disease, or cold.

Escape from Captive Situations (unintentional)

ESCAPES FROM AQUACULTURE

Escapes from aquaculture facilities occur when the outflow pipes lack suitable screening, the ponds overflow during flood events, or birds transport animals and drop them in a nearby water body. In recent years, more than 100,000 Atlantic salmon have escaped net pens in Puget Sound either during rough seas or because otters tore the nets (Goldburg and Triplett 1997). Shelton and Smitherman (1984) candidly state, "For whatever purpose an exotic fish is used [in aquaculture], escape is virtually inevitable; thus, this eventuality should be considered" (p. 262). Three of the Chinese carp (grass carp, bighead carp, and silver carp) have become established in large river systems in the United States largely through aquaculture escapes. A fourth species of Chinese carp, the black carp, has also escaped into the wild and poses a significant threat to the native mussel and snail populations if it becomes established (Nico and Williams 1996). Better facility design could reduce the chance of escape, and the use of certified triploids would reduce the chance of such escapees from becoming established and perpetuating biological damage. Triploids have the ability to revert to diploids and reproduce; therefore, triploidy does not completely eliminate the possibility of establishment.

ESCAPE FROM FUR FARM/GAME FARM

Nutria and the South American capybaras were both imported for their fur. Nutria and, to a lesser extent, capybaras have become established as a result of escape or release from private game or fur farms. Capybaras are believed to be established in Florida and possibly North Carolina as a result of escapes from game farms; their impacts are as yet unknown. Nutria have been reported in forty states (Hygnstrom et al. 1994) and are now established in many areas of the country, including the Southeast, Great Lakes, East Coast, and Pacific Northwest. These introductions were a result of farm escapes and of deliberate releases by fur farmers of unwanted animals when the fur prices declined.

Nutria were imported and released several times in the early 1930s, but never became established. The first successful introduction was in 1938 when between 12 and 20 were brought into the country for fur. They were kept on an island off the coast of Louisiana. They multiplied rapidly and many escaped. In 1940, a hurricane that passed through released about 150 animals (Griffo 1957). Many made it to the mainland where, combined with animals that had been stocked for vegetation control, they started a new population and reproduced very quickly. Within twenty years, the population was estimated at 20 million (Lowery 1974, Anonymous 1998).

Nutria cause significant damage to *Spartina* salt marshes by eating vegetation and tunneling into dikes and levees. For example, in Jefferson Parish, Louisiana, nutria damaged canal banks costing an estimated $2 million per year (Chabreck and Nyman 1995). In Maryland, the Department of Interior is spending $2.9 million per year to control the population because of erosion and marsh damage (Anonymous 1998).

Stowaways (unintentional)

Some species enter the country hidden in imported commodities. The greenhouse frog (*Eleutherodactylus planirostris*) and the Cuban tree frog (*Osteopilus septentrionalis*) are both believed to have entered the country on goods imported from the Caribbean (Neill 1951, Ashton and Ashton 1988). Both species are now established in the wild in Florida. The Cuban tree frog is known to eat native tree frogs (*Hyla* spp.) (Ashton and Ashton

1988) and also has the annoying trait of causing power outages by short-
ing out electrical transformers, to which they are apparently attracted by
the insect-like buzzing (Fichter 1970). A single tessellated water snake
(*Natrix tesselata*) collected in Norfolk, Virginia, at the navy shipyard is
believed to have come over on a navy ship from Europe (Mitchell 1994).

Canals (unintentional)

Canals were constructed in the late 1800s and early 1900s to give ships
easier ways around land obstacles. Two of the largest canals connecting
major water bodies are the Panama Canal that connects the Atlantic and
Pacific Oceans and the Suez Canal that connects the Mediterranean Sea to
the Red Sea. Another example is the Erie Canal in the United States, which
connects the Hudson River to the Great Lakes, saving ships the northward
journey to the St. Lawrence Seaway in Canada.

Artificial canals can connect two drainages that would otherwise not be
connected, or can bypass a natural barrier that would have prevented animal movement. These canals can be the primary method of introduction of
a species to a new drainage, such as the sea lamprey (*Petromyzon marinus*)
in the upper Great Lakes via the Welland Canal, or they can aid in dispersal
once a species is introduced to a new area. An example of the latter is the
round goby (*Neogobius melanostomus*), which was introduced into the
Great Lakes. It made its way through the Chicago Shipping and Sanitary
Canal to the Mississippi basin and spread throughout the country. The
zebra mussel used this same route in its invasion of the northern United
States. Other species have used this thoroughfare to go the other direction,
from the Mississippi to the Great Lakes—for example, the skipjack herring
(*Alosa chrysochloris*) (Fago 1993), the river darter (*Percina shumardi*)
(Becker 1983), and possibly the gizzard shad (*Dorosoma cepedianaum*)
(Miller 1957, Becker 1983). Mills et al. (1999) and Crossman and Cudmore
(1999) provided an overview of the spread of nonindigenous species via
canals. Although new navigation canals are no longer being built, organisms will continue to use existing canals to expand their ranges.

Ballast Water Release (unintentional)

Many animals survive transport in ship ballast and are introduced into the
United States from around the world. The round goby, tubenose goby
(*Proterorhinus marmoratus*), European flounder (*Platichthys flesus*), and
ruffe (*Gymnocephalus cernuus*) are examples of species that arrived in the
Great Lakes by this means. Four gobies—the yellowfin goby (*Acanthogo-
bius flavimanus*), the Chameleon goby (*Tridentiger trigonocephalus*), the

Shimofuri goby (*T. bifasciatus*), and the Shokihaze goby (*T. barbatus*)—have all been introduced into San Francisco Bay via ballast water (Brittan et al. 1970, Meng et al. 1994, Matern and Fleming 1996). Even a live school of mullet was found in a ballast tank of a ship during sampling (Wonham et al. 2000).

Released Research Animals (intentional; unauthorized)

A few introductions are known to be research animal releases after termination of the researcher's experiment. The pike killifish (*Belonesox belizanus*) was released into South Florida in 1957 after medical experiments (Belshe 1961). It is now established in that area of the state.

DISCUSSION

How the Likelihood of Success of an Introduction Is Related to the Pathway

Three attributes of the pathway are related to the potential success of an introduction: the number of individuals introduced, the frequency of the introductions from that pathway, and the likelihood that the pathway will deliver healthy individuals. Pathways that deliver hundreds, or even millions, of individuals are more likely to result in the establishment of a species than a pathway that delivers only single individuals. For example, when species are stocked, thousands or tens-to-hundreds of thousands of individuals are often introduced at one time. In addition, ballast water discharge can deliver millions of planktonic individuals at one time. If one multiplies this by the number of ships in port at any given time and the number of trips in a year, it is readily apparent why ports are being taken over by foreign species. Escapes of large groups of individuals from aquaculture or other captive situations also carry a fairly high risk of establishment. On the other hand, releasing an individual pet is far less likely to result in the establishment of that species. Exceptions to this density-dependence may exist for parthenogenetic species, as well as gravid or pregnant females, especially if the species is very fecund.

Frequent inoculations, such as those from ballast water releases or yearly fish stocking, are more likely to become established. Repeated releases of single individuals in a localized area can build to a population of potential mates. This can happen with pet releases in urban or suburban areas near large population centers.

Pathways that deliver healthy individuals are also more likely to lead to establishment of a species. Liberated sick or dying pets are not likely to survive, let alone reproduce. Well-fed hatchery fish may have extra reserves to get them through the initial adjustment period. However, hatchery fish that are raised in crowded conditions with poor water quality or are fed a poor diet are not likely to survive.

Recommendations Concerning Pathways for Freshwater Aquatic Vertebrates

Most vertebrate introductions are intentional and therefore preventable. Intentional introductions can be either authorized or unauthorized. Most authorized introductions are conducted by state personnel. Public education is key to preventing future introductions, because the public is often unaware of the problems resulting from introductions. The number of authorized introductions could be reduced if changes in policy dictated that state stocking efforts use only native species or, at least, did not introduce any new species that are not already established.

Although most states have laws prohibiting the release of exotic species, these animals continue to turn up all across the nation. Sometimes these introductions occur through ignorance, other times through deliberate intent to establish a new species in an area. The fact that the releases continue indicates that legislation alone is not enough to prevent the illegal introduction of species. Grass carp is one such example of this failure (Guillory and Gasaway 1978). Although many states either prohibit their use or require only sterile triploids to be stocked, grass carp are now established in large rivers in the central portion of the country.

In order to determine how best to prevent introductions in a specific region, each region should be specifically analyzed for origins and pathways of introduced species. For example, the management approaches for reducing the introduction of nonindigenous species in Hawaii, Nebraska, and San Francisco Bay will all be very different. Once the pathways and entry points are identified and ranked in terms of importance, a management plan can be implemented to reduce, prevent, or intercept new introductions.

Intentional, Authorized Introductions

Preventing further intentional, authorized introductions is a matter of education and changing state agency stocking policies. Stocking is the prime pathway for fish introductions. Stocking with native species rather than nonnative species is encouraged to prevent further introductions. The state of Colorado is undertaking such a plan, entitled "Bring back the natives."

Intentional, Unauthorized Introductions

Preventing further intentional, unauthorized introductions is a matter of educating the public. Unauthorized pet release is the primary pathway for reptile introductions, and is responsible for at least 75 percent of the species introduced. It also accounts for approximately 25 percent of fish introductions, and 28 percent of amphibian introductions (USGS 2000). These introductions are preventable with education. Implementing a clean list or dirty list approach, so that each species is reviewed for potential threats to an area before it is allowed in, appears to be a good solution. Unfortunately, this approach is far less practical on a broad scale and is highly controversial, particularly with the pet industry.

Bait release accounts for approximately 16 percent of the fish species introduced and a few of the amphibian species (USGS 2000). Often anglers do not realize the consequences of releasing unused bait. They also do not know if a species is native to the drainage where they are fishing. Public education would serve to inform the anglers of the consequences of dumping live bait. Allowing only native species to be used as bait would further reduce the chance of a new introduction. Some areas already experiencing problems with bait introductions prohibit the use of any live bait.

Unintentional Introductions

Preventing further unintentional introductions is not as easy. Some, such as those resulting from ballast water release, may require technological remedies. Although research is being conducted to find a way to prevent these releases, a feasible, safe, and cost-effective solution has not yet been found. Ballast water introductions are also being targeted through growing legislation. Certain ports and regions now require open-water ballast exchange when it is safe to do so. High seas ballast exchange reduces the problem, but does not eliminate it because the tanks can never be completely emptied or fully exchanged. When a feasible solution is found to treat ballast water, it will have a significant role in reducing the invasion of marine species.

Staging more inspectors at ports of entry could prevent other unintentional introductions. Because of the large volume of goods imported into this country and the small staff responsible for inspecting these goods, only a small percentage is actually examined. Increasing the number of inspectors at entry points would allow for closer examination, and could intercept stowaways or prevent misidentified species from coming into the country. These inspectors need to be well trained to recognize the myriad species they may face.

Unintentional introductions, resulting from stock contamination, are preventable by thoroughly examining all stock prior to release for species identity, stock contaminants, parasites, and general health.

Ensuring that baitfish shipments and sales are of all one species, rather than a mixture with contaminants (unintended species), will reduce the likelihood of unintentionally introducing a nonnative (though the release itself is an intentional act). Some areas in the Great Lakes are training bait dealers to inspect and recognize problematic contaminants such as round goby and ruffe.

Unintentional introductions resulting from aquaculture escapes are largely preventable, as follows.

• The most significant effect would result from requiring better design standards in aquaculture facilities. Currently, no national design standards exist (U.S. Congress 1993:150). The facilities should be designed to withstand a 100-year flood. The first and most obvious way to achieve this is to carefully choose the location for the facility and not place it in a floodplain. Once a suitable location is found, the facility can be further secured by placing tall berms around the farm (at the 100-year flood stage) and surrounding the entire facility with perimeter fencing.

  Although this may be effective for freshwater aquaculture facilities, this can't be applied to mariculture facilities, such as open-ocean salmon net pens. All outfalls should be screened to prevent accidental discharge of animals. Better yet, a closed circulation system could be used so that there is no water discharged. Practicing bird control, using nets or scare tactics, would reduce or prevent birds from transporting fish to nearby waterways. It will also reduce farmers' losses to predation.

• Implement measures that decrease the chances of species becoming established should they escape. Options include allowing only triploids, stocking only males, or using sterile hybrids for potentially problematic species. On a more general scale, encourage the culture of native species and discourage the culture of exotic species.

Other captive situations (game farms, zoos, etc.) should also have properly designed facilities to contain the animals that will be housed there. The facility design needs to take into account natural disasters that may arise.

Canals that allow species to expand their ranges into nonnative areas will continue to exist and be accessible to aquatic animals. Recent research is aimed at designing electrical barriers that can be installed along these waterways to prevent this (Hauser 1998, Keppner and Theriot 1998).

Introductions and spread through recreational boating or fishing can be decreased by thoroughly cleaning boats and other equipment used in waters with nonindigenous species to prevent their spread to the next water body where the equipment will be used.

In general, reducing the likelihood of further freshwater aquatic introductions requires vigilance, planning, and public education.

ACKNOWLEDGMENTS

The U.S. Geological Survey is indebted to the vast number of biologists and interested public parties who take the time to make us aware of these introductions.

REFERENCES

Anonymous. 1998. $2.9 million from feds going toward getting rid of nutria. *The Star Democrat Online.* Available from <http://www.talb.lib.md.us/stardemocrat/1998/october/1016nutria.html>. Accessed 13 October 1998.

Ashton, R. E., and P. E. Ashton. 1988. *Handbook of reptiles and amphibians of Florida. Part Three: The amphibians.* Miami: Windward Publishing, Inc.

Baird, S. F. 1874. *Report of the Commissioner of Fish and Fisheries for 1872–1873.* Washington, D.C.: U.S. Commission of Fish and Fisheries.

Baird, S. F. 1876. *Report of the Commissioner of Fish and Fisheries for 1873–1874.* Part III. Washington, D.C.: U.S. Commission of Fish and Fisheries.

Barse, A. M., and D. H. Secor. 1999. An exotic nematode parasite of the American eel. *Fisheries* 24(2):6–10.

Becker, G. C. 1983. *Fishes of Wisconsin.* Madison: University of Wisconsin Press.

Belshe, J. F. 1961. Observations of an introduced tropical fish (*Belonesox belizanus*) in southern Florida. M.S. thesis, University of Miami, Coral Gables.

Brittan, M. R., J. D. Hopkirk, J. D. Conners, and M. Martin. 1970. Explosive spread of the oriental goby *Acanthogobius flavimanus* in the San Francisco Bay–Delta region of California. *Proceedings of the California Academy of Sciences* 38:207–214.

Burkhead, N. M., and J. D. Williams. 1991. An intergeneric hybrid of a native minnow, the golden shiner, and an exotic minnow, the rudd. *Transactions of the American Fisheries Society* 120:781–795.

Carter, J., A. Lee Foote, and L. A. Johnson-Randall. 1999. Modeling the effects of nutria (*Myocaster coypus*) on wetland loss. *Wetlands* 19(1):209–219.

Chabreck, R. H., and J. A. Nyman. 1995. *Environmental assessment of alternatives for controlling nutria damage in the drainage canal system of Jefferson Parish.* <http://jeffparish.net/departments/nutria.html>. Page accessed 7 March 2000.

Conniff, R. 1989. Keeping an immigrant in check. *National Wildlife* 27:43–44.

Courtenay, W. R. Jr., D. A. Hensley, J. N. Taylor, and J. A. McCann. 1984. Distribution of exotic fishes in the continental United States. In *Distribution, biology, and management of exotic fishes,* W. R. Courtenay, Jr., and J.R. Stauffer, Jr., eds., p. 41–77. Baltimore: Johns Hopkins University Press.

Courtenay, W. R. Jr., and G. K. Meffe. 1989. Small fishes in strange places: a review of introduced poeciliids. In *Ecology and evolution of livebearing fishes (Poeciliidae)*, G. K. Meffe and F. F. Snelson, Jr., eds., p. 319–331. Englewood Cliffs: Prentice-Hall.

Courtenay, W. R. Jr., and C. R. Robins. 1989. Fish introductions: good management, mismanagement, or no management? *CRC Critical Reviews in Aquatic Sciences* 1:159–172.

Crossman, E. J., and B. C. Cudmore. 1999. Summary of fish introductions through canals and diversions. In *Nonindigenous freshwater organisms: vectors, biology, and impacts,* R. Claudi and J. H. Leach, eds., pp. 393–395. Boca Raton: CRC Press LLC.

Danielsen, T. L. 1968. Differential predation on *Culex pipiens* and *Anopheles albimanus* mosquito larvae by two species of fish (*Gambusia affinis* and *Cyprinodon nevadensis*) and the effects of simulated reeds on predation. Doctoral dissertation, University of California, Riverside.

Deacon, J. E. 1988. The endangered woundfin and water management in the Virgin River, Utah, Arizona, Nevada. *Fisheries* 13:18–29.

Dill, W. A., and A. J. Cordone. 1997. *History and status of introduced fishes in California, 1871–1996.* Fish Bulletin of the California Department of Fish and Game 178.

Fago, D. 1993. Skipjack herring, *Alosa chrysochloris,* expanding its range into the Great Lakes. *Canadian Field Naturalist* 107:352–353.

Fichter, G. 1970. The new nature of Florida. *Florida Wildlife* (December):10–15.

Ford, D. M. 1981. The hobby of ornamental fish keeping. In *The diseases of ornamental fishes,* D. M. Ford, ed., pp. 317–322. Waltham symposium no. 3. Waltham-on-the-Wolds, Leicestershire: Pedigree Pet Foods Animal Studies Centre.

Fuller, P. L., L. G. Nico, and J. D. Williams. 1999. *Nonindigenous fishes introduced into inland waters of the United States.* Bethesda: American Fisheries Society Special Publication 27.

Goldburg, R., and T. Triplett. 1997. *Murky waters: environmental effects of aquaculture in the United States.* Washington, D.C.: Environmental Defense Fund.

Griffo, J. V. Jr. 1957. The status of the nutria in Florida. *Quarterly Journal of the Florida Academy of Sciences* 20:209–215.

Guillory, V., and R. D. Gasaway. 1978. Zoogeography of the grass carp in the United States. *Transactions of the American Fisheries Society* 107:105–112.

Haramis, M., and R. Colona. 1999. *The effect of nutria (Mycocaster coypus) on marsh loss in the lower eastern shore of Maryland: an exclosure study.* Available from <http://www.pwrc.nbs.gov/resshow/nutria.htm>. Page accessed 16 December 1999.

Hauser, M. 1998. Champlain Canal fish barrier study. *Aquatic Nuisance Species Digest* 2:26–27.

Hygnstrom, S. E., R. M. Timm, and G. B. Larson. 1994. *Nutria: damage prevention and control methods.* Port Allen: USDA, APHIS, Animal Damage Control.

Jackson, D. D. 1990. Orangetooth is here to stay. *Audubon* 92:88–94.

Keppner, S. M., and E. A. Theriot. 1998. Controlling round gobies in the Illinois waterway system. *Aquatic Nuisance Species Digest* 2:20–22.

King, W. 1968. As a consequence many will die. *Florida Naturalist* 41:99–103, 120.

Lachner, E. A., C. R. Robins, and W. R. Courtenay, Jr. 1970. Exotic fishes and other aquatic organisms introduced into North America. *Smithsonian Contributions to Zoology* 59:1–29.

Lafferty, K. D., and C. J. Page. 1997. Predation on the endangered tidewater goby, *Eucyclogobius newberryi*, by the introduced African clawed frog, *Xenopus laevis*, with notes on the frog's parasites. *Copeia* 1997:589–592.

Lowery, G. H. Jr. 1974. *The mammals of Louisiana and its adjacent waters.* Baton Rouge: Louisiana State University Press.

Matern, S. A., and K. J. Fleming. 1996. Invasion of a third Asian goby species, *Tridentiger bifasciatus*, into California. *California Fish and Game* 81:71–76.

McKeown, S. 1996. *A field guide to reptiles and amphibians in the Hawaiian Islands.* Los Osos, CA: Diamond Head Publishing.

Meng, L., P. B. Moyle, and B. Herbold. 1994. Changes in abundance and distribution of native and introduced fishes of Suisun Marsh. *Transactions of the American Fisheries Society* 123:498–507.

Miller, R. R. 1957. Origin and dispersal of the alewife, *Alosa pseudoharengus*, and the gizzard shad, *Dorosoma cepedianum*, in the Great Lakes. *Transactions of the American Fisheries Society* 86:97–111.

Mills, E. L., J. R. Chrisman, and K. T. Holeck. 1999. The role of canals in the spread of nonindigenous species in North America. In *Nonindigenous freshwater organisms: vectors, biology, and impacts*, R. Claudi and J. H. Leach, eds. pp. 347–379. Boca Raton: CRC Press LLC.

Mitchell, J. C. 1994. *The reptiles of Virginia.* Washington, D.C.: Smithsonian Institution Press.

Modin, J. 1998. Whirling disease in California: a review of its history, distribution, and impacts, 1965–1997. *Journal of Aquatic Animal Health* 10:132–142.

Moyle, P. B. 1976. *Inland fishes of California.* Berkeley: University of California Press.

Moyle, P. B., F. W. Fisher, and H. Li. 1974. Mississippi silversides and log perch in the Sacramento–San Joaquin River system. *California Fish and Game* 60:145–147.

Myers, G. S. 1967. *Gambusia*, the fish destroyer. *Australian Zoology* 13:102.

Nehring, B. 1999. Stream fisheries investigations: whirling disease investigations. *Federal Aid in Fisheries Progress Report F-237-R6.* Ft. Collins: Colorado Division of Wildlife.

Neill, W. T. 1951. Florida's air-plants and their inhabitants. *Florida Naturalist* 24:61–66.

Nico, L. G., and P. L. Fuller. 1999. Spatial and temporal patterns of nonindigenous fish introductions in the United States. *Fisheries* 24:16–27.

Nico, L. G., and J. D. Williams. 1996. *Risk assessment on black carp (Pisces: Cyprinidae). Final report to the Risk Assessment and Management Committee of the Aquatic Nuisance Species Task Force.* Gainesville: U.S. Geological Survey, Biological Resources Division.

Patuxtent Wildlife Research Center. 1999. *South American nutria destroys marshes. PWRC Fact Sheet 1999–01.* Patuxtent, MD: U.S. Geological Survey.

Power, P. 1996. Direct and indirect effects of floating vegetation mats on Texas wildrice (*Zizania texana*). *The Southwestern Naturalist* 41:462–464.

Radonski, G. C., N. S. Prosser, R. G. Martin, and R. H. Stroud. 1984. Exotic fishes and sport fishing. Page 313–321 in W. R. Courtenay, Jr., and J. R. Stauffer, eds. *Distribution, Biology, and Management of Exotic Fishes.* Baltimore: Johns Hopkins University Press.

Shafland, P. L. 1995. Introduction and establishment of a successful butterfly peacock fishery in southeast Florida canals. *American Fisheries Society Symposium* 15:443–445.

Shelton, W. L., and R. O. Smitherman. 1984. Bacteria, parasites, and viruses of aquarium fish and their shipping waters. Exotic fishes in warmwater aquaculture. In *Distribution, biology, and management of exotic fishes,* W. R. Courtenay, Jr., and J. R. Stauffer, Jr., eds., pp. 262–301. Baltimore: Johns Hopkins University Press.

Shotts, E. B., and J. B. Gratzek. 1984. Bacteria, parasites, and viruses of aquarium fish and their shipping waters. In *Distribution, biology, and management of exotic fishes,* W. R. Courtenay, Jr., and J. R. Stauffer, Jr., eds., p. 215–232. Baltimore: Johns Hopkins University Press.

Skinner, W. F. 1984. Oreochromis aureus (Steindachner; Cichlidae), an exotic fish species, accidentally introduced to the lower Susquehanna River, Pennsylvania. *Proceedings of the Pennsylvania Academy of Science* 58:99–100.

Skinner, W. F. 1986. Susquehanna River tilapia. *Fisheries* 11:56–57.

Smith, H. M. 1896. A review of the history and results of the attempts to acclimatize fish and other water animals in the Pacific states. *Bulletin of the U.S. Fish Commission for 1895* 40:379–472.

St. Amant, J. A., F. G. Hoover, and G. R. Stewart. 1973. African clawed frog, *Xenopus laevis laevis* (Daudin), established in California. *California Fish and Game* 59:151–153.

Stauffer, J. R. Jr., S. E. Boltz, and J. M. Boltz. 1988. Cold shock susceptibility of blue tilapia from the Susquehanna River, Pennsylvania. *North American Journal of Fisheries Management* 8:329–332.

Stewart, J. E. 1991. Introductions as factors in diseases of fish and aquatic invertebrates. *Canadian Journal of Fisheries and Aquatic Sciences* 48(Suppl. 1):110–117.

U.S. Congress, Office of Technology Assessment. 1993. *Harmful non-indigenous species in the United States.* Washington, D.C.: U.S. Government Printing Office OTA-F-565.

U.S. Geological Survey (USGS). 1982. A U.S. Geological Survey data standard, codes for the identification of hydrologic units in the United States and the Caribbean outlying areas. *U.S. Geological Survey Circular* 878-A.

U.S. Geological Survey. 2000. *Nonindigenous aquatic species database.* Gainesville, FL: USGS.

Welcomme, R. L. 1984. International transfers of inland fish species. In *Distribution, biology, and management of exotic fishes,* W. R. Courtenay, Jr., and J. R. Stauffer, Jr., eds., pp. 22–40. Baltimore: Johns Hopkins University Press.

Wonham, M. J., J. T. Carlton, G. M. Ruiz, and L. D. Smith. 2000. Fish and ships: relating dispersal frequency to success in biological invasions. *Marine Biology* 136:1111–1121.

Chapter 7

In Ships or on Ships? Mechanisms of Transfer and Invasion for Nonnative Species to the Coasts of North America

Paul W. Fofonoff, Gregory M. Ruiz,
Brian Steves, and James T. Carlton

A critical parameter in both interpreting historical patterns of invasions and predicting future invasions is an understanding of how, when, and where specific vectors operate to deliver species to new regions. In marine and estuarine (brackish-water) environments, shipping has been the predominant vector of human transport of nonindigenous species around the world (Carlton 1985, Cohen and Carlton 1995, Hewitt et al. 1999, Reise et al. 1999, Ruiz et al. 2000). Organisms were historically and unintentionally transported in and on ships by (in association with) hull fouling, holes in wooden vessels, solid ballast, cargo, and deck and bilge debris. By 1900, another mechanism of transport associated with shipping had appeared, with most new long-range ships using ballast water (instead of ballast rocks or ballast sand) for stability (Carlton 1985).

Despite the clear role of ship-mediated transfer in many coastal invasions, the relative importance of transfer on the outside of ships (hull fouling communities) or the inside of ships (ballast water communities) is poorly resolved. Lacking even a first-order resolution limits both retrospective and prospective analyses of marine bioinvasions. Ballast water has received much attention as a vector since the late 1980s, because of dramatic invasions associated with ballast water around the world. These include the invasion of the Laurentian Great Lakes by zebra and quagga mussels (*Dreissena polymorpha* and *D. bugensis,* respectively) and other invertebrate and fish species from Eurasia (Mills et al. 1993), the invasion of southern Australia by Japanese dinoflagellates (Jones 1981), and the invasion of the Black Sea by the American ctenophore *Mnemiopsis leidyi* (Harbison and Volovick 1994).

In contrast, it is often assumed that hull fouling may now be a less important vector than it was for centuries or millennia, in part as a result of the focus on ballast water. There are multiple reasons to believe that this may be the case (Carlton et al. 1995): (1) the use of increasingly effective antifouling paints has vastly expanded over the twentieth century; (2) ships spend far less time in ports than they did historically (and thus there would be less opportunity for colonization of ships' hulls); and (3) vessel speeds have increased considerably, with the concomitant probability that fewer organisms remain on the hulls because of increased sheer forces.

On the other hand, there are also reasons to believe that ship fouling may still be an important vector (Carlton et al. 1995, Gollasch 2002). The extent of fouling communities on most oceangoing vessels has not been well characterized in recent years, and certain species continue to appear in new regions for which hull fouling is the most likely, or only, vector. The faster speeds of modern ships may have in fact operated to enhance survival among certain taxa, such as oligohaline (low-salinity) species that would not survive as well under longer exposures to full-strength seawater (e.g., Roos 1979). Further, many slow-moving vessels and craft, such as semisubmersible drilling platforms and floating dry docks, still regularly cross the world's oceans. Despite the biocidal action of fouling paints, some taxa have now evolved resistance to copper-based antifouling paints (Hall 1981), a phenomenon that is probably well underestimated. In addition, the recent decreased use of certain antifouling compounds, such as paints containing tributyltin (TBT), may have led or may lead to the increase in fouling communities on certain vessels (Nehring 2001). Finally, increased water quality in recent years in many harbors and ports around the world may also lead to more abundant and diverse fouling communities that could colonize the hulls of visiting ships (Carlton 1996).

In this chapter, we characterize the state of knowledge about the relative importance of ships' hulls and ballast water to temporal patterns of invasion in coastal ecosystems. As a model system, we have used data on established nonnative species of invertebrates, algae, and fish in North America. First, we identify the invasions attributed to shipping. Second, we classify ship-associated invasions according to life-history characteristics, habitat(s) of occurrence, and mobility. Third, using this classification scheme, we assign the ship-mediated invasions to either hull fouling or ballast water, or both, to evaluate the relative strength of each component. Our intent is to provide a first, albeit coarse, analysis at a continental scale, using general biological attributes of the species to develop a "decision tree" approach to classification. We consider the merits and limitations to this approach in the discussion.

METHODS

For this analysis, we used our database (NEMESIS 2002) that identifies and characterizes 316 nonnative species of invertebrates and algae that are established in coastal marine habitats (including brackish waters and tidal freshwaters of estuaries) of continental North America, including the United States and Canada. For each coast of North America (Atlantic, Pacific, and Gulf), the database includes information on the geographic origin, date of first record, and vector(s) of introduction. We have also included 5 species of fishes introduced by shipping. Of about 73 fish species introduced and established in North America estuaries, predominantly in freshwater reaches, only these 5 species are considered ship-related introductions (NEMESIS 2002). ·

We published a previous analysis of invasion patterns for 298 of the 316 nonnative invertebrate and algal species (Ruiz et al. 2000). This earlier analysis identified those invasions associated with shipping and other vectors. In this chapter, adding the additional records for fish and newly documented invertebrates and algae, we sought to further characterize the specific modes (or subvectors) of shipping that may have led to ship-mediated invasions.

Our approach was to classify each of the nonnative species associated with shipping, using multiple traits, to assess the likelihood of transport by each of the subvectors. More specifically, we examined life history, ecology, invasion history (especially time of reported invasion), salinity tolerance, and habitat utilization, in order to attribute the invasions to one or more particular shipping subvectors. We used these data to create a "decision tree," or flowchart, to classify the species by subvector, as outlined below.

Adult Form and Habitat Utilization

Adult characteristics greatly influence the potential for different modes of shipping transport. Sessile and sedentary forms, which occur on hard or soft substrate of bottom communities, are often associated with ships' hulls. Holoplanktonic forms occur in the water column, are readily transported in ballast water, and have a low probability of being carried by other shipping subvectors. If an organism tolerates air exposure, transport by dry ballast or in cargo is a possibility.

We classified adult characteristics of the nonnative species using the following categories:

- *Sessile.* Species attached to hard surfaces, with little or no adult movement. Examples include macroalgae, sponges, hydroids, sea anemones, serpulid and spirorbid tubeworms, mussels, kamptozoans, bryozoans, and tunicates.
- *Sedentary.* Species residing in attached tubes or burrows in wood, but also capable of occasional movement outside of tubes or burrows. Examples include spionid polychaetes, wood-boring and tube-dwelling crustaceans, and their protozoan and crustacean associates.
- *Mobile.* Species capable of actively moving or staying in suspension in water, or actively moving in or on sediment. Subcategories include the following:

    1. *Holoplanktonic.* Organisms that occur predominantly or frequently in the water column. Examples include planktonic diatoms, planktonic copepods, cladocerans, and mysids.
    2. *Epibenthic with eggs attached to hard surfaces.* Organisms that live on the surface of hard or soft substrates, but only occasionally in the water column, and attach eggs or egg masses to hard substrates. Examples include some prosobranch gastropods, most opisthobranch gastropods, gobiid fishes, and parasites of these forms.
    3. *Epibenthic without attached eggs.* Organisms that live on the surface of hard or soft substrates, but only occasionally in the water column, and do not attach eggs to hard substrates. Examples include free-living epibenthic polychaetes, free-living amphipods, isopods, decapods, and some fishes.
    4. *Infauna.* Organisms that burrow in soft sediment. Examples include many foraminiferans, annelids, bivalves, and crustaceans.
    5. *Intertidal.* Organisms with some tolerance to air exposure and desiccation. Examples include littoral snails (e.g., *Littorina littorea, Myosotella myosotis*), upper intertidal amphipods and isopods (*Transorchestia engimatica* and *Ligia exotica,* respectively), and crabs (*Carcinus maenas, Hemigrapsus sanguineus*).
    6. *Intertidal–supratidal.* Organisms with some life stages that are essentially terrestrial, with great tolerance to air exposure and desiccation. Examples include terrestrial insects with aquatic larvae, and a semiterrestrial crab (*Platchirograpsus spectabilis*).

Planktonic Life Stages (and Reproductive Mode)

Although organisms may be transferred in their adult form, the whole life history of the species must be considered in regard to possible modes of shipping transport. Many species have planktonic larval stages of varying duration. The length of time in the water column, or development time, can determine the potential for long-distance dispersal by ballast water. We estimated characteristics of planktonic stages, using literature for individual species. For most species, literature on larval development was not available. For these, we used literature for broader taxonomic groups (e.g., Barnes 1974) and applied these generalized characteristics to species within the groups. Using this approach, we classified individual species among the following plankton life stage categories:

- *None.* Species that have direct development and little or no active swimming.
- *Short larval duration (0–5 days).* Species that have short-duration larval periods are primarily lecithotrophic larvae. For 13 species examined in this category with specific, quantitative measures, the mean larval duration was 0.9 days (standard deviation = 1.2 days).
- *Long larval duration (>5 days).* Species that have long-duration larval periods are primarily planktotrophic larvae. For 18 species examined in this category with quantitative measures, the mean larval duration was 24.1 days (standard deviation = 16.5 days).
- *Sporadic (occasional swimming).* These species, mostly peracarid crustaceans (e.g., amphipods and isopods), have direct development, but may swim and even swarm. Many of these species swarm in the plankton at night or during reproductive periods (Bourdillon 1958, Menzies 1961, Grabe 1996).
- *Holoplanktonic.* Organisms that occur predominantly or frequently in the water column, as larval and adult forms (note that some of the above categories are included here as well).
- *Unknown.* Species or taxonomic groups for which insufficient information existed to characterize planktonic life stages.

Other Characteristics

Other characteristics can also affect the potential for transfer by the various shipping subvectors. Some of these characteristics are directly related to the organisms; others have to do with the vector operation. The following are associated with organisms:

- *Mobility and size.* Nektonic species and large, mobile epibenthic species (e.g., benthic fishes, shrimps, swimming crabs) are unlikely to remain associated with the hull of a moving ship, due to speed and sheer forces, and are more likely to be transferred in ships' ballast tanks. Small species that may lack planktonic stages and live at or near the sediment surface (e.g., oligochaetes, certain mollusks, foraminifera) may easily be suspended in the water column ("tychoplankton") by turbulence, especially by a ship's propellers or by wave action, and taken up in a ship's ballast water. For these species, interstices associated with other vectors (such as solid ballast, oysters, and aquatic plant shipments) also provide opportunities for transfer.
- *Salinity tolerance.* Most epibiotic species found at low salinities (including freshwater) cannot survive long exposure to seawater in a ship's fouling community, and so cannot be transported across ocean basins on ships' hulls. Such species are more likely transported in ballast water, or possibly solid ballast or wet dunnage (packing material).
- *Tolerance to air exposure.* Species found in the upper intertidal zone, in or on sediment and stones, have the potential for transport in solid ballast. Since transport with solid ballast is an historical vector, we judged this in part based upon either the date of discovery (that is, if during a time period when solid ballast was being used) or the probability of being overlooked (that is, the potential for the species having been introduced long ago, but only detected in more modern times; see example below). Species with nonaquatic life stages (e.g., flying insects with aquatic larvae; fauna associated with emergent intertidal plants) or semi-terrestrial habitats (e.g., spray-zone insects and crustaceans) have a potential for transport in cargo, on deck, or in containers, in plant material used as packing. Reeds, rushes, marsh grasses, and rice straw were once widely used for packing fragile goods. Living nonindigenous invertebrates (snails, insects, etc.) have been found in discarded packing material (Mills et al. 1993).

Invasion history (and especially the timing of establishment) is also highly relevant to the likelihood of transfer among vectors. The use of ballast water by ships became increasingly common by 1900, largely replacing dry ballast. Thus, the likelihood of transport by these two modes shifted through time, and is also influenced by the date of colonization and discovery. For example, one likely solid-ballast introduction, the Southern Hemisphere littoral amphipod *Transorchestia engimatica*, was not discovered in San Francisco Bay until 1962, although possibly introduced there as early as the nineteenth century (Cohen and Carlton 1995).

A number of wood-boring organisms (e.g., the shipworm *Teredo navalis* and the isopods *Sphaeroma* spp. and *Limnoria* spp.) were transported around the world in the hulls of wooden ships commencing at least by the fifteenth century, but many of these wood-borers, as well as a wide range of their commensal and parasitic associates, were only first discovered in the twentieth century (Carlton 1979, Cohen and Carlton 1995, Kensley and Schotte 1999). This is a clear artifact of the onset of taxonomic interest in these small animals. In our analysis, we assumed that these particular organisms were introduced with their hosts, in boring communities of ships' hulls.

Classification and Analysis

For all of the species included in our analysis, our database (NEMESIS 2003) identifies those species invasions attributed to shipping as either the sole vector or one of multiple possible vectors (see Ruiz et al. 2000 for further discussion of this process and results for 298 of these species). For species attributed to the shipping vector, either solely or in combination with other possible vectors, we identified here one or more possible shipping subvectors (ballast water, fouling, dry ballast, cargo/packing material), using a multiple-step process. First, we classified all species according to three character sets: adult form, planktonic life stages, and year of first record. Second, using this classification matrix along with some additional character sets (e.g., size, tolerance to high salinity and air exposure, timing of vector operation), we assigned invasions to one or more subvectors as below.

For each species, we used the date of first report on the coastlines of continental North America, north of Mexico (i.e., excluding Hawaii and the Caribbean Islands). Only the first invasion of a species to the continent was considered. For example, the first report of the Southern Hemisphere barnacle *Balanus amphitrite* in North America was on the Pacific coast (California) in 1921 (Carlton 1979). First records on the Atlantic and Gulf coasts were in 1955 (Zullo et al. 1972, Wells 1966, Henry and McLaughlin 1975), but we used only the earlier 1921 record in our analysis. Invasions of species native to a coast or a part of a coast of North America were included in this analysis. An example is the dark false mussel *Mytilopsis leucophaeata*, native from Chesapeake Bay to Mexico, but found north of the Chesapeake in the Hudson River in 1937, and later found at several other sites in New England in the 1990s (Smith and Boss 1995). In such cases, the earliest date (e.g., 1937) outside the native range was used. For 7 other species that have invaded coastal bays and estuaries, the first reported introduction was to the Great Lakes from Europe; the Great Lakes date of first record was used in the analysis (but we do not otherwise consider invasions to the Great Lakes, or any other nontidal freshwater locations, in the present analysis).

We considered ships to shift predominantly from use of solid ballast to ballast water in 1900. Obviously, this date is somewhat arbitrary. The transition was not abrupt at the turn of the century, as ballast water was used experimentally as early as the mid–nineteenth century; some sailing cargo ships, using dry ballast, persisted into the twentieth century. Nonetheless, we chose a specific transition point to estimate the growing importance of ballast water around this time; thus, a few species attributed to ballast water may have arrived by dry ballast, but the species pool sampled by each subvector differed substantially.

To evaluate possible modes of transport, information about adult and larval life histories must be integrated, using knowledge specific to each taxonomic group. For example, hydrozoans with sessile polyp stages and planktonic medusae were treated as "sessile" organisms with a "long" planktonic stage. These species potentially can be transported either by ballast water or by fouling. Other taxa that sometimes reach adult (and presumably reproductive) sizes in internal ballast fouling communities include barnacles, bryozoans, and mussels. The zoospores of macroalgae are mostly very short-lived (Santelices 1990, Reed et al. 1992), but spores of a few species (e.g., the brown kelp *Undaria pinnatifida*, and the green alga *Enteromorpha intestinalis*) can remain planktonic for up to 14 days in light (Jones and Babb 1968, Forrest et al. 2000). Even for these species, planktonic periods are likely to be much shorter in darkness, since these cells are photosynthetic (Jones and Babb 1968). Thus, planktonic stages (zoospores) of macroalgae were generally considered to have short planktonic periods. Nonetheless, since detached fragments of adult plants suspended or floating in the water column are also potential propagules (Santelices 1990), we considered both ballast water and hull fouling to be mechanisms of transport for macroalgae.

For each species associated with shipping, we considered one or more subvectors as possible mechanisms for transfer as follows:

1. Hull fouling was considered a possible subvector for species with (a) sessile or sedentary forms, or (b) attached eggs to hard substrates. For species that cannot tolerate high salinities for multiple days, however, hulls were not considered a viable transfer mechanism.
2. Dry ballast was considered a possible subvector for species that (a) likely colonized before 1900, (b) occur in the intertidal or superlittoral zone (i.e., relatively high on the shore), and (c) can tolerate prolonged periods in damp environments out of water.

3. Ballast water was identified as a possible subvector for species that colonized after 1900 and met one of the following criteria: (a) occur commonly in the water column for postlarval stages, (b) occur in the water column as larval stages with relatively long duration (>5 days), or (c) are small surface-dwelling organisms that may be resuspended and entrained during ship operations.
4. Cargo or packing materials were considered possible vectors for species (a) known to be associated with particular commodities (or packing materials) and (b) tolerant of long periods of air exposure.

Using this approach, we summarized the relative importance of these subvectors, focusing especially on temporal changes for invasions attributed to ballast water and hulls. We presented the results of these analyses in a partitioned fashion, showing (a) the number of invasions for which the subvector is the sole mechanism of introduction and (b) the additional number of invasions for which the subvector is one of multiple possible mechanisms. These partitioned results were intended to highlight not only the potential importance of the respective subvectors but also the proportion of cases in which both ballast water and hulls were viable mechanisms.

In addition, for invasions where other nonshipping vectors were a possible mode of introduction, we indicated the number of species for which oyster transplantation was an alternate vector versus those for which it was not. We based this determination primarily on specific criteria (salinity range, habitat, life history, etc.), as compiled, analyzed, and reported by Carlton (1979) and Cohen and Carlton (1995).

Assumptions and Further Considerations

Overall, our goal in this analysis is to provide a first approximation at estimating the relative importance of the respective subvectors over time. It is worth noting the coarse nature of some criteria. One such criterion is larval duration. We used larval duration as a key criterion to attribute certain species as transfers to North America by ballast water. Most initial marine invasions (i.e., not including secondary spread) to North America have occurred from across ocean basins or across the continent (Ruiz et al. 2000), where voyage duration is routinely 7–10 days (Carlton and Geller 1993, Smith et al. 1999), suggesting that larval duration is an important limitation in the dispersal process.

It is important to note, however, that boundaries between long and short larval durations are by no means rigid. Some lecithotrophic species can have planktonic periods prolonged by low temperatures or absence of suitable substrata for settlement. A few species have poecilogonic life histories, and can shift development patterns either genetically or in response to varying environmental conditions (e.g., the nudibranch *Tenellia adspersa,* and the spionid polychaete worm *Streblospio benedicti;* Chester 1996, Levin 1984). There are relatively little data available for most individual species concerning larval duration and variation (including capacity for delayed metamorphosis) under the complex conditions found in the wide variety of ballast tanks and ballasted cargo holds. In addition, in some cases, larvae of sessile organisms may settle out on debris in ballast tanks, and this debris may subsequently be discharged. Thus, our estimates serve as general approximations for some groups.

We also created two discrete decision criteria associated with ballast water, where in reality a continuum exists. The first of these involved the transition from dry ballast to ballast water (see "Classification and Analysis"). The second involves our distinction between the hull and ballast water subvectors. For simplicity, we considered here ships' hulls to consist of everything external to the ship, whereas we included in our ballast water category sea chests, internal piping that supplies water to ballast tanks and engines, and all contents of ballast tanks and ballasted cargo holds. In future analyses, we will examine these individual components, to gain a more detailed understanding of how each may operate.

Thus, a vessel offers many surface or near-surface environments for marine life to colonize. Shipworms (teredinid bivalve mollusks) and gribbles (limnoriid isopod crustaceans) bored into the hulls of wooden boats, creating complex micro- and macrogalleries and abandoned holes in which secondary organisms could nestle. Both wood and steel-hulled vessels had or have numerous habitats on which organisms can settle. These include the hull (which varies in physical characteristics from waterline to keel), the rudder, the propeller and propeller shaft, through-hull fittings (for both the intake of water and the discharge of waste materials), and the grates and screens covering various intakes or discharges. Vessels taking in water for engine cooling or ballast water may do so via a boxlike chamber (opening) in the hull known as a "sea chest," in which certain fouling organisms (that might otherwise be washed away on the outer "skin" of the vessel) may be transported successfully. In and among all of the fouling organisms that could colonize these substrates may be a large number of crevicolous dwellers deep within the fouling matrix interstices. These boring, fouling, and secondary nestling species (including sedentary, sessile, and free-living taxa) are broadly characterized as "fouling" organisms.

Evidence from a number of studies suggests that many of these various ship microhabitats may favor distinct taxa.

In a similar way, the ballast systems of ships provide a complex array of environments that can be colonized by organisms delivered by ballast water. As noted above, fouling organisms can colonize sea chests, piping, and the interior of tanks (Hülsmann and Galil 2002). Beyond those organisms in the water itself, organisms and sediments entrained in tanks can form bottom communities, which may persist for months to years, creating the opportunity to "reseed" the water column and be discharged at subsequent ports of call (National Research Council 1996; Colautti et al., this volume).

RESULTS

Overall Strength of the Shipping Vector

Of 316 nonnative species of invertebrates and algae established in North America, 164 (52%) are attributed solely to shipping, and another 87 (27.5%) include shipping as one of multiple possible vectors. Shipping was considered a possible vector for all but 65 species (20.5%). This dominance of the shipping vector is illustrated in Figure 7.1, showing the cumulative number of species invasions attributed solely or possibly to shipping.

Shipping is associated with established invasions across 16 different phyla, either as the sole vector or one of multiple possible vectors (Fig. 7.2). Crustaceans and molluscs are the most numerous taxa, accounting for 32% and 15% of species, respectively. Most of the shipping-associated invasions have occurred on the Pacific coast, comprising 66% of the total across coasts, compared to 31% on the Atlantic coast and 4% on the Gulf coast (Fig. 7.3).

Of invasions associated with shipping on the Pacific coast, a relatively large proportion (40%) included shipping as one of multiple possible mechanisms of introduction, compared to that observed for the Atlantic coast (27%) (Fig. 7.3). This difference results primarily from the widespread introductions of the Pacific oyster (*Crassostrea gigas*) and Atlantic oyster (*C. virginica*) to the Pacific coast, whereby many species (especially fouling organisms) could have been transported with oysters or by ships. Both vectors were in operation from the same geographic source regions, and overlap in time, such that the contributions of these two vectors are impossible to separate for many species (see Cohen and Carlton 1995 for further discussion). By comparison, there have been relatively few translo-

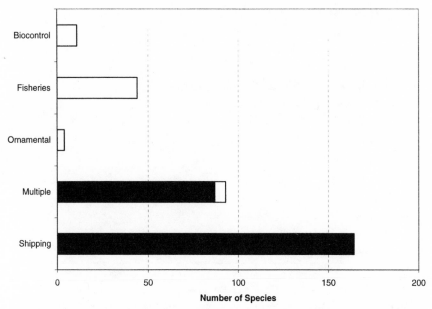

FIGURE 7.1. Transfer mechanism (vector) responsible for the introduction of nonnative coastal marine species established in North America (continental United States and Canada). Shown are the number of invertebrate and algal species associated with each vector; species for which more than one vector was possible were included under the vector category "Multiple." Species for which shipping is a possible vector are shown in black; those for which shipping is not a possible vector are indicated by white bars. (Species shown are primarily from Ruiz et al. [2000] with some additional, recent records [NEMESIS 2002]. Total number of species = 316.)

cations or introductions of oysters to the Atlantic and Gulf coasts. The most numerous oyster movements on these other coasts have been transplants of Atlantic oysters from southern bays to depleted oyster beds on the northeast coast of the United States, resulting in the introduction of a few invertebrates from the Gulf and southern Atlantic coasts (Carlton and Mann 1996).

Temporal Changes in Life-History Patterns

From the nineteenth century to the twentieth century, the overall numbers and types of life-history characteristics of introduced species attributed solely to shipping changed drastically (Fig. 7.4). The numbers of recorded invasions that are attributed only to shipping were much greater after 1900 than before. Although increases were observed for both sessile and mobile species, the percentage increase was greatest for mobile taxa.

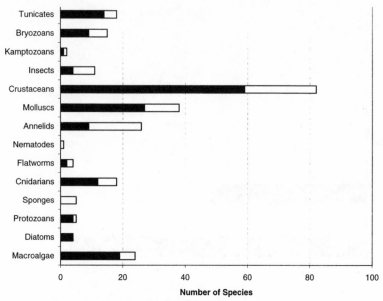

FIGURE 7.2. Taxonomic composition of nonnative coastal marine species established in continental North America for which shipping was a possible mechanism of introduction. The species included correspond to the lower two bars in Figure 7.1 for invertebrates and algae associated with shipping (shown in black), plus 5 additional species of fish. Those species introductions attributed solely to shipping as the vector are shown in black, whereas those for which "multiple" vectors were possible (including shipping) are indicated as white bars.

For sessile and sedentary organisms, there was a 475% increase (pre-1900, 16 species; post-1900, 76 species), while the increase was 733% for mobile taxa (pre-1900, 9 species; post-1900, 66 species). For the possible shipping introductions (i.e., those for which shipping was one of multiple possible vectors), there were even greater increases: 1,700% for sessile taxa (pre-1900, 2 species; post-1900, 34 species); and 2,600% for mobile taxa (pre-1900, 2 species; post-1900, 52 species).

Sessile/sedentary and mobile species differed in some life-history features consistently across time periods (Fig. 7.4). The frequency of species with larvae of short planktonic duration (≤5 days) was greatest for the sessile/sedentary organisms, occurring in 56% of pre-1900 introductions attributed solely to shipping (9 of 16 species) and 52% (40 of 76 species) of those post-1900 introductions. By comparison, short planktonic life histories occurred in none of the pre-1900 mobile species introductions attributed solely to shipping, and in only 1% of those for post-1900 (0 of 9 and 1 of 67 species, respectively). This pattern is only further amplified

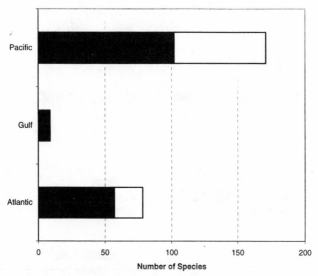

FIGURE 7.3. Number of nonnative coastal marine species established in continental North America by coast (Pacific, Gulf, and Atlantic coasts), for which shipping was a possible mechanism of introduction. The species included correspond to the lower two bars in Figure 7.1 for invertebrates and algae associated with shipping (shown in black), plus 5 additional species of fish. Those species introductions attributed solely to shipping as the vector are shown in black, whereas those for which "multiple" vectors were possible (including shipping) are indicated as white bars.

when considering those species for which shipping is one of multiple possible vectors (Fig. 7.4).

Species lacking planktonic life stages occurred almost exclusively among the mobile organisms (Fig. 7.4). Of the 128 sessile/sedentary species attributed to shipping, as either the sole vector or one of multiple possible vectors, only 1 species is known to lack any planktonic stages: the Asian orange-striped sea anemone *Diadumene lineata* (which can have planktonic larvae) appears to engage only in asexual fission in North America populations (Ting and Geller 2000). In contrast, 24% (30 of 127 species) of all mobile species associated with shipping, either as the sole vector or as a possible vector, had no planktonic stages.

Associated with the large increase in the proportion of mobile biota after 1900, a striking shift occurred in the proportion of these taxa spending long periods of time in the plankton (Fig. 7.4). For mobile biota invasions attributed solely to shipping, the proportion of species that were holoplanktonic or had long larval durations increased dramatically from 11% before 1900 (1 of 9 species) to 55% after 1900 (36 of 67 species).

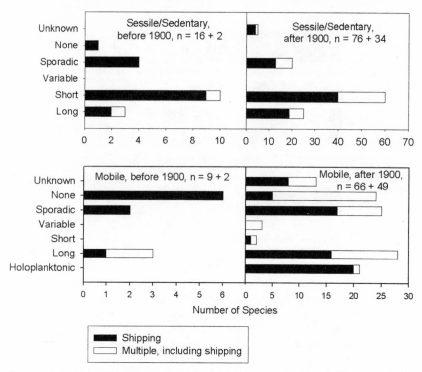

FIGURE 7.4. Number of nonnative coastal marine species established in continental North America, associated with shipping, as a function of adult form, planktonic life stages, and date of arrival. The numbers of species (invertebrates, algae, and fish) are divided between those with sessile/sedentary forms (top) and those with mobile forms (bottom); each of these groups is further divided between those established before 1900 (left) and after 1900 (right). For each of these four groupings, the species are classified according to planktonic life stages for those (a) attributed to shipping as a sole vector (black bar) and (b) those attributed to shipping as one of multiple possible vectors (white bar); these categories correspond to the lower bars of Figure 7.1 (see text for further description).

When including those species possibly introduced by shipping (among other vectors), the same general shifts in favor of long-duration and holoplanktonic organisms are evident but less pronounced (27% before 1900 versus 39% after 1900; Fig. 7.4).

Because we set 1900 as the date for the onset of ballast water transport, our analyses would therefore show no shipping vector introductions of holoplankton before this date. Although this would appear to be a bias of our vector time boundary, in fact no clear holoplankton invasions in North America were noted for another 50 years. The first record is the surf diatom *Attheya armata* (previously *Chaetoceros armatus*), of uncertain origin, which appeared on the Oregon coast around 1950 (Lewin 1973). In the last half century, 21 species of introduced holoplankton have appeared in North American estuaries and coastal waters.

Differences across time periods were smaller for sessile organisms. For those species attributed to shipping as the sole vector, the proportion of species with long (>5 days) planktonic larval duration increased from 12.5% before 1900 (2 of 16 species) to 26% after 1900 (20 of 76 species), while the proportion of species with short-duration larvae (≤5 days) remained about the same, 56% before 1900 and 53 percent afterward (Fig. 7.4).

Modes of Shipping Transport (subvectors)

Here we summarize our assignment of subvectors to the invasions associated with shipping, using the approach described earlier (see "Methods"). Details of subvector classification for each species are available in the database (NEMESIS 2002).

Mobile Biota Invasions since 1900

Table 7.1 summarizes the classification of subvectors for mobile species introduced after 1900 that were attributed to shipping as a sole vector or as a possible vector among others (i.e., lower right panel of Fig. 7.4). This table shows the results by adult form for this subset of the total, illustrating the general approach used across all species. The first number, outside of parentheses, in each column refers to the number of species attributed to the respective subvectors, of those invasions attributed solely to the shipping vector. The numbers within parentheses indicate the additional number of species for which this subvector is a possible source of invasion, along with other nonshipping vectors, divided into two categories (number for which oysters were also a possible vector, number for which oysters were not a possible vector) (see Table 7.1 for further details).

Of mobile biota introduced by shipping (as sole vector) since 1900, 32 of 66 species were attributed only to ballast water. This included 20 holoplanktonic species and 12 benthic mobile species (8 infaunal forms with planktonic larvae, 3 epibenthic mobile species, and 1 upper-intertidal mobile species). Another 26 species were attributed to either ballast water or hulls as the subvector, whereas only 1 species in this group was attributed solely to hull fouling. The only mobile species found after 1900 that was considered a clear-cut fouling introduction was the New Zealand commensal isopod *Iais californica*, described in 1904, but most likely introduced before 1900, with its isopod host *Sphaeroma quoyanum*. Taken

TABLE 7.1. Number of nonnative "mobile" species established after 1900 in coastal marine environments of continental North America, associated with different shipping subvectors.

	Vectors					
Adult Form/Habitat Utilization	Ballast Water	Ballast Water/ Hull Fouling	Hull Fouling	Ballast Water/ Other Shipping Vectors	Other Shipping Vectors	Total
Holoplankton	20 (0, 1)	0 (0,0)	0 (0,0)	0 (0,0)	0 (0,0)	20 (0, 1)
Epibenthic mobile w/attached eggs	0 (0,0)	12 (3, 0)	0 (0,0)	0 (0,0)	0 (0,0)	12 (3, 0)
Epibenthic mobile	3 (0, 11)	8 (8, 0)	0 (0,0)	0 (0,0)	0 (0,0)	11 (8, 11)
Infauna	8 (7, 1)	2 (1, 0)	1 (0,0)	1 (6, 4)	0 (0,0)	12 (14, 5)
Upper-intertidal mobile	1 (0,0)	0 (0,0)	0 (0,0)	0 (0 0)	1 (0, 1)	2 (0, 1)
Intertidal–supratidal mobile	0 (0,0)	0 (0,0)	0 (0,0)	2 (0, 1)	2 (0, 6)	4 (0, 7)
Unknown	0 (0,0)	4 (0,0)	0 (0,0)	0 (0,0)	0 (0,0)	4 (0, 0)
TOTAL	32 (7, 13)	26 (12, 0)	1 (0, 0)	3 (6, 5)	3 (0, 7)	65 (25, 25)

The number of species associated with each category of shipping subvector(s), as a possible mechanism of the introduction, is shown by adult form and habitat utilization. The first number in each category corresponds to species for which shipping was the sole vector. The numbers in parentheses correspond to species for which shipping was one of multiple possible vectors; these numbers are divided, showing those that include oysters as an alternate vector and those that do not include oysters. (This table corresponds to the lower right panel of Figure 7.4; see text for description of sole and multiple vectors.)

together, the combination of ballast water and/or hull fouling was considered responsible for 95% of these invasions (63 of 66 mobile species since 1900 attributed only to shipping).

For mobile species that included other possible nonshipping vectors, ballast water was considered a possible mechanism for 20 species. For 7 of these species, oysters were also a possible mode of introduction; for the other 13 species, vectors such as transport with stocked fishes, ornamental aquatic plant shipments, or aquarium escapes were also considered possible. Another 12 species were associated with both ballast water and hull fouling as a possible vector, along with oysters; most of these were epibenthic mobile species that have attached eggs. Thus, altogether, 44 of 50 species (88%) that included shipping as one of multiple possible vectors were associated with ballast water and hull fouling, either alone or in combination.

Sessile, Sedentary, and Mobile Biota

SHIPPING AS THE SOLE VECTOR

Across all biota and time periods, ballast water and fouling alone accounted for 90% of the 168 species attributed solely to shipping, and were included as a possible subvector (along with other shipping subvectors) for 94% of these species (Table 7.2). Of the 167 species, 36% (60 species) were classified as resulting from hull fouling alone (Table 7.2). These included 57 sessile and sedentary species and 3 mobile species. The most numerous of the sessile taxa are tunicates (14 species), commensals and parasites of wood-boring crustaceans and shipworms (9 species, including protozoans, flatworms, crustaceans), bryozoans (8 species), and wood-borers (6 crustaceans, 3 shipworms). The 3 mobile species are all isopods (*Sphaeroma quoyanum, Synidotea laevidorsalis, Iais californica*) introduced to the Pacific coast between 1890 and 1904 (Carlton 1979).

For the same group of 168 species, 35 species (20%) were classified as resulting from the ballast water subvector alone. This includes 33 mobile species that colonized after 1900, as described above (see previous section), and 2 sessile species (the zebra mussel *Dreissena polymorpa* [Mellina and Rasmussen 1994; Mills et al. 1997]; the quagga mussel *D. bugensis* [Mills et al. 1996]).

Another 34% (57 of these 168 species) were classified as transfers from either ballast water or hull fouling, whereby the contribution of each subvector could not be determined (Table 7.2). These species include epibenthic mobile species, with eggs attached to hard surfaces (11 species, mostly nudibranchs and gobiid fishes); smaller mobile epibenthic fauna, capable of remaining in the fouling community on a moving ship (15 species); and

TABLE 7.2. Number of nonnative coastal marine species established in continental North America associated with different shipping subvectors' as a function of adult form, for all time periods.

Vectors	Sessile/Sedentary	% of Total	Mobile	% of Total	Total	% of Total
Ballast water	2 (0,0)	1 (0,0)	33 (7,13)	44 (13,25)	35 (7,13)	20 (8,15)
Ballast water/Hull fouling	31 (18,0)	33 (50,0)	26 (12,0)	35 (23,0)	57 (30,0)	34 (34,0)
Hull fouling	57 (18,0)	62 (50,0)	3 (1,0)	4 (2,0)	60 (19,0)	36 (21,0)
Ballast water/Dry ballast	0 (0,0)	0 (0,0)	1 (6,4)	1 (12,8)	1 (6,4)	1 (7,5)
Hull fouling/Dry ballast	2 (0,0)	1 (0,0)	1 (1,1)	1 (2,2)	3 (1,1)	1 (1,1)
Hull fouling/Dry ballast/Ballast water	1 (0,0)	1 (0,0)	0 (0,0)	0 (0,0)	1 (0,0)	1 (0,0)
Dry ballast	0 (0,0)	0 (0,0)	2 (0,1)	3 (0,2)	2 (0,1)	1 (0,1)
Ballast water/Cargo or packing material	0 (0,0)	0 (0,0)	2 (0,1)	2 (0,2)	2 (0,1)	1 (0,1)
Dry ballast/Cargo or packing material	0 (0,0)	0 (0,0)	5 (0,0)	7 (0,0)	5 (0,0)	3 (0,0)
Cargo or packing material	0 (0,0)	0 (0,0)	2 (0,6)	3 (0,12)	2 (0,6)	1 (0,7)
TOTAL	93 (36,0)	98 (100,0)	75 (27,26)	100 (52,51)	168 (63,26)	99 (71,30)

Column 1 shows the different shipping subvectors, alone or in combination. The numbers in each of the subsequent columns 2, 4, and 6 are as follows: Column 2 shows the number of sessile and sedentary species associated with each subvector category; column 4 shows the number of mobile species associated with each subvector category; column 6 shows the combined total for sessile/sedentary + mobile species associated with each subvector category. The first number in each category of these columns corresponds to species for which shipping was the sole vector; the numbers in parentheses correspond to species for which shipping was one of multiple possible vectors; these numbers are divided, showing those that include oysters as an alternate vector and those that do not include oysters. The percentages in columns 3, 5, and 7 are derived from the previous numerical columns as follows: (A) The first number in each category (outside of parentheses) represents the percent of total species from the previous column that were associated with that subvector and had shipping as a sole vector. For example, 31 species in column 2 were associated with ballast water/hull fouling of 93 total species for the column, yielding 33% in column 3. (B) The second and third numbers in each category (within parentheses) represent the percent of total species from the previous column that were associated with that subvector, and had shipping as one of multiple vectors; these are divided between species associated with oysters as an alternate vector and those associated with another vector. For example, 18 species in column 2 were associated with ballast water/hull fouling + oysters of 36 total, yielding 50% of all species that had a possible nonshipping vector.

sessile fauna with planktonic life stages (31 species). Major taxa in the latter category include 12 species of macroalgae, which could be transported either by fouling or by floating fragments; 6 species of cnidarians, with polyp/medusa life cycles; 4 species of barnacles, with planktonic larvae; and 4 species of mussels, also with planktonic larvae.

The relatively dry modes of ship transport (including dry ballast, cargo/packing material) were possible vectors for 16 species that were attributed solely to shipping, of which 9 species included ballast water or shipping as a possible vector (Table 7.2). Major groups included 6 marine oligochaetes, possibly transported by oysters, fouling, solid ballast, or ballast water; 4 freshwater oligochaetes (dry ballast, ballast water, aquatic plant shipments); and 5 insects associated with introduced marsh plants (*Phragmites australis, Typha angustifolia*) (cargo, aquatic plant shipments). This is in contrast to the relatively large number of plant invasions associated with dry ballast (e.g., Ruiz et al. 1999; Mack, this volume).

SHIPPING AS ONE OF MULTIPLE VECTORS

Another 89 species included shipping as a possible vector, among other nonshipping vectors ("multiple" category of Fig. 7.1; see Table 7.2). Of these species, 69% included ballast water as a possible vector, and 56% included hull fouling as a possible vector; 34% included both ballast water and hull fouling as possible vectors. Moreover, ballast water or hull fouling were considered a possible vector for 82 of the 89 species (92%). When combined with species attributed solely to shipping (above), either ballast water or hull fouling was a possible vector (alone or in combination with other shipping and nonshipping vectors) for 94% of the 257 species (Table 7.2). Of the 89 species that had shipping as one of multiple vectors, 63 species (71%) included oysters as a possible vector, and another 26 species (29%) included vectors others than oysters. For the "oyster + shipping" species, the ballast water and hull fouling were both considered possible subvectors for 48% (30 of 63 species), ballast water was considered the only possible shipping subvector for 11% (7 of 63 species), and hull fouling was considered the only possible shipping subvector for 30% (19 of 63 species). The distribution of sessile/sedentary versus mobile taxa among these various subvector categories was roughly similar to that described for those 168 species with shipping as the sole vector (Table 7.2).

Species whose alternate vectors were modes other than oyster plantings tended to be freshwater species, which included as possible vectors aquatic plant shipments, fish stockings, and aquaculture. For these organisms, half (13 of 26) were possible ballast water introductions. Other possible subvectors included dry ballast and packing material.

Temporal Changes in the Shipping Vector

The rate of invasions attributed to shipping has increased exponentially over time, from the early 1800s to the present. Figure 7.5 shows the number of newly detected invasions per time interval that were attributed solely to shipping (solid line) and the additional number when including species attributed to shipping as one of multiple vectors (dashed line).

Although shipping was the single largest vector for marine species introductions in North America (Fig. 7.1; see also Cohen and Carlton 1995; Ruiz et al. 2000), the full contribution of shipping to invasions remains obscured by the existence of multiple possible vectors for 89 (28%) of the 321 species considered here. Most (71%) of these species with multiple vectors included oysters, introduced to the Pacific coast in

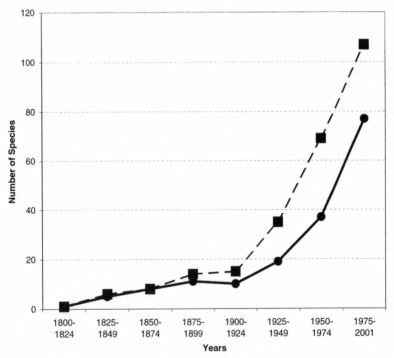

Figure 7.5. Rate of invasion for nonnative coastal marine species established in continental North America and associated with shipping. The numbers of species (invertebrates, algae, and fish) that were newly detected in each time interval are shown for (a) those invasions for which shipping was the sole vector (black line) and (b) those invasions for which shipping was the sole vector or one of multiple possible vectors (dashed line); these categories correspond to the lower black bars of Figure 7.1 (see text for further description).

the late nineteenth and early twentieth centuries, as an alternate vector (as above).

As we shift focus to examine the shipping vector in greater detail, the relative contribution of individual subvectors over the same time periods is difficult to discern (Figs. 7.6 and 7.7). The combination of ballast water and hull fouling account for nearly all invasions attributed to shipping, and this shows a strong pattern of increase over time (Fig. 7.5, Table 7.2). However, our present ability to attribute most of these invasions to one or the other subvector is confounded by the multiple life stages of many species that could be transferred by either hull fouling or ballast water, as well as other nonshipping vectors that may have transferred some species. This is reflected in Figures 7.6 and 7.7, indicating the number of newly detected invasions over time that are attributed to ballast water or hull fouling as the sole vector (solid lines in Figs. 7.6 and 7.7, respectively), and the additional number of invasions that may have resulted from either vector (dashed line; this last group includes species for which shipping is the sole vector as well as one of multiple vectors). Thus, although the num-

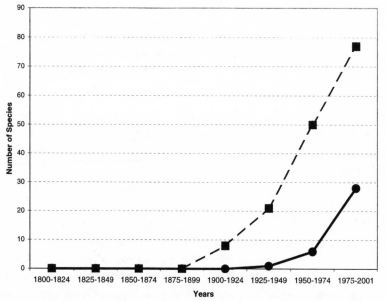

FIGURE 7.6. Rate of invasion for nonnative coastal marine species established in continental North America and associated with ships' ballast water. The number of species (invertebrates, algae, and fish) that were newly detected in each time interval are shown for (a) those invasions for which ballast water was the sole vector (black line) and (b) those invasions for which ballast water was the sole vector or one of multiple possible vectors (dashed line); see text for further description.

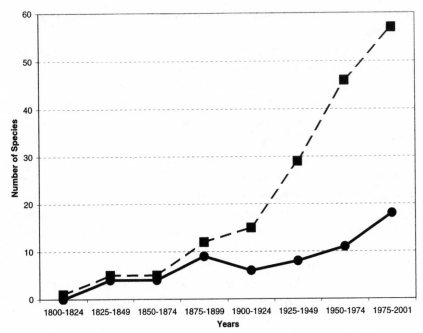

FIGURE 7.7. Rate of invasion for nonnative coastal marine species established in continental North America and associated with ships' hulls. The number of species (invertebrates, algae, and fish) that were newly detected in each time interval are shown for (a) those invasions for which ships' hulls were the sole vector (black line) and (b) those invasions for which ships' hulls were the sole vector or one of multiple possible vectors (dashed line); see text for further description.

ber attributed to each subvector has increased over time, the potential magnitude of this increase (dashed lines) and the relative contribution of each of these two subvectors are largely unresolved for North American invasions.

DISCUSSION

Shipping is a major driver of invasions by coastal marine organisms in North American estuaries. Shipping has historically contributed the majority of established nonnative species in coastal North America, and invasions attributed to shipping alone have exhibited the greatest rate of change through time among the various vectors (Cohen and Carlton 1995, 1998, Ruiz et al. 2000). A similar pattern has emerged in other regions of the world, suggesting this pattern is widespread (Hewitt et al. 1999, Reise et al. 1999). Although is it evident that two primary subvectors—ballast water and hull fouling—are responsible for most shipping invasions, it remains a challenge to distinguish or quantify the relative importance of each.

Both hull fouling and ballast water clearly transfer many taxa, but the characteristics of these organisms and the transfer process are not fully resolved. Although there are many cases of individual species reported from ship hulls, surprisingly few are comprehensive assessments of ship fouling communities, identified to species (e.g., Visscher 1927, Pyefinch 1950, Skerman 1960, Carlton and Hodder 1995, Gollasch 2002). Emphasis in these studies is usually on larger, sessile organisms, rather than on mobile species living in association with hull fouling communities. The frequency of transport for free-living, walking, crawling, and sporadically swimming epibenthic organisms, such as polychaetes, isopods, amphipods, and crabs, on hulls of modern ships is poorly understood.

In addition, while numerous studies have examined the biota of ballast water, underscoring the general scale and magnitude of species transfer, these analyses have focused primarily on small holoplankton, using 20–100 μm mesh nets (i.e., meso- and microplankton; Carlton 1985, Locke et al. 1991, Carlton and Geller 1993, Chu et al. 1997, Harvey et al. 1999, Lavoie et al. 1999, Smith et al. 1999). Some studies have also begun to examine microorganisms (McCarthy and Crowder 2000, Drake et al. 2001, Hülsmann and Galil 2002). However, relatively few studies of organisms in ballast tanks have been designed to examine infaunal or epibenthic organisms, such as ostracods, mysids, amphipods, and benthic fishes, which occur at some frequency (Carlton 1985, Locke et al. 1991, Carlton and Geller 1993, Smith et al. 1999, Wonham et al. 2000). Although these organisms appear to be relatively rare compared to the holoplankton, significant populations may be present when considering the entire ballast tank, and few studies have been designed to sample for such organisms. Similarly, fragments of macroalgae were collected in only one extensive survey of ballast water (Carlton and Geller 1993), but the potential for significant transport of potentially reproductive pieces of seaweeds cannot be excluded.

Both shipping subvectors are clearly delivering a diverse array of organism types on massive scales, resulting in the observed increases in invasion. Commercial vessels arrive to U.S. ports alone at an estimated rate of 50,000 arrivals per year from outside the country (Ruiz et al. 2001). Carlton et al. (1995) estimated that these vessels delivered >70 million metric tons of ballast water in 1991. More recently, Ruiz et al. (2003) estimated commercial vessel arrivals amount to >300 million m^2 of submerged hull surface arriving each year to U.S. ports. Thus, both mechanisms are in play, capable of extensive transfer, and are plausible mechanisms for the invasion of many different types of taxa and multiple life stages—making it extremely difficult to discern the relative contribution of each subvector for many species.

Further, the operation of these subvectors is changing through time. The number, size, speed, and source regions of ships are in flux, and each of these attributes may influence transfer of species (Carlton 1996; Ruiz and Carlton, this volume). Ships are being asked to undertake ballast water management, in order to limit the transfer of organisms and opportunities of invasion (e.g., Ruiz et al. 2001, Taylor et al. 2002); this management may take the form of open-ocean exchange of water (i.e., flushing of tanks) or treatment by various physical or chemical means. In addition, bottom coatings of ships are changing through time and may have important consequences on operation of this subvector (Nehring 2001).

Our analysis in this chapter focuses on the current state of knowledge about vectors of marine invasions, and especially the relative importance of ballast water and hull fouling, based on life-history characteristics of the invading organisms and the operation of vectors. Although we believe the results are fairly robust, we emphasize that this is a first, coarse-level analysis. In particular, the resolution of this analysis is limited at the present time by knowledge about life histories for some taxa, and also the extent to which some taxa are transferred by each subvector. For example, some infaunal species can inhabit interstices of fouling communities. Carlton and Hodder (1995) found infaunal capitellid polychaetes on the hull of a slow-moving sailing ship. Infaunal forms, such as capitellids and oligochaetes, frequently occur on stationary surfaces, in pockets of accumulated sediment, and among dense growths of sessile fouling organisms. These microhabitats may be less likely to remain stable and intact on a fast-moving modern ship. Also, we have treated ballast water as a potential vector for all macroalgae, assuming that algal fragments are potential propagules. However, we consider transport of macroalgae in the fouling community to be a more likely vector, since algae on a ship's hull are exposed to light and more likely to reach reproductive maturity in an exotic port.

Throughout our analysis, we have made assumptions about particular organisms and generalized shipping as a vector with two primary subvectors, the exterior outer surface (hull fouling) and interior compartments (ballast water). This classification of subvectors includes many different environments, which may in fact operate to selectively transfer (i.e., favor) particular types of organisms. For example, we have included the sea chest as part of the interior compartment, considering it here as part of ballast water. Although the sea chest is part of the ballast system and exists as a recessed compartment, it is also primarily exterior to the ship and in constant contact with surrounding waters. Being somewhat sheltered from the slipstream of the moving ship, extensive fouling communities may develop more regularly in the sea chest than in ballast tanks or ballasted cargo holds. Unlike ballast tanks, sea chests are also subject to more rapid changes of temperature and salinity during voyages. Some examinations of sea chest communities have been made (Carlton 1999, Ruiz et al., unpublished data). Richards (1990) describes the transport of an Indo-Pacific muricid snail (*Thais blandfordi*) to England in the sea chest of a cargo ship. This snail, lacking planktonic larvae, successfully reproduced in the sea chest, to the point of blocking the intake pipes (Richards 1990).

In general, the assumptions in our analyses serve to underestimate the number of possible shipping subvectors for organisms. For example, if the sea chest is taken as part of the hull fouling subvector, organisms that appear unlikely to persist on hulls would be included in both the hull fouling and ballast water subvectors (e.g., large mobile organisms such as crabs and fish, infaunal organisms). In addition, we have not considered the importance of other possible shipping subvectors, such as sediments and associated organisms that can be transported with anchor chains.

Despite the uncertainty surrounding the relative importance of shipping subvectors to contemporary invasions, management efforts to reduce ballast water have emerged as a high priority at state, national, and international levels (Ruiz and Carlton, this volume). We believe such actions are appropriate and highly desirable. It is clear that many invasions are occurring through ballast water, and some of these are having very significant ecological and economic impacts. Although the relative roles of hull fouling or ballast water may be unclear for many species introduced over the past 100 years or more, the functional nature of species inoculations by ships' ballast systems versus hull fouling differs sufficiently to warrant significant focus on ballast management. It is conceivable that a vessel could enter and leave a port without leaving a trace of its fouling community (either because the fouling species did not depart the hull or reproduce), but ballast water release is a "guaranteed" mechanism of inoculation—scores or hundreds of species, and sometimes millions of individuals, are discharged into the receiving environment by nature of the vector operation.

In this regard, then, the rapid (since the late 1990s) global expansion and interest in ballast water management, more so than the expansion of ship fouling management, may provide a compelling "management-based experiment" in invasion science. We predict that if there is a differential reduction of "ship in" versus "ship out" subvectors there will be a shift in invasions toward species that are more easily linked to ship fouling versus ship ballast. The existing level of uncertainty underscores the necessity of tracking changes in invasion rates coincident with new management actions, as well as improving our knowledge base about the operation of specific shipping subvectors and the specific biological attributes of associated organisms. Only with this approach can we understand vectors of marine invasions and develop effective management to reduce future invasions and to limit the unwanted consequences of nonindigenous species.

Acknowledgments

We wish to thank Richard Everett, Whitman Miller, Fred Dobbs, Lisa Drake, Bella Galil, Chad Hewitt, Anson Hines, and the ICES/IMO Working Group on Ships Ballast Water and Other Shipping Vectors for numerous discussions over multiple years about ship-mediated invasions, contributing substantively to our present perspectives. For support of our research in this area, we also thank the U.S. Department of Defense Legacy Program, Maryland Sea Grant, National Sea Grant, Pew Fellows Program in Marine Conservation, Pew Oceans Commission, and U.S. Fish and Wildlife Service.

References

Barnes, R. D. 1974. *Invertebrate zoology.* Philadelphia: Saunders.

Bourdillon, A. 1958. La dissémination des crustacés xylophages *Limnoria tripunctata* Menzies et *Chelura terebrans* Philippi. *Année Biologique* 34: 437–463.

Carlton, J. T. 1979. *History, biogeography, and ecology of the introduced marine and estuarine invertebrates of the Pacific Coast of North America.* Davis: University of California.

Carlton, J. T. 1985. Transoceanic and interoceanic dispersal of coastal marine organisms: the biology of ballast water. *Oceanography and Marine Biology, an Annual Review* 23: 313–371.

Carlton, J. T. 1996. Pattern, process, and prediction in marine invasion ecology. *Biological Conservation* 78: 97–106.

Carlton, J. T. 1999. Molluscan invasions in marine and estuarine communities. *Malacologia* 41: 439–454.

Carlton, J. T. and J. B. Geller. 1993. Ecological roulette: the global transport of nonindigenous marine organisms. *Science* 261: 78–82.

Carlton, J. T. and J. Hodder. 1995. Biogeography and dispersal of coastal marine organisms: experimental studies of a replica of a 16th-century sailing vessel. *Marine Biology* 121: 721–730.

Carlton, J. T. and R. Mann. 1996. Transfers and world-wide introductions. In *The eastern oyster,* V. S. Kennedy, R. I. E. Newell, and A. F. Eble, eds., p. 691–706. College Park: Maryland Sea Grant.

Carlton, J. T., D. M. Reid, and H. van Leeuwen. 1995. *The role of shipping in the introduction of nonindigenous aquatic organisms to the coastal waters of the United States (other than the Great Lakes) and an analysis of control options.* Washington, DC: U.S. Coast Guard.

Chester, C. M. 1996. The effect of adult nutrition on the reproduction and development of the estuarine nudibranch, *Tenellia adspersa* (Nordmann 1845). *Journal of Experimental Marine Biology and Ecology* 188: 113–130.

Chu, K. H., P. F. Tam, C. H. Fung, and Q. C. Chen. 1997. A biological survey of ballast water in container ships entering Hong Kong. *Hydrobiologia* 352: 201–206.

Cohen, A. N. and J. T. Carlton. 1995. *Nonindigenous aquatic species in a United States estuary: a case study of the biological invasions of the San Francisco Bay and Delta.* Washington, DC: U.S. Fish and Wildlife Service and National Sea Grant College Program (Connecticut Sea Grant).

Cohen, A. N. and J. T. Carlton. 1998. Accelerating invasion rate in a highly invaded estuary. *Science* 279: 555–558.

Drake, L. A., K. H. Choi, G. M. Ruiz, and F. C. Dobbs. 2001. Global redistribution of bacterioplankton and virioplankton communities. *Biological Invasions* 3: 193–199.

Forrest, B. M., S. N. Brown, M. D. Taylor, C. R. Hurd, and C. H. Hay. 2000. The role of natural dispersal mechanisms in the spread of *Undaria pinnatifida* (Laminariales, Phaeophyceae). *Phycologia* 39: 547–553.

Gollasch, S. 2002. The importance of ship hull fouling as a vector of species introductions into the North Sea. *Biofouling* 18: 105–121.

Gollasch, S. and 12 other authors. 2000b. Fluctuations of zooplankton taxa in ballast water during short-term and long-term ocean-going voyages. *International Revue of Hydrobiology* 85: 597–608.

Grabe, S. A. 1996. Composition and seasonality of nocturnal peracarid zooplankton from coastal New Hampshire (USA) waters, 1978–1980. *Journal of Plankton Research* 18: 881–894.

Hall, A. 1981. Copper accumulation in copper-tolerant and non-tolerant populations of the marine fouling alga *Ectocarpus siliculosus* (Dillw.) Lyngbye. *Botanica Marina* 24: 223–228.

Harbison, G. R. and S. P. Volovick. 1994. The ctenophore, *Mnemiopsis leidyi,* in the Black Sea: a holoplanktonic organism transported in the ballast water of ships. In *Nonindigenous and introduced marine species,* pp. 25–36. Washington, DC: Government Printing Office.

Harvey, M., M. Gilbert, D. Gauthier, and D. M. Reid. 1999. A preliminary assessment of risks for the ballast water–mediated introduction of nonindigenous marine organisms in the estuary and Gulf of St. Lawrence. *Canadian Technical Report of Fisheries and Aquatic Sciences* 2268: 1–35.

Henry, D. P. and P. A. McLaughlin. 1975. The barnacles of the *Balanus amphitrite* complex (Cirripedia, Thoracica). *Zoologische Verhandelingen* 141: 1–203.

Hewitt, C. L., M. L. Campbell, R. E. Thresher, and R. B. Martin, eds., 1999. *Marine biological invasions of Port Phillip Bay, Victoria.* Hobart: CSIRO Marine Research.

Hülsmann, N. and B. S. Galil. 2002. Protists—a dominant component of the ballast transported biota. In *Invasive aquatic species of Europe: distributions, impacts, and management,* E. Leppäkoski, S. Gollasch, and S. Olenin, eds., pp. 20–26. Dordrecht: Kluwer Academic Publishers.

Jones, M. M. 1981. Marine organisms transported in ballast water: a review of the Australian scientific position. *Bureau of Rural Resources Bulletin* 11: 1–48.

Jones, W. E. and M. S. Babb. 1968. The motile period of swarmers of *Enteromorpha intestinalis. British Phycological Bulletin* 3: 525–528.

Kensley, B. and M. Schotte. 1999. New records of isopods from the Indian River Lagoon, Florida (Crustacea: Peracarida). *Proceedings of the Biological Society of Washington* 112: 695–713.

Lavoie, D. M., L. D. Smith, and G. Ruiz. 1999. The potential for intracoastal transfer of non-indigenous species in the ballast water of ships. *Estuarine, Coastal, and Shelf Science* 48: 551–564.

Levin, L. A. 1984. Multiple patterns of development in *Streblospio benedicti* Webster (Spionidae) from three coasts of North America. *Biological Bulletin* 166: 494–508.

Lewin, J. 1973. Blooms of surf-zone diatoms along the coast of the Olympic Peninsula, Washington. III. Changes in the species composition of the bloom since 1925. *Beihefte zur Nova Hedwigia* 45: 251–256.

Locke, A., D. M. Reid, W. G. Sprules, J. T. Carlton, and H. C. Van Leeuwen. 1991. Effectiveness of mid-ocean exchange controlling freshwater and coastal zooplankton in ballast water. *Canadian Technical Report of Fisheries and Aquatic Sciences* 1822: 1–93.

McCarthy, H. P. and L. B. Crowder. 2000. An overlooked scale of global transport: phytoplankton species richness in ballast water. *Biological Invasions* 2: 321–322.

Mellina, E. and J. B. Rasmussen. 1994. Occurrence of zebra mussel (*Dreissena polymorpha*) in the intertidal region of the St. Lawrence Estuary. *Journal of Freshwater Ecology* 9: 81–84.

Menzies, R. J. 1961. Suggestion of night-time migration by the wood-borer *Limnoria. Oikos* 12: 170–172.

Mills, E. L., J. H. Leach, J. T. Carlton, and C. L. Secor. 1993. Exotic species in the Great Lakes: a history of biotic crises and anthropogenic introductions. *Journal of Great Lakes Research* 19: 1–54.

Mills, E. L., G. Rosenberg, A. P. Spidle, M. Ludyanskiy, and Y. Pligin. 1996. Review of the biology and ecology of the quagga mussel (*Dreissena bugensis*), a second species of freshwater dreissenid introduced to North America. *American Zoologist* 36: 271–286.

Mills, E. L., M. D. Scheuerell, J. T. Carlton, and D. Strayer. 1997. Biological invasions in the Hudson River: an inventory and historical analysis. *New York State Museum Circular* 57: 1–51.

National Research Council. 1996. *Stemming the tide: controlling introductions of nonindigenous species by ships' ballast water.* Washington, DC: National Academy Press.

Nehring, S. 2001. After the TBT era: alternative anti-fouling paints and their ecological risks. *Senckenbergiana Martima* 31: 341–351.

NEMESIS (National Exotic Marine and Estuarine Species Information System). 2002. Unpublished database, partly accessible at Smithsonian Environmental Research Center's Web site <http://invasions.si.edu>.

Pyefinch, K. A. 1950. Notes on the ecology of ship-fouling organisms. *Journal of Animal Ecology* 19: 29–35.

Reed, D., C. D. Amsler, and A. Ebel. 1992. Dispersal in kelps: factors affecting spore swimming and competency. *Ecology* 73: 1577–1585.

Reise, K., S. Gollasch, and W. J. Wolff. 1999. Introduced marine species of the North Sea coasts. *Helgoländer Meeresunters* 52: 219–234.

Richards, A. 1990. Muricids: a hazard to navigation? *Hawaiian Shell News* 39 (5): 10.

Roos, P. J. 1979. Two-stage life cycle of a *Cordylophora* population in the Netherlands. *Hydrobiologia* 62: 231–239.

Ruiz, G. M., P. W. Fofonoff, J. T. Carlton, M. J. Wonham, and A. H. Hines. 2000. Invasion of coastal marine communities in North America: apparent patterns, processes, and biases. *Annual Review of Ecology and Systematics* 31: 481–531.

Ruiz, G. M., P. Fofonoff, A. H. Hines, and E. D. Grosholz. 1999. Non-indigenous species as stressors in estuarine and marine communities: assessing invasion impacts and interactions. *Limnology and Oceanography* 44: 950–972.

Ruiz, G. M., A. W. Miller, K. Lion, B. Steves, A. Arnwine, E. Collinetti, and E. Wells. 2001. Status and trends of ballast water management in the United States. First biennial report of the National Ballast Information Clearinghouse. Submitted to U.S. Coast Guard. 48 pp.

Ruiz, G. M., A.W. Miller, B. Steves, and R. A. Everett. 2003. Global shipping patterns and marine bioinvasions: the hull story? In review.

Santelices, B. 1990. Patterns of reproduction dispersal and recruitment in seaweeds., *Oceanography and Marine Biology, an Annual Review* 28: 177–276.

Skerman, T. M. 1960. Ship-fouling in New Zealand waters: a survey of marine fouling organisms from vessels of the coastal and overseas trade. *New Zealand Journal of Science* 3: 620–648.

Smith, D. G. and K. J. Boss. 1995. The occurrence of *Mytilopsis leucophaeta* (Conrad 1831) (Veneroidea: Dreissenidae) in southern New England. *Veliger* 39: 359–360.

Smith, L. D., M. J. Wonham, L. D. McCann, G. M. Ruiz, A. H. Hines, and J. T. Carlton. 1999. Invasion pressure to a ballast-flooded estuary and an assessment of inoculant survival. *Biological Invasions* 1: 67–87.

Taylor, A., G. Rigby, S. Gollasch, M. Voigt, G. Hallegraeff, T. McCollin, and A. Jelmert. 2002. Preventive treatment and control techniques for ballast water. In *Invasive aquatic species of Europe: distributions, impacts, and management*, E. Leppäkoski, S. Gollasch, and S. Olenin, eds., pp. 484–507. Dordrecht: Kluwer Academic Publishers.

Ting, J. H. and J. B. Geller. 2000. Clonal diversity in introduced populations of an Asian sea anemone in North America. *Biological Invasions* 2: 23–32.

Visscher, J. P. 1927. Nature and extent of fouling of shi"s bottoms. *Bulletin of the Bureau of Fisheries* 43: 193–252.

Wells, H. W. 1966. Barnacles of the northeastern Gulf of Mexico. *Quarterly Journal of the Florida Academy of Sciences* 29: 81–95.

Wonham, M. J., J. T. Carlton, G. M. Ruiz, and L. D. Smith. 2000. Fish and ships: relating dispersal frequency to success in biological invasions. *Marine Biology* 136: 1111–1121.

Zullo, V. A., D. B. Beach, and J. T. Carlton. 1972. New barnacle records (Cirripedia, Thoracica). *Proceedings of the California Academy of Sciences* 39: 65–74.

Chapter 8

Pathways of Biological Invasions of Marine Plants

María Antonia Ribera Siguan

Less is known about exotic species and biological invasions in the marine than in the terrestrial environment, particularly for marine plants. The scarcity of information can be explained as follows: (1) There is a lack of historic knowledge of marine flora on most coasts of the world. Most floristic studies on marine vegetation are recent, limiting comparisons of flora composition over time. (2) The sea remains a cryptic medium, where the presence of new species may often go undetected. (3) In most cases, the consequences of marine plant invasions only appear over long periods, sometimes as a result of changes in biological and environmental factors. (4) In general, the direct repercussions of marine plant invasions on human activity are much lower than those of invasions on land or even in freshwater. As a result, interest in the study, evaluation, and control of marine plant invasions has been relatively low.

It is difficult to obtain a clear picture of spatial and temporal patterns for marine plant invasions even though there are many sources of information. Floristic studies have been carried out in many countries and geographical regions (Abbott and Hollenberg 1976, Adams 1994, Lawson and John 1982, Noda 1987, Scagel et al. 1989, South and Tittley 1986, Taylor 1957, 1960, Verlaque 1994, 2001, Womersley 1984, 1987, 1994). Many papers and reports focus on particular exotic species (Boudouresque et al. 1994, Cranfield et al. 1998, Eno et al. 1997, ICES 1981, 1987, 1989, 1991, 1997, Munro et al. 1999, Ribera and Boudouresque 1995, see also reference section). In addition, information is increasingly available through electronic databases (e.g., Britain [Eno et al. 2000], Hawaii [Anonymous 2000a], United States [Anonymous 2000b, Invasive Spartina Project 2002]). However, given limitations associated with the historic record, most floristic studies are not explicit as to whether a species is native or exotic, and there remains a lack of data for many parts of the world.

To present an overview of the known world distribution of nonnative marine plants, I have grouped, compiled, and examined records from the following geographical regions: the Mediterranean Sea, the European Atlantic coast, the North American Atlantic coast, the North American Pacific coast, and the Australian and New Zealand coasts. The exotic species from islands or countries that do not belong to any of the geographic regions cited, such as China, the Canary Islands, the Azores, and the Hawaiian Islands, are grouped as "others." Because of the relative lack of information published about marine plant invasions in Central and South America, these regions are not included in this chapter (although some sporadic information about the use of exotic species for their commercialization does exist). Moreover, this overview considers both long- and short-distance transport—the former allowing a species to colonize a new continent or geographical area; the latter allowing a species, native or introduced, to extend its distribution range to neighboring countries. Therefore, within any one region—for example, the European Atlantic coasts—the total number of introduced species includes both invasions from outside Europe and introductions of species within and between European countries. For example, *Sargassum muticum* is considered an exotic species for the Atlantic and has subsequently colonized the coasts of several European countries via diffusion (natural or not). Likewise, although *Mastocarpus stellatus* is native to the European coastline, it was introduced by scientific experiments to the Helgoland area (one of the North Frisian Islands), where it did not previously occur.

This assessment of marine plants includes 189 exotic species (Fig. 8.1, Appendix 8.1). The list does not include taxa recorded at supraspecific levels or doubtful species. The composition of the known exotic marine flora is as follows: 100 red algae (53%), 42 brown algae (22%), 21 green algae (11%), 15 phytoplanktonic species (8%), and 11 higher plants (6%). All these data are probably underestimates. In the case of macroalgae, the number of invaders in each group (red, brown, and green) is roughly proportional to the total number of species known globally in each category. However, following this line of thought, the number of exotic microalgae is likely to be underestimated with respect to the total number of global taxa (possible reasons for this are discussed on page 188). In contrast, the percentage of successful exotic higher plants is elevated (6%) relative to

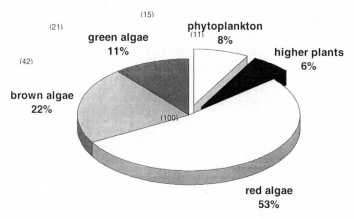

FIGURE 8.1. Number and percentage of nonnative species of marine plants, presented by systematic groups.

the total number of known marine phanerogams in the world. This observation suggests that risk of invasions is relatively great for higher marine plants as a taxonomic group (Ribera and Boudouresque 1995).

The main aim of this chapter is to review the pathways by which species are introduced into different environments, and to relate these pathways to taxa. However, before presenting the results, I wish to emphasize some overarching considerations about the relation between species and pathways, and about interpreting my analyses:

1. It is difficult, and sometimes impossible, to identify the dispersal mechanism (or vector) of exotic marine species. There are few species for which a single vector can be clearly identified. The possible vector is determined on the basis of known associations and life-history characteristics. For instance, ships' hulls are a possible vector for species that (a) appear in a port and (b) are known to occur on ships' hulls in the same port or in nearby waters. The same reasoning can be used for ships' ballast water as a vector. Of course, finding a species on a ship's hull does not exclude it from ballast water transport, and vice versa. For some species, many transfer mechanisms are possible and the likelihood of any one is difficult to estimate; I have considered the pathway of these species (55 total) to be uncertain at the present time.

2. The same taxon may also be involved in various pathways among regions. For example, depending on the region, invasion by *Codium fragile* subsp. *tomentosoides* has been associated with ships' deballasting, transfer by oysters, packing materials used for baits, fishing nets,

ships' hulls, and ships' propeller shafts. Given that one of the main points of interest here is the relationship between the transport systems of different species, it is important to emphasize and adequately reflect the possible occurrence of multiple pathways for the same species. In the analyses (see Conclusions section, p. 202), this explains why the total number of nonnative species distributed in distinct pathways is always greater than the total number of nonnative species, and also why some species are cited more than once in Appendix 8.1.

3. For most species it is very difficult to define species-specific relationships with particular pathways or vectors. Most of the main pathways are not usually highly selective, making it impossible to draw up a profile of risk organisms for each one. Instead, factors such as the propagule supply (pressure) from a particular region, resulting from the extent and frequency of maritime transport and the abundance of a species in a given source region, are likely to determine the dispersal vector from that region.

DESCRIPTION OF MAIN PATHWAYS

The following review of the patterns of invasion for marine plants associated with particular pathways is based on my synthesis and analyses of records for focal geographic regions. This summary excludes species records for which the pathway is considered uncertain (55 total).

Maritime Transport

Transport by vessels (ships) may be the greatest mechanism of global dispersal of marine organisms. This vector includes both the intentional and accidental transport of organisms both outside and inside the vessel (Carlton 1994). Intentional transport aboard ships mostly concerns animals, so the focus here is only on unintentional pathways. Marine macrophytes (including higher plants, red algae, brown algae, and green algae) have been detected most frequently on the outside of vessels, on or in various structures or parts of the ship, such as hulls, rudders, propeller shafts, and anchors. These organisms are usually sessile—that is, fixed on the surface of these structures (fouling)—but they can also be vagile species (clinging) or species that live in holes (boring) (Zibrowius 1991). Inside the vessel, organisms can be found in seawater pipe systems, in the anchor chain locker, and in other places, but in general these "habitats" do not provide suitable survival conditions for marine plants; in contrast, animals and unicellular plants (phytoplankton) are more abundant in the ballast tanks of vessels (Carlton 1994).

Transport on Ships' Hulls

Dispersal of at least 39 marine plants is attributed to hull fouling (Fig. 8.2), and a high number of those nonnative species considered to have an uncertain pathway may also have been transported via this vector (Ribera and Boudouresque 1995). Of these 39 species, most (31) were red algae, and it is interesting to note that higher plants are not involved in this pathway (Appendix 8.1).

Species transported by fouling usually correspond to small specimens that have a flexible and filamentous vegetative structure and a high growth rate, such as species belonging to Acrochaetiales, Ceramiaceae, Ectocarpales, and *Cladophora*. Furthermore, very large sporophytes of the big brown alga *Undaria pinnatifida*, which measure more than a meter, have been observed on the hull of a boat in a New Zealand harbor after sailing hundreds of kilometers (Hay 1990). Therefore, this species may have reached New Zealand through the fishing fleets of Korea and Japan (Adams 1994). In addition, this vector may disseminate large plants whose life cycles include a microscopic stage, such as the microscopic gametophytes of *U. pinnatifida*. Fletcher and Manfredi (1995) predicted that coastal traffic would be responsible for the significant spread of *U. pinnatifida* in northern European waters. Equally, species with a wide range of ecological requirements can survive successive submersion in waters with

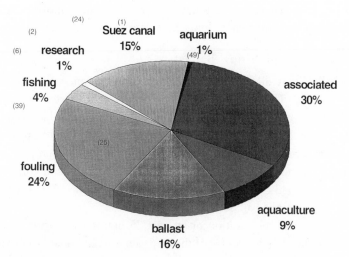

FIGURE 8.2. Number and percentage of nonnative species of marine plants introduced by each pathway.

different physical and chemical characteristics, especially temperature and salinity. In some cases these differences may act as biological barriers, impeding the introduction of species into a new area. This is the case of the Baltic Sea, where the water is virtually fresh (Leppäkoski 1994).

In the Mediterranean Sea, many established exotic plant species belong to the group Ceramiaceae, such as *Acrothamnion preissii*, *Antithamnion amphigeneum*, or *Antithamnionella spirographidis*. Most of these species produce blooms in the initial naturalization phase or periodically, depending on the environmental conditions. These fouling species have sometimes become dominant in invaded communities. For example, in the northwestern Mediterranean Sea, *Asparagopsis armata* covers as much as 100 percent of the available hard substratum in winter but virtually disappears for the rest of the year (Ribera and Boudouresque 1995). In addition, *Polysiphonia harveyi* is now the dominant small seaweed during much of the year at many locations on the North Sea; it has a wide range of biological features that have given it an advantage as an immigrant (Maggs and Stegenga 1999).

The transport of fouling organisms on ships' hulls is the most probable mechanism of transoceanic dispersal of *Codium fragile* subsp. *tomentosoides* (Carlton and Scanlon 1985). The direct observations of this species on the hulls of vessels traveling from Europe to North America (Loosanoff 1975) lend strong support to the hypothesis that the population of this Asian species on North American coasts originated in Europe. Similarly, Hillson (1976) suggested that this species might have been transported to Virginia on a ship's hull. However, according to Carlton and Scanlon (1985), other mechanisms may have contributed to the movement of *C. fragile* subsp. *tomentosoides* along the Atlantic coast of North America.

Transport through Ships' Ballast

Ships' ballast may also make a significant contribution to the transfer of marine plants. The volume of ballast water transported globally is immense; it is estimated that 79 million metric tons of ballast water are released annually into U.S. waters (Carlton et al. 1995) and up to 3,000 million tons worldwide (Gollash 1996). Ballast water frequently contains sediment, which is an ideal medium for the transport of certain organisms. In a survey of 343 cargo vessels, Hallegraeff and Bolch (1992) found that 65 percent contained ballast tank sediments.

I attribute the colonization of at least 25 nonnative marine plants to this vector (Fig. 8.2). Most (16) of these species are phytoplankton (Appendix 8.1). I believe these data are underestimates, as the scarcity of floristic studies on the phytoplankton of geographical regions and the numerous problems in the taxonomy of planktonic microalgae make it difficult to detect nonnative species. Recognition of the presence of a new species is often restricted to the appearance of algae blooms (red tides), suggesting that many rare species may go undetected. Improved methods used in taxonomic analyses, especially on dinoflagellates, based on new techniques of ultrastructural chemical composition and genetic molecular analysis should increase understanding in these areas and allow detection of new species. For example, the reports of *Gymnodinium catenatum* in northern European waters do not refer to a recently introduced exotic species but constitute a misidentification of a native overlooked in these waters: *G. nolleri* (Elbrächter 1999). This author refers to this kind of species as pseudoexotic species.

The percentage (20%) of higher plants that have colonized new areas in this way is surprising; this may be attributable to the production of fruits or seeds that are adapted to survive for long periods under constant humidity (water or sediment) and low light.

Carlton and Geller (1993) note that ships' ballast is among the least selective means of species transfer, from an ecological and taxonomic point of view. As an example, they pointed out that 317 species were recorded in the ballast of Japanese ships on arrival in Oregon (USA). In spite of this complexity, I can attempt to identify the characteristics of plants that are susceptible to this mode of transport:

1. Planktonic species. Ballast is probably the most feasible method of transport for microscopic organisms living free in a mass of water.
2. Plants that can develop structures of resistance during their life cycle, such as cysts, zygotes, spores of resistance, or fruits. This strategy can be present in both micro- and macrophytes, and can guarantee survival in ballast tanks (normally in sediment ballast), and especially viability in later life in the new area. Thus diatoms, with their siliceous frustule, and dinoflagellates, with their cysts, may be best adapted to ballast water or sediments. One single ballast tank was estimated to contain more than 300 million dinoflagellate cysts, which could germinate into confirmed toxic cultures (Hallegraeff and Bolch 1991).
3. Plants with vegetative reproduction. Ballast can be the pathway for macrospecies that form specific propagules or fragments of plants, like cuttings. Some large brown algae, such as *Sargassum muticum*, fragment easily and can even form large floating masses; these fragments could easily withstand ballast tank conditions for several days. The red alga *Furcellaria lumbricalis* and the brown alga *Fucus serratus* were probably introduced to the area of the Gulf of St. Lawrence (Canadian

Atlantic coast) in the late nineteenth century, and ships' ballast has been suggested as the likely mechanism (Novaczek and McLachlan 1989).

Ballast introductions are dependent on the survival of organisms during transit; the specific conditions in ballast tanks may be a barrier to invasion for some organisms or may enhance the survival of others (Galil and Hülsmann 1997). The darkness of ballast tanks, combined with continued zooplankton grazing pressure, result in a considerable decrease in algal densities, thereby precipitating a decline in the light-driven food web (Galil and Hülsmann 1997). A study of ballast carried by cargo vessels entering Australian ports reported that 50 percent of sediment samples contained dinoflagellate cysts and 5 percent were cysts of the toxic dinoflagellates *Alexandrium catenella* and *A. tamarense* (Hallegraeff and Bolch 1992). Similarly, MacDonald and Davidson (1998) detected *Alexandrium* cysts (including *A. minutum* and *A. tamarense*) in 17 percent of the ships arriving in Scottish ports (European Atlantic coast) after short voyages through regional European seas. In contrast, no cysts of potentially harmful dinoflagellates were detected in a study of ballast tanks of vessels on the Mediterranean coast of Israel (Galil and Hülsmann 1997).

These microalgae, whether they are naturalized (i.e., established for a relatively long period of time) or recent arrivals, can give rise to blooms that have a large impact on the environment or the human population. For example, the introduction of the diatom *Coscinodiscus wailesii* to the North Sea has caused the clogging of fishing nets by extensive mucus production (Boalch and Harbour 1977). The dispersal of toxic dinoflagellates can be more dangerous because they pose a serious threat to aquaculture and public health. The human illness Paralytic Shellfish Poisoning (PSP) results from the consumption of products contaminated with alkaloid toxins from any of 11 species of planktonic dinoflagellates (Hallegraeff et al. 1995), such as *Alexandrium catenella, A. minutum, A. tamarense,* and *Gymnodinium catenatum.* Although in a strict sense PSP is a natural phenomenon, well known to native North American Indian tribes, until 1970 poisoning records were confined to temperate waters of Europe, North America, and Japan (Hallegraeff 1993). However, by 1990, PSP outbreaks were documented throughout the Southern Hemisphere, including South Africa, Australia, New Zealand, and Papua New Guinea, and the Northern Hemisphere, including India, Thailand, and the Philippines (Hallegraeff 1998). The introduction of Southeast Asian dinoflagellates of the genera *Gymnodinium* and *Alexandrium* into Australian waters in the 1980s caused PSP, and as a result there was a ban on the sale of cultured and wild shellfish, mussels, oysters, and scallops, which had a catastrophic economic impact (Pollard and Hutchings 1990). Similarly, blooms of *Gymnodinium catenatum* were first recognized in waters of southern Tasmania in late 1985 (Hallegraeff and Summer 1986). The most likely vector of this introduction is ballast water discharge either from northeast Asian ports or, less likely, from European ports (McMinn et al. 1997).

Transport by Other Parts of a Ship

There are other means of plant transfer by maritime transport, for example, adhesion to the propeller or anchor. This kind of vector may be more important in marginal rather than remote dispersal of marine species. The dissemination of *Caulerpa taxifolia* across the Mediterranean was probably caused by successive anchorage, often over long distances, by pleasure boats plying between the French coast and the Balearic Islands or the Italian coast (Meinesz 1992). By following the most frequent pleasure routes in the Mediterranean (Meinesz et al. 1998), I can reconstruct the progress of *C. taxifolia* from the French Mediterranean coast. Ecophysiological studies on *C. taxifolia* confirm that this species survives long periods in the prevailing conditions of an anchor locker (Sant et al. 1996). Therefore, Knoepffler-Péguy et al. (1985) suggested that small boats caused the expansion of *Sargassum muticum* outside the Thau Lagoon (France). Species transported by this method develop rapidly and can regenerate from vegetative fragments. Moreover, if the species are tolerant to desiccation, a wide range of temperatures, and the dark, their potential survival in new areas is enhanced.

Aquaculture Activities

The introduction of 64 exotic marine plant species has been associated with aquaculture, accounting for 39 percent of the total (Fig. 8.2, Appendix 8.1). Aquaculture activities are comprised of three separate pathways: the controlled culture of algae (deliberate pathway); the unintentional escape of individuals from controlled cultures; and the transfer of marine plants associated with imported aquaculture species (unintentional pathway).

RELEASE OF NONNATIVE SPECIES INTO THE OPEN SEA FOR AQUACULTURE PURPOSES

At present, 15 nonnative species of marine plants (11 red algae, 4 brown algae) are used in aquaculture as experimental cultures or as exploited resources (Fig. 8.2). Only red and brown algae are transported by this pathway, since the greatest economic profit to be obtained from algae today is from their use as raw material for the extraction of colloids (algi-nates from brown algae, and agar and carrageenan from red algae). These two groups are also the algae most extensively used in the food industry.

Laminaria japonica is the alga used most frequently in marine aqua-culture throughout the world, with a global production of 3,521,108 tons fresh weight in 1993 (Perez 1997). China, where the alga was intentionally introduced in 1925, is now the world's largest producer of *L. japonica*, esti-mated in 1996 as 3,900,000 tons fresh weight (Perez 1997). This species has also been introduced for the same purpose in Korea (Perez 1997). *Undaria pinnatifida* was also deliberately introduced for farming to the coast of Brittany (French Atlantic coasts) from the Thau Lagoon (French Mediterranean Sea), where it had been accidentally introduced with Japa-nese oysters (Perez et al. 1984).

In an attempt to produce more economically viable species, the intro-duction of exotic species in close proximity to natives is very common. For example, since the 1980s *Porphyra yezoensis* has been one of the nonna-tive species with better prospects for industrial cultivation off the coasts of Canada and the United States (ICES 1997). Numerous studies have been done on its biology and the impact of natural populations on indigenous communities.

By crossing small geographical barriers, sometimes a nonnative species is imported from neighboring countries, where the species is native. This may modify natural distribution. Equally, the transfer of species within a country with a long coastline means new introductions: although a species is included in the national floristic lists, a native species should be consid-ered nonnative in a geographical region where it did not exist previously. In Chile, for example, the transfer of *Gracilaria* species from one region to another for aquaculture activity has altered the geographical range of this genus (Santelices 1989). Finally, the transfer of individuals of a native species (potentially representing a distinct genome) from a different geo-graphical area must also be considered an introduction. For example, strains of *Hypnea musciformis* from Senegal were cultured in Corsica (Mediterranean Sea) in the 1980s, because they appeared to be more pro-ductive than the Mediterranean strains (Mollion 1984). The introduction of geographically isolated strains of the same species or of close relatives to the native ones may lead to a new kind of biological contamination, genetic mixing, the consequences of which are unpredictable (Ribera and Boudouresque 1995).

Most countries have no legislation on the management of introduction and transfer of marine plants. Normally the importation, control, and quarantine measures of these organisms are governed by regulations on agriculture or fisheries. Consequently, in most cases, aquaculture projects in the open sea, which do not have sufficient knowledge of the species or the appropriate measures for control, are permitted.

Escapes from Controlled Cultures

Most nonindigenous cultured species do not reach reproductive maturity and, therefore, do not complete their life cycle in the new territory. To control the culture of nonnative species, the presence of naturalized individuals of these species is searched for in neighboring areas. Of all the exotic species cultivated in Europe, only *Undaria pinnatifida* has colonized regions near its farming site on the French Atlantic and Mediterranean coasts (ICES 1987). When this species was introduced in Brittany (France) in 1984, experts claimed that it was not possible for the gametophytes to mature and produce sporophytes in situ because of the water temperature prevailing in this region (these claims are noted by Wallentinus 1994). But the full life cycle of this species could be completed (Floc'h et al. 1991) and at present *U. pinnatifida* forms wild populations along the French Atlantic coast (Castric-Frey et al. 1993, 1999). This escape started a controversy about whether cultivation should be allowed to continue, but after several evaluations of the risks, these were considered too small to stop further development of other farms (ICES 1989). However, not all countries have applied the same criteria. For example, proposals to introduce the farming of this species in Ireland were rejected (Eno et al. 1997).

Attempts to profit from naturalized nonindigenous species are habitual, and sometimes even used as a possible way to control spread. However, if economic benefits are optimal, this may lead not only to exploitation of natural populations but also to their culture and spread. These cases could be considered marginal dispersal pathways. For example, *Laminaria japonica* and *Undaria pinnatifida*, naturalized in the interior of the Thau Lagoon in the French Mediterranean Sea, are now cultivated in the open sea along the nearest coasts (Perez et al. 1984).

In the industrial production of algae, certain phases of development (or a whole generation) may be attained in controlled cultivation conditions away from the open sea. This method is often used because many plants cannot complete their life cycle with an economically viable growth rate in the open sea. Many decisions about whether to import exotic species for cultivation are based on this premise of containment. However, for some species this is not always true, and they escape to form natural populations.

All projects to cultivate nonnative species, whether contained or open cultivation, should first evaluate formally the opportunities for escape, the viability in situ, and the possible consequences of target species and associated organisms. The results of these studies may not support the establishment of cultures. For example, ICES did not allow continuation of the culture of the brown alga *Macrocystis pyrifera*, started in the 1970s off the Atlantic coast of France, because young sporophytes were allowed to grow for about seven months to 13 meters, and they reached the fertile stage with sporangia.

In 1981 specimens of the red alga *Eucheuma spinosum* from the Philippines were immersed in the open sea in the French Antilles (Guadaloupe), forty-eight hours after being collected, to compare their growth rate with native specimens of the same species. No quarantine measures were applied. Two months later the introduced specimens were all dead, victims of necroses that the authors of the experiment attributed to a disease known in southern Asia as ice-ice (Barbaroux et al. 1984). The authors noted that the indigenous plants were resistant to this disease, and therefore assumed that they might have introduced the ice-ice disease in the Caribbean Sea.

Further, the literature contains several examples of unsuccessful experiments of algal culture. *Gracilaria tenuistipata* from China was cultivated in land-based fishponds on the Baltic Sea coast (1989–90) where it did not survive a water temperature below 7°C (Haglund and Pedersen 1992). Growth experiments with *Alaria esculenta* have been carried out along the Helgoland coast, which is within its range of distribution in European waters, but where the species does not grow naturally because of the high summer temperatures (Munda and Lüning 1977).

ASSOCIATED SPECIES WITH AQUACULTURE SPECIES TRANSFER

The accidental transport of epibiont or endobiont organisms associated with the intentional transport of target species is one of the main introduction pathways for marine plants nowadays, especially in association with the importation of oyster spat (seed) for aquaculture. Further, transfers of the Japanese oyster *Crassotrea gigas* have been the most important, both for the volume of imports and for its impact. This species has been progressively transferred to several global regions, including Canada, the United States, Tasmania, New Zealand, Europe, and China (Grizel 1994).

This "associated" pathway is the largest single pathway for naturalized (established) populations, both within aquaculture activities (49 of 64 species) and among all pathways (30% of 189 species considered here; Fig. 8.2). Among these, red algae predominate with 29 species (59%). However, brown algae, with 14 species (29%), have exhibited the most invasive behavior, in terms of the spread, abundance, and negative impacts. For example, *Sargassum muticum* has colonized the Atlantic coasts of Europe (Scandinavia, the Netherlands, Belgium, Britain, France, Spain), the Mediterranean Sea (France), and the Pacific (Canada, the United States, Mexico, China, Japan, Korea, Philippines, Russia) (Critchley et al. 1990). It was first discovered in European waters off the Isle of Wight (Britain) in 1973 (Farnham et al. 1973) and has spread extensively in European waters from Portugal to Scandinavia, and even into the Mediterranean. Since its introduction this species has become dominant in many lower littoral areas with standing waters and in the sublittoral fringe, covering a broad range of conditions and habitats (Farnham 1994). In some cases it has formed large masses of floating algae, thanks to its air vesicles, which have had serious impacts on marine transport, especially in the English Channel. According to Farnham (1994), *S. muticum* has generally performed as an opportunist "gap-grabber," without significantly displacing or outcompeting native benthic plants, as had originally been feared. Although new small populations still appear, the clearly invasive aspect of its initial spread has declined more recently, and some initial populations (which survived for several years) have disappeared. This species has been the object of many eradication and evaluation studies, although these have been undertaken on a limited spatial scale (Belsher 1991).

History demonstrates that sites of shellfish culture carry a high risk, as locations of shellfish farming are often "hot spots" in terms of the number of exotic species. For example, the Thau Lagoon at Sète on the French Mediterranean coast is an area of very active shellfish cultivation where up to 45 nonnative species have been identified, 43 of them from the Pacific region (Verlaque 2001). Other aquaculture sites in the Mediterranean Sea are now developing, such as the Venice Lagoon in the Adriatic, where the number of nonnative marine plants is increasing and includes *Sargassum muticum*, *Undaria pinnatifida*, and *Grateloupia doryphora* (Verlaque 1994). These aquaculture farms may be potent sources of infection (spread) to new areas for two reasons:

1. Once established, naturalized species can increase their distributions in the region by other vectors, such as maritime transport activities (sometimes in close proximity to aquaculture) or by natural dispersion. Some nonnative species in the Thau Lagoon, such as *Porphyra yezoensis*, *Solieria chordalis*, *Leathesia difformis*, *Lomentaria hakodatensis*, *Undaria pinnatifida*, and *Sargassum muticum*, have colonized part of the open sea (Verlaque 1994).

2. From these sites, spat or adult shellfish are transferred to other aquaculture farms. This occurs not only with Japanese oysters but also with other bivalves, such as the oysters *Crassotrea virginica* and *Ostrea edulis*, which can become another vector for these exotic plant species. In the Thau Lagoon, for example, exotic species, such as *Cladosiphon zosterae*, *Agardhiella subulata*, and *Chondria coerulescens*, may have been introduced along with oyster transfers from Atlantic farms (Verlaque 2001). Similarly, the worldwide distribution kinetics of the *Sargassum muticum* follow the routes of oyster importation (Ribera and Boudouresque 1995).

Outside the European coast, other cases of nonnative species carried by this vector are known. The transport on shells of commercial oysters moved along the coast for relocation may be responsible for the initial establishment of *Codium fragile* subsp. *tomentosoides* on the south shore of Cape Cod and Boothbay Harbor on the eastern coast of North America (Malinowski and Ramus 1973). In 1996 *C. fragile* subsp. *tomentosoides* was detected in Canadian Atlantic waters and was believed introduced with shellfish from the United States (Campbell 1997). The presence of *Polysiphonia subtilissima* seems to be associated with oyster and mussel farms in New Zealand (Ribera and Boudouresque 1995).

To date, no microalgae are involved in this vector. A unicellular alga that has recently been reported on the Atlantic coast of France, *Fibrocapsa japonica* (Raphidophyceae), may have arrived with ballast water rather than with the spat of the oyster *Crassostrea gigas*, as Billard (1992) had hypothesized. Within this group we can include the endobionts of the oysters, which are less visible but very dangerous if they are toxic. The faeces and digestive tracts of bivalves can be loaded with viable *Alexandrium* cells (Bricelj et al. 1991) and can also contain resistant resting cysts (Hallegraeff 1993).

This vector also transported a higher plant. The sea grass *Zostera japonica* was first reported in 1957 from the state of Washington (North American Pacific coast) probably carried by imported Japanese oysters (Harrison and Bigley 1982). Its later occurrence to the north and south of Washington State may have the same origin or result from spread (natural or human-aided) along the northern Pacific coast (Munro et al. 1999).

Research Activities

Scientific research has been an uncommon pathway for introduced marine plants. We have recorded only two species associated with this pathway

(Fig. 8.2), but nonnative species or strains are increasingly being used in ecological and physiological studies. I am aware of numerous laboratory experiments on exotic marine plants in the past and present, but in most cases we do not know whether they are still in use. For example, *Gracilaria lemanaeformis* from Florida, *G. verrucosa* from Puerto Rico, and *G. tenuistipitata* from China were used for laboratory tests to produce protoplasts in Sweden (Björk et al. 1990).

Escapes from Laboratories

Normally, research is carried out with specimens maintained in controlled conditions in laboratory tanks, and their use is usually temporary. In most cases, the behavior and potential ecological impacts of these exotic species (if they became naturalized in a new environment) is not known. Protective measures are usually taken to prevent the escape of these species from their cultures to the environment. Although no case of exotic marine plant species having escaped from laboratories has been reported to date, the risk of this is very high in research centers having an open seawater system; therefore, any drainage network for wastewater can be a mode of transport to the sea. Nowadays, the threat is increasing because research with toxic species is being carried out.

Accidental Release of Discards

In most cases, previous knowledge of the ecological requirements of the species is sufficient to establish the means by which to safely dispose of live material. The main problem occurs with the fate of the individuals of a species at the end of studies. A total destruction of the material is always advisable, and waste should never be allowed to enter the main drains, as these always lead to the sea.

According to Koch (1951), the populations of the red alga *Bonnemaisonia hamifera* on Helgoland (European Atlantic coast) may have originated from discarded material sampled on scientific expeditions to Norway. However, its presence along other Atlantic coasts in Europe, such as in England, can be related to shellfish cultures (Tittley, pers. comm.; Eno et al. 1997).

Scientific Equipment (diving suits, objects installed in the water for a period of time)

There is no evidence that this vector is the specific cause of the naturalization of any marine plant, but it may play a role in marginal dispersion. How many times have small traces of algae appeared on diving suits? Therefore, all scientific equipment used in environmental research should be thoroughly cleaned, decontaminated, and dried. Fixed installations in the sea for an extended time act as artificial reefs, which allow the rapid colonization of opportunist algae.

Release of Nonnative Species into the Environment

Scientists carry out ecophysiological studies with nonindigenous species in the environment to compare them with native species or for their possible exploitation. Only one case of naturalized species by this vector is known. In the late 1970s, as an experiment, the red alga *Mastocarpus stellatus* was transplanted onto a rock on the island of Helgoland in Germany, on which it does not grow naturally. By the early 1990s it had colonized the whole western coast of the island (Munro et al. 1999). In this case, naturalization was predictable because this intertidal species is widely distributed along North Atlantic coasts.

In some cases, if the plants transferred to the open sea showed signs of adapting to the new environment in order to complete their life cycle, researchers have removed them. For example, hybridization experiments between the European brown alga *Laminaria saccharina*, the Canadian *L. longicruris* from British Columbia, and the Japanese *L. ochotensis* were carried out at Helgoland. The hybrids were cultivated in the sea but removed before they became fertile (Bolton et al. 1983).

Aquarium Trade

The aquarium industry involves the constant and uncontrolled commerce of exotic species, both animals and plants. This activity could be regulated by national legislation and also by the Washington Convention (CITES), but the latter does not include plants. For example, in 1993 the Europrix Company catalogs offered 102 seaweed species, natives or otherwise, which included *Caulerpa taxifolia* and *Sargassum muticum* (Boudouresque and Ribera 1994). Since the invasion of *C. taxifolia*, some Mediterranean countries (France, Italy, and Spain) have national or regional legislation that prohibits the transport, trade, and possession of these species (Boudouresque et al. 1996).

To date, in contrast to freshwater plants, relatively few nonnative marine plants are associated with this pathway. There is one clear example (Appendix 8.1, Fig. 8.2). Despite the low signal, the impacts associated with this one case have been extreme (see below).

Escapes from Public or Private Aquaria

Many public aquaria have water supply systems that run on an open circuit with the sea, and in numerous cases, normally for economic reasons, treatment of these waters is incomplete (Ribera and Boudouresque 1995). Under these conditions the risk of species escaping from an aquarium is high, but there has been only one known example that involves marine vegetation.

In the Mediterranean Sea, the tropical alga *Caulerpa taxifolia* has been one of the most spectacular marine invasions and was probably introduced via this vector. Its presence in the Mediterranean has been on record since 1984 on the coasts of Monaco, and it probably escaped from an aquarium at the Oceanographic Centre (Meinesz and Hesse 1991). The Mediterranean population has different morphological and physiological characteristics from the tropical populations, exhibiting more invasive behavior such as rapid spread and overgrowth of other species (Boudouresque et al. 1996). Today *C. taxifolia* is found along the coasts of five countries in the northern Mediterranean (Monaco, France, Italy, Spain, and Croatia) and about 6,000 hectares have been affected by this invasion (Meinesz et al. 1998). Sixteen years after the appearance of this alga, it is still advancing with the same vitality and shows no sign of decline. Recently, a new population of this species has been detected on the Tunisian coast in the southern Mediterranean (Langar et al. 2000).

Genetic–molecular studies have been used to confirm the aquarium origin of the Mediterranean population of *C. taxifolia*. The individuals of this species used in private or public aquaria in Europe seem to have the same origin: the Wilhelma Zoologish-Botanischer Garden of Stuttgart (Germany). At the beginning of the 1980s, this strain was given to the Tropical Aquarium of Nancy (French Atlantic coast) and from here to the Aquarium of Monaco (Mediterranean coast) (Meinesz and Boudouresque 1996). *Caulerpa taxifolia* individuals from several European aquaria and Mediterranean populations belong to a single strain, based on DNA sequence identity (Jousson et al. 1998).

During the spring of 2000, new settlements of *C. taxifolia* were discovered in the Pacific and Atlantic Oceans:

1. *Caulerpa taxifolia* was reported on the east coasts of Australia, about 30 kilometers south of Sydney (A. Miller, pers. comm., Algae-L [Algae-Listserver is a "Forum for marine, freshwater, and terrestrial algae"], May 30, 2000). Although the species grows naturally in some islands in the Tasman and Coral Seas, it was apparently new to this area.

2. Similarly, the species was reported in southern Florida, in the Atlantic Ocean (E. C. Oliveira, pers. comm., Algae-L, May 31, 2000). Oliveira proposed two theories about this new species: either the invasive strain of *C. taxifolia* from the Mediterranean was introduced into Florida, or the local strain of *C. taxifolia* (native in the Caribbean) is

blooming because pollution in the area is decreasing. Carlton (pers. comm.) pointed out that the Florida population of *C. taxifolia* is now considered the native stock of pantropical *C. taxifolia*.

3. In the Pacific Ocean, two populations of this species have been detected on the California coast, one in the San Diego Lagoon (Agua Hedionda) and the other in Huntington Harbor near Los Angeles (Kaiser 2000).

Deliberate Release of Aquarium Content

The presence in nature of freshwater exotic species from the release of the content of private aquariums is well documented (Welcomme 1992). In contrast, this method of introduction is very uncommon for marine plants. This pathway could explain the appearance of *Caulerpa taxifolia* in Les Lecques Harbor on the coast of Var (French Mediterranean). The presence of a few blocks of tropical coral reef just beside the population of *C. taxifolia* suggests that the content of a personal aquarium had been emptied into the sea (Laborel 1992). Similarly, Hoffmann (in Colmer, pers. comm., Algae-L, July 6, 2000) believes that the presence of this species in the San Diego Lagoon is the result of emptying the contents of a private aquarium into a storm drain or directly into the lagoon. The release of the contents of private aquaria to the environment most likely results from (the avoidable) limited public information about associated risks and possible safeguards. Alternatively, in the case of *C. taxifolia* it may be the reaction to the fear of having a dangerous or illegal species at home.

Fishing Activities

This group includes activities as diverse as the use of plants as packing for bait, fish, and shellfish, and the transport by fishing nets. While packing is a long-distance transport pathway, transport by fishing nets is more common as a marginal dispersion system. This pathway involves 6 species (2 Rhodophyceae, 2 Phaeophyceae, 1 Chlorophyceae, 1 higher plant), 5 of which have shown invasive behavior (Fig. 8.2).

Unintentional Release of Packing Material

Marine plants are often used as packing materials to keep marine animals (used for bait, direct human consumption, or aquaculture) cool and damp during transport. These plants are usually medium-sized, branched or foliaceous, with a fleshy consistency, and they can retain humidity for a long time.

Seaweeds used for packing bait are routinely released into aquatic environments, where they can naturalize. Two examples are known in the Mediterranean Sea. The Atlantic brown alga *Fucus spiralis* forms a small population in the Gruissan Lagoon (Sancholle 1988), and the red alga *Polysiphonia nigrescens* in the Prévost Lagoon (Verlaque and Riouall 1989); both sites are in the French Mediterranean Sea. Along the Atlantic coast of the United States the use of the green alga *Codium fragile* subsp. *tomentosoides* to wrap fishing baits has probably contributed to its spread. Carlton and Scanlon (1985) suggested that this species may have been transported to the Delaware–Maryland–Virginia region with shipments of the fucoid alga *Ascophyllum nodosum,* which are sent to that region from Massachusetts with bait worms and are then discarded.

Various species are used for packing fish and shellfish, but only two examples of plants established in a new area by this pathway are known. The sea grass *Zostera japonica* was first reported in 1957 in Willapa Bay in Washington State, and from then it continued to spread along the coasts (Harrison and Bigley 1982). This sea grass also occurs along the U.S. coast from Oregon to British Columbia. Although Z. *japonica* has not replaced the native Z. *marina,* largely because the former occurs higher in the intertidal zone, the previously barren mudflats in those areas are drastically changed by its spread (Posey 1988).

Transport by Nets

The production of very extensive biomass, either floating or easily detachable, of nonnative plants may permit their interfacing with other dispersal mechanisms, such as fishing nets. Nets have played a key role in the transport of filamentous algae such as *Womersleyella setacea* and *Acrothamnion preissii.* The filamentous red algae (*W. setacea*), introduced by ship fouling, is spreading in the Mediterranean Sea and produces temporal blooms, which clog up fishing nets (Verlaque 1989). In this case, nets may be the most direct cause (vector) of range expansion for this species. The same is true of *A. preissii,* which Italian fishermen refer to as *pelo* in the Liguria area (Cinelli et al. 1984). On the Atlantic coast of the United States, Carlton and Scanlon (1985) reported that vessels sailed from Nantucket Sound to the fishing grounds north of Cape Cod with *Codium fragile* subsp. *tomentosoides* in their nets, possibly promoting its spread.

Opening Maritime Canals

The best example of this pathway is the opening in 1869 of the Suez Canal, linking two biogeographical marine provinces that had been separated for several million years: the Mediterranean Sea and the Red Sea. The presence of highly saline lakes along the course of the canal had been an obstacle to the transfer of species for many years. Today, progressive desalination of these lakes has permitted the exchange of organisms between the two seas. According to Por (1990), an estimated 200–300 species from the Red Sea have colonized the Mediterranean. Sometimes called Lessepsian immigrants, in reference to the engineer Ferdinand de Lesseps who designed the canal, Por (1989) further considered this wholesale transfer of species to be the largest biogeographic phenomenon witnessed in contemporary oceans. To date, there are 24 known Lessepsian marine plants in the Mediterranean (15 red algae, 4 brown algae, 4 green algae, 1 higher plant) (Fig. 8.2).

Most of the Lessepsian immigrants are localized in the eastern Mediterranean (Boudouresque 1994), a region referred to by Por (1990) as the Lessepsian province. Few of these species reach the occidental basin. Examples include the higher plant *Halophila stipulacea* and the red alga *Sarconema filiforme,* located respectively on the coasts of Italy and France (Ribera 1994). Dispersal of most of the Lessepsian plants from the mouth of the Suez Canal has likely resulted from natural dispersal, as suggested by the linear progression spread. Although spread of some species is multidirectional, as in the simultaneous northern and southern spread of the brown alga *Stypopodium schimperi* (Verlaque and Boudouresque 1991), range expansions most frequently follow the dominant anticlockwise current into the eastern basin (Ribera 1994).

In contrast, the Panama Canal, which links the Pacific and the Atlantic Oceans, has resulted in the passage of a very limited number of euryhaline marine species, and there is no evidence that any plant is involved (Por 1978). The primary barrier to successful transport and colonization of the Caribbean species may not be the freshwater lakes, which could disrupt some transfers, but herbivore activity and the lack of reef-generated refuge areas on the Pacific coast (Hay and Gaines 1984).

CONCLUSIONS

Importance and Geographical Distribution of the Different Pathways

On a worldwide scale, shellfish transport is the main dispersal pathway for exotic species (49 species, 30%). This is followed by ship fouling (39 species, 24%), after which the following vectors are, in order of importance, ballast (25 species, 16%), the Suez Canal (24 species, 15%), and importation for aquaculture activity (15 species, 9%) (Fig. 8.2). The other vectors (scientific research, fishing, aquaria) represent 6 percent of the total. Overall, maritime transport accounts for 40 percent (including ship fouling and ship ballast), and the combination of intentional and unintentional transfers associated with aquaculture activities accounts for 39 percent (Fig. 8.2). Intentional (partially aquaculture and partially research) and accidental vectors account for 10 percent and 90 percent of marine plant invasions, respectively. As noted earlier, these estimates exclude cases in which vector was considered uncertain.

The high number of exotic plant species involved with shellfish transfers is determined almost exclusively by the importance of this vector in the Mediterranean and European Atlantic coasts (Fig. 8.3A). This result may reflect partially the frequency of studies on the aquaculture installations in Europe, but it also reflects the present-day increase in the number of aquaculture farms in European countries. A quarter of the marine plant invasions of Europe are associated with oyster transfer. In some countries this proportion is greater: half the nonnative marine algae found in Britain are believed introduced via this pathway (Eno et al. 1997).

In the other geographic regions, fewer marine plant invasions are associated with shellfish transfers, but it should be emphasized that aquaculture still leads to important invasions around the world, such as *Sargassum muticum*, *Grateloupia doryphora*, and *Codium fragile* subsp. *tomentosoides*. Moreover, I wish to emphasize that almost all species dispersed by shellfish transport on the North American coasts are also found on the European coasts, introduced by the same vector (Appendix 8.1). This underscores that aquaculture installations are often points of dispersal of exotic species. Thus, although most of the exotic species came directly from the Pacific or Indian Oceans at the beginning of the development of this kind of industry, the Mediterranean and the European Atlantic have now become exporters of Indo-Pacific species.

The transport of fixed organisms on ships' hulls can be considered the oldest transport vector of marine plants. This vector is undoubtedly partly responsible for the present-day distribution of many species of algae, some of which have a worldwide distribution. The data in the literature are clearly incomplete since there are very few references of species dispersed by this pathway on North American coasts. Genetic studies may help resolve cryptic invasions. For example, studies of DNA hybridization in the genus *Cladophora* (van den Hoeck et al. 1990) on both sides of the Atlantic showed that populations of *C. albida* have only recently diverged, and those of *C. sericea* are identical. In both cases a transfer by fouling across the Atlantic in one direction or another is suggested (Ribera and

Boudouresque 1995). My data show that transport on ships' hulls is most important in New Zealand (19 species), followed by the Mediterranean Sea (12 species) and the European Atlantic coast (8 species) (Fig. 8.3B). The reduced number of Australian exotic species involved in this vector could be an artifact, since the high number of species with an uncertain pathway may correspond to this vector.

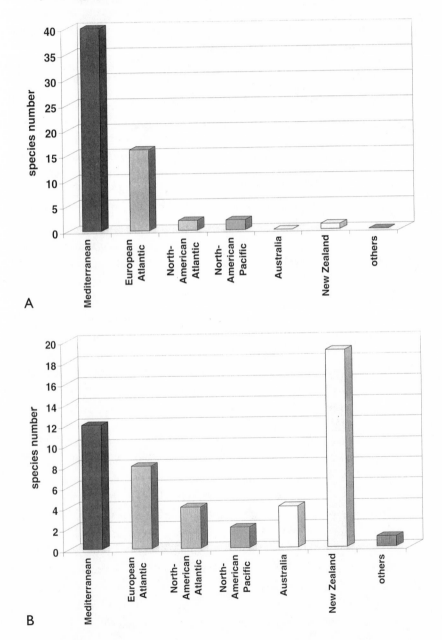

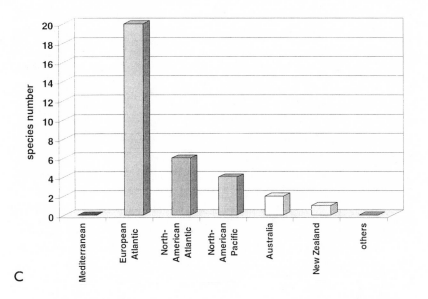

C

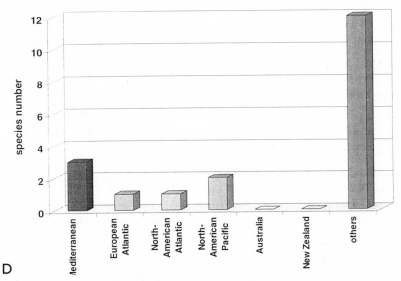

D

FIGURE 8.3. Importance of different pathways in the introduction of nonnative marine plant species in each geographical area. (A) Oyster transfers. (B) Ships' hulls. (C) Ships' ballast. (D) Aquaculture purposes.

Ships' ballast is the dominant pathway in the European Atlantic coast (20 species), making a relatively small contribution on the Atlantic (6 species) and Pacific coasts of America (4 species) (Fig. 8.3C). As suggested earlier, these data likely reflect the lack of knowledge about nonnative microalgae species, which are among the most common organisms in ballast tanks. For example, in the Mediterranean Sea, where maritime traffic is intense, no nonnative planktonic species have been cited. Occasional or periodic blooms of some dinoflagellates, both toxic and nontoxic forms, are known in the Mediterranean, but there are very few studies available to evaluate adequately any association to nonindigenous species. Similarly, in areas like Australia, New Zealand, and the Hawaiian Islands, the number of exotic species involved in this pathway could be expected to be higher because there is a large amount of maritime traffic. It is perhaps particularly surprising that there are so few nonnative plant species recognized as ballast introductions in Australia, where this vector has had a strong ecological and economic impact on the shellfish farms.

The cultivation of nonnative species of seaweed is not common at the moment for the geographical areas studied, except Hawaii (Fig. 8.3D in others). Nine species, most of them red algae of the genera *Gracilaria,* *Kappaphycus,* and *Eucheuma,* have been introduced to this region for commercial purposes, especially to obtain different classes of carrageenan (Russell 1992) (Appendix 8.1). In contrast, the three species introduced to China for aquaculture, used for the production of alginates and as a food source, are the large brown algae *Laminaria japonica, Macrocystis integrifolia,* and *Macrocystis pyrifera* (Appendix 8.1). As mentioned before, some of the nonnative species used now in European aquaculture were introduced earlier by other vectors.

The introduction of species as a result of fishing is known only from the Mediterranean Sea (3 species) and from Atlantic (2 species) and Pacific (2 species) waters of North America. Although the examples of introductions with packed baits may be well documented, marginal dispersion by nets has not been studied in depth and may therefore be underestimated.

Discharges (intentional or accidental) from aquaria, although responsible for the presence of a large number of exotic species (especially animals) in freshwaters have contributed <1 percent of the plant species introduced to marine waters. However, the biological invasion of *Caulerpa taxifolia* in the Mediterranean Sea underscores the potential risks associated with any species and the ability of certain organisms to adapt to the artificial or novel conditions found in aquaria. In such cases, the strain that best adapts may be a unique, genetically modified strain that may not survive in nature. Alternatively, genetic modifications may occur in the populations subjected to stress. In addition, in this environment, plants usually reproduce only vegetatively, which can give rise to a clone that spreads by fragmentation. To date, the common origin of various populations of *C. taxifolia* in the Mediterranean has been identified with those of the Stuttgart Center. If the same origin is confirmed for the new populations from the Pacific and Atlantic coasts of the United States, we will be facing a new biological invasion at a worldwide scale with a single origin, similar to the *Sargassum muticum* spread.

Presence of Exotic Species in the Different Geographical Regions of the World

The comparison of geographical areas shows that the Mediterranean Sea contains the highest number of reported exotic marine plant species (83 species), followed by the European Atlantic coast with 49 species (Fig. 8.4). The other areas have fewer reported exotic species, having roughly similar numbers among regions: 26 species on the Australian coast, 21 on the New Zealand coast, 19 on the North American Pacific coast, 17 on the North American Atlantic coast, and 22 on the coasts of the other regions considered (Hawaii, China, Canary Islands, the Azores) (Fig. 8.4). Therefore, the Mediterranean appears to be an exceptional nucleus, or hot spot, for nonnative species of marine plants.

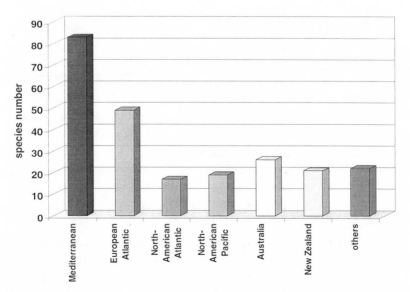

FIGURE 8.4. Diversity of nonnative species of marine plants in each geographical area.

Multiple mechanisms, operating alone or in combination, may explain the observed geographic pattern, and especially the extent of invasions in the Mediterranean. The number of pathways in the area, and their activity level (e.g., number of ships, amount of oyster movement, etc.), may be relatively great in this region. Alternatively, environmental and biological characteristics of the area may influence invasion patterns. Other regions referred to as Mediterranean-like, such as California, Chile, South Africa, and South Australia, are often thought to have climatic conditions that favor colonization of exotic species. Moreover, other features in the Mediterranean may enhance opportunities for colonization, relative to other regions, including (a) a wide range of water temperatures, allowing the acclimation of both cold-adapted and tropical species, (b) a great variety of biotopes, and (c) a scarcity of both large perennial algae and herbivores.

In the Mediterranean Sea the dominant vector is shellfish transfer (40 species) followed by the Suez Canal (24 species) (Fig. 8.3). The other pathways are of limited importance in terms of the number of species. Although a minor contribution to the overall number, it is noteworthy that the unintentional release from a public aquarium of *Caulerpa taxifolia* has had arguably the greatest ecological impact among biological invasions in the Mediterranean Sea. As mentioned before, the absence of species related to ballast transfer might be attributable to the lack of studies on this subject in the Mediterranean area.

The transport of nonnative species by the Suez Canal, which until the beginning of the 1990s was the largest for both flora and fauna in the Mediterranean (Ribera 1994), has now been relegated to second place, at least for flora, because of the alarming growth of species related to the importation of *Crassostrea gigas*. And in this sea, France is the leading importer of exotic species (Ribera and Boudouresque 1995), perhaps because the relevant legislation is deficient and not fully enforced. In particular, especially in the last twenty years, a giant shellfish farm has been developed in the Thau Lagoon. This French lagoon is quickly becoming the most important site of introduction for macroalgae, not only in the Mediterranean Sea but also in European waters (Verlaque 2001). Thus, weighted heavily by oyster imports, and Lessepsian immigrants, the Indo-Pacific Ocean has been the main source of nonnative species found in the Mediterranean.

On the European Atlantic coasts the dominant vector is ballast (20 species), closely followed by oyster transfer (16 species) (Fig. 8.3). France may be considered the country that has imported the largest number of species associated with shellfish aquaculture (Ribera and Boudouresque 1995), while the North Sea contains the largest number of species transported by ballast (Reise et al. 1999). As occurs in other geographic areas, the nonnative species classified as having an uncertain pathway could correspond to transport by ships' hulls, since the role of the fouling in these European coasts may be underestimated (see below).

On both the Atlantic and Pacific coasts of North America, the number and relative contribution of the different pathways are similar (Fig. 8.3). I emphasize that in these two regions, in contrast to the other areas, all the main pathways have contributed a comparable number of species, including aquaculture (including associated species), ballast, fouling, and fishing. These two areas also have a relatively high number of species with uncertain pathways that, as mentioned above, may correspond to ship fouling (Appendix 8.1). It is interesting to point out that only 5 species are found on both coasts of North America. Among the exotic species with the widest distribution, I can highlight the absence of *Grateloupia doryphora* from the North American Pacific coast, while *Sargassum muticum* is not reported on the North American Atlantic coast (Appendix 8.1). If we compare the exotic species on both the eastern and western Atlantic coasts, the number of common species is equally low (6 species). Concerning the nonnative marine flora on the coasts of Australia and New Zealand, my data only permit me to point out that fouling seems to be the most important vector, with 19 species for New Zealand and 4 species in Australia; again, the relatively high number of species with uncertain pathways for Australia (20 species) may obscure the importance of ship fouling. There is a need for deeper studies to determine the contribution of deballasting of ships and the species associated with species transfers in aquaculture for these regions.

Rate and Prediction of the Different Pathways

In general, it can be argued that the rates of both intentional and accidental introduction of marine plants may be increasing, owing to the growing importance of marine products for human activities, along with the increasing volume of marine trade and transport. To assess the rate of introductions, it is necessary to identify the date of arrival of each species, which is sometimes difficult or even impossible. The results presented here should be interpreted with caution because there could be many artifacts related to the information. It is clear, for example, that the data for the last century and the beginning of this one are underestimations. It should also be borne in mind that the date of the first report of an exotic species could in some cases be much later than the date of arrival.

The rate of reported introductions for nonnative marine plants appears to confirm this temporal increase in Europe (Fig. 8.5). I have selected

Europe, including Atlantic and Mediterranean coasts (in this case), as an illustrative example, because (a) I am most familiar with records from this region and (b) it includes the largest number of species for any geographic region. This graph reveals three periods in the rate of increase: up to the beginning of the 1900s, from 1920 to 1960, and from 1970 to the present. In the first two periods the increase is linear, although the number of species in the second period is double that in the first. In contrast, in the third period, the increase is exponential. This increase may be due to greater interest in the subject, to increased pressures to document invasions, or of course to actual increased numbers of invasions. However, this increase does not appear to occur at the same rate in different geographic areas. Since little has been published on the subject, it is very difficult to identify where rates of introductions are increasing and where they are decreasing. Among the European coasts, for example, the exotic species number presents a linear increase in Britain, whereas in the Mediterranean this increase is exponential.

The rate of new introductions does not follow the same pattern for each pathway. The Suez Canal, for instance, is a vector that is likely to lose force in the Mediterranean as long as environmental conditions are not modified, but in any case the eastern Mediterranean is still a very receptive area for new introductions, owing to its biological characteristics.

In contrast, aquaculture activities (and especially the marine culture industry for algae, as well as other organisms) are flourishing for two major reasons. First, demand for algae products is increasing as a result of (a) an increase of Oriental populations and foods in Europe and North

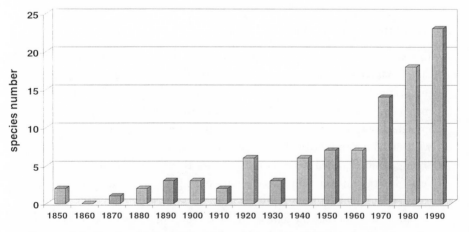

FIGURE 8.5. Number of nonnative marine plant species that were newly reported on the European coast in each decade (noncumulative data).

America, and (b) increased use of natural products for cosmetics, medicines, and diet foods (Perez 1997). Second, additional countries are entering the market to emulate the profitable algal colloids industries developed in China, Japan, and Korea. Both factors promote the importation and naturalization of the most economically competitive algal species despite their geographical origin.

Likewise, the development of fish and shellfish cultures is causing an indiscriminate transport of associated species that may be increasing. The arrival of exotic species associated with shellfish aquaculture on the European coast over time shows a marked increase since the 1970s; in the last ten years the number of exotic algae involved in this pathway has increased fivefold (Fig. 8.6A). According to Eno et al. (1997), before 1960 there were substantial introductions of marine species in association with imports for aquaculture; however, since then quarantine regulations have halted this method of entry into Britain. These measures do not seem to have the same effect in all countries. Verlaque (2001) indicates that failure in decontamination processes and/or quarantine of these authorized or nonauthorized imports has resulted in an increasing number of species introductions to Thau Lagoon. In general, we can argue that the current development of fish and shellfish farming will produce a dramatic increase in the range of species that are cultivated and consequently in the number of introductions (Ribera and Boudouresque 1995).

The temporal pattern of marine plant invasions associated with ships is less clear. For example, the frequency of toxic algae blooms on the Atlantic coasts of Canada has tripled over the past fifteen years, to which ship-mediated invasions may have contributed, but a definitive connection with ballast release has yet to be established (Smith and Kerr 1992, Subba Rao et al. 1994). The level of risk of introductions also remains to be quantified (Forbes 1994). Explanations for the apparent global increase of toxic dinoflagellate blooms include increased scientific awareness caused by the developing aquaculture industry and stimulation of dinoflagellate blooms by increased coastal eutrophication. However, in a limited number of cases, dispersal of nonindigenous estuarine dinoflagellate species across oceanic boundaries—via either ships' ballast water or transport of shellfish products—is more probable (Hallegraeff and Bolch 1991).

On the other hand, the quantity of transoceanic shipping has increased greatly, and the tendency of modern vessels to move faster through the water may increase the survival of both fouling species and those carried in ballast (Eno et al. 1997). Consequently, despite the data on fouling for the European coast suggesting relatively few recent plant introductions associated with ships (Fig. 8.6B), I predict that maritime transport will be

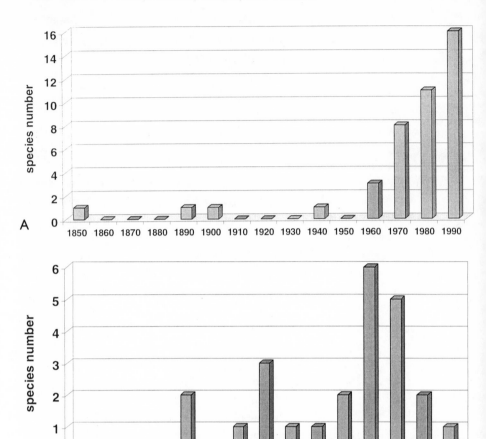

FIGURE 8.6. Number of nonnative marine plant species that were newly reported on the European coast by decade, according to different pathways (noncumulative data). (A) Nonnative species associated with shellfish transfer. (B) Nonnative species introduced on ships' hulls.

one of the main vectors of exotic species introductions in the future. Moreover, we should not forget that any change in marine transport routes could lead to the arrival of new species, like a change in the location of the fishing grounds, or a new tourist industry on a small island that previously had few visitors.

We have seen that the intensity of each pathway (as number of introduced nonnative species) is different for each country and geographical region. The temporal sequence and development of pathways may also depend on the region. For example, the same pathways shown for the European coast can operate at different timescales in another region. This is one of the major risks associated with isolated regions that are undergoing, or will undergo, rapid economic development, which can swiftly shift or increase human activities known to transfer organisms. Islands, and small developing states, may be especially vulnerable to such rapid change.

The conclusion, realistic and pessimistic, is that the increasing rate of species introductions is linked, directly or indirectly, to economic interests. These may be in conflict with other economic interests (introduction of diseases or parasites, competition with commercially exploited indigenous species, nuisances, other uses of coastal areas) and also with ethical values that are difficult to evaluate in economic terms (Boudouresque 1994).

ACKNOWLEDGMENTS

I would like to thank Dr. Amelia Gómez Garreta for her continuous help during the preparation of this manuscript, Mr. Luca Lavelli for the graphical design, and Mr. Robin Rycroft for reviewing the English text.

REFERENCES

Abbott, I. I. and G. J. Hollenberg. 1976. *Marine algae of California.* Stanford: Stanford University Press.

Adams, N. M. 1994. *Seaweeds of New Zealand. An illustrated guide.* New Zealand: Canterbury University Press.

Anonymous. 2000a. *Marine invasives in Hawai'i.* <http://www.botany.hawaii. edu/invasive/default.htm>.

Anonymous. 2000b. *Nonindigenous aquatic algae within the USA.* <http://nas. er.usgs.gov/algae/algaelist.htm>.

Barbaroux, O., R. Perez, and J. P. Dreno. 1984. L'algue rouge *Eucheuma spinosum.* Possibilités d'exploitation et de culture aux Antilles. *Science et Pêche* 348: 2–9.

Belsher, T. 1991. Sargassum muticum *(Yendo) Fensholt sur le littoral français.* IFREMER, Brest, DEL-91.25.

Billard, C. 1992. *Fibrocapsa japonica* (Raphidophyceae), algue planctonique nouvelle pour les côtes de France. *Cryptogamie, Algologie* 13: 225–231.

Björk, M. P., P. Ekman, A. Wallin, and M. Pedersen. 1990. Effects of growth rates and other factors on protoplast yield from four species of *Gracilaria* (Rhodophyta). *Botanica Marina* 33: 433–439.

Boalch, G. T. and D. S. Harbour. 1977. Unusual diatom off south-west England and its effect on fishing. *Nature* 192: 279–280.

Bolton, J. J., I. Germann, and K. Lüning. 1983. Hybridisation between Atlantic and Pacific representatives of simplices section of *Laminaria* (Phaeophyceae). *Phycologia* 22: 133–140.

Boudouresque, C. F. 1994. Les espèces introduites dans les eaux côtières d'Europe et de Méditerranée: état de la question et conséquences. In *Introduced species in European coastal waters,* C. F. Boudouresque, F. Brian, and C. Nolan, eds., p. 8–27. Luxembourg: European Commission.

Boudouresque, C. F., E. Ballesteros, F. Cinelli, Y. Henocque, A. Meinesz, D. Pesando, F. Pietra, M. A. Ribera, and G. Tripaldi. 1996. Synthèse des résultats du programme CCE-LIFE Expansion de l'algue verte tropicale *Caulerpa taxifolia* en Méditerranée. In *Second International Workshop on Caulerpa taxifolia*, M. A. Ribera, E. Ballesteros, C. F. Boudouresque, A. Gómez, and V. Gravez, eds., pp. 11–57. Publicacions Universitat de Barcelona.

Boudouresque, C. F., F. Brian, and C. Nolan, eds. 1994. *Introduced species in European coastal waters.* Luxembourg: European Commission.

Boudouresque, C. F. and M. A. Ribera. 1994. Les introductions d'espèces végétales et animales en milieu marin—Conséquences écologiques et économiques et problèmes législatifs. In *First International Workshop on Caulerpa taxifolia*, C. F. Boudouresque, A. Meinesz, and V. Gravez, eds., pp. 29–102. GIS Posidonie.

Bricelj, V. M., M. Greene, J. H. Lee, and A. D. Cembella. 1991. *Growth response and fate of dinoflagellate cells and paralytic shellfish poisoning (PSP) toxins in mussels,* Mytilus edulis. Newport: Abstracts Fifth International Conference on Toxic Marine Phytoplankton.

Campbell, M. 1997. *National Report for Canada.* Report of the WGITMO, ICES CM 1997: 24–30.

Carlton, J. T. 1994. Biological invasions and biodiversity in the sea: the ecological and human impacts of nonindigenous marine and estuarine organisms. In *Nonindigenous Estuarine and Marine Organisms (NEMO), Proceedings of the Conference and Workshop, Seattle, Washington, April 1993.* Keynote Address, pp. 5–11. U.S. Department of Commerce, National Oceanic and Atmospheric Administration, Office of the Chief Scientist, 125 pp. (September 1994). Government Document No. C55.2:N73, Government Printing Office No. 0208-C-04.

Carlton, J. T. and J. B. Geller. 1993. Ecological roulette: the global transport of non-indigenous marine organisms. *Science* 261: 78–82.

Carlton, J. T., G. M. Ruiz, D. Smith, and A. Hines. 1995. Ballast waters. *Aliens* 1: 18–19.

Carlton, J. T. and J. A. Scanlon. 1985. Progression and dispersal of an introduced alga: *Codium fragile* ssp. *tomentosoides* (Chlorophyta) on the Atlantic coast of North America. *Botanica Marina* 28: 155–165.

Castric-Frey, A., C. Beaupoil, J. Bouchain, E. Pradier, and M. T. L'Hardy-Halos. 1999. The introduced alga *Undaria pinnatifida* (Laminariales, Alariaceae) in the rocky shore ecosystem of the St. Malo area: growth rate and longevity of the sporophyte. *Botanica Marina* 42: 83–96.

Castric-Frey, A., A. Girard, and M. T. L'Hardy-Halos. 1993. The distribution of *Undaria pinnatifida* (Phaeophycea, Laminariales) on the coast of St. Malo (Brittany, France). *Botanica Marina* 36: 351–358.

Cinelli, F., U. Salghetti, and F. Serena. 1984. Nota sull'areale di *Acrothamnion preissii* (Sonder) Wollaston nell'Alto Tirreno. *Quaderni del Museo di Storia naturale di Livorno* 5: 57–60.

Cranfield, H. J., D. P. Gordon, R. C. Willan, B. A. Marshall, C. N. Battershill, M. P. Francis, W. A. Nelson, C. J. Glasby, and G. B. Read. 1998. *Adventive marine species in New Zealand. NIWA Technical Report 34.*

Critchley, A. T., W. F. Farnham, T. Yoschida, and T. A. Norton. 1990. A bibliography of invasive alga *Sargassum muticum* (Yendo) Fensholt (Fucales; Sargassaceae). *Botanica Marina* 33: 551–562.

Elbrächter, M. 1999. Exotic flagellates of coastal North Sea waters. *Helgoländer Meeresuntersuchungen* 52: 235–242.

Eno, N. C., R. A. Clark, and W. G. Sanderson, eds. 1997. *A review and directory of non-native marine species in British waters.* Joint Nature Conservation Committee, U.K.

Eno, N. C., R. A. Clark, and W. G. Sanderson, eds. 2000. *Directory of non-native marine species in British waters.* <http://www.jncc.gov.uk/marine/dns/de>.

Farnham, W. F. 1994. Introduction of marine benthic algae into Atlantic European waters. In *Introduced species in European coastal waters,* C. F. Boudouresque, F. Brian, and C. Nolan, eds., pp. 32–36. Luxembourg: European Commission.

Farnham, W. F., R. L. Fletcher, and L. M. Irvine. 1973. Attached *Sargassum muticum* found in Britain. *Nature* 243: 231–232.

Fletcher, R. L. and C. Manfredi. 1995. The occurrence of *Undaria pinnatifida* (Phaeophyceae, Laminariales) on the south coast of England. *Botanica Marina* 38: 355–388.

Floc'h, J. Y., R. Pajot, and I. Wallentinus. 1991. The Japanese brown alga *Undaria pinnatifida* on the coast of France and its possible establishment in European waters. *Journal Conseil International Exploration Mer* 47: 379–390.

Forbes, R., ed. 1994. Proceedings of the Fourth Canadian Workshop on Harmful Marine Algae. *Canadian Technical Report Fisheries Aquatic Sciences 2016.*

Galil, B. S. and N. Hülsmann. 1997. Protist transport via ballast waters—Biological classification of ballast tanks by food web interactions. *European Journal of Protistology* 33: 244–253.

Gollash, S. 1996. *Untersuchungen des Arteintrages durch den internacionales Schiffsverskehr unter besonderer Berücksichtigung nichtheimischer Arten.* PhD Thesis, Dr. Kovac, Hamburg.

Grizel, H. 1994. Réflexions sur les problèmes d'introduction de mollusques. In *Introduced species in European coastal waters,* C. F. Boudouresque, F. Brian, and C. Nolan, eds., pp. 50–55. Luxembourg: European Commission.

Haglund, K. and M. Pedersen. 1992. Outdoor cultivation of the marine red alga *Gracilaria tenuistipitata* var. *liui* Zhang et Xia (Gracilariales, Rhodophyta) in brackish water in Sweden. Growth, nutrient uptake, cocultivation with rainbow trout and epiphyte control. In *Photosynthesis and growth of some marine algae, with emphasis on the Rodophyte* Gracilaria tenuistipitata. Ph.D. Thesis, Department of Physiological Botany, Uppsala University.

Hallegraeff, G. M. 1993. A review of harmful algal blooms and their apparent global increase. *Phycologia* 32: 79–99.

Hallegraeff, G. M. 1998. Transport of toxic dinoflagellates via ships' ballast water: bioeconomic risk assessment and efficacy of possible ballast water management strategies. *Marine Ecology Progress Series* 168: 297–309.

Hallegraeff, G. M. and C. J. Bolch. 1991. Transport of toxic dinoflagellates cysts via ships' ballast water. *Marine Pollution Bulletin* 22: 27–30.

Hallegraeff, G. M. and C. J. Bolch. 1992. Transport of dinoflagellate cysts in ships' ballast water: implications for plankton biogeography and aquaculture. *Journal of Plankton Research* 14: 1067–1084.

Hallegraeff, G. M., M. A. McCausland, and R. K. Brown. 1995. Early warning of toxic dinoflagellate blooms of *Gymnodinium catenatum* in southern Tasmania waters. *Journal of Plankton Research* 17: 1163–1176.

Hallegraeff, G. M. and C. E. Summer. 1986. Toxic plankton blooms affect shellfish farms. *Australian Fish* 45: 15–18.

Harrison, P. G. and R. E. Bigley. 1982. The recent introduction of the seagrass *Zostera japonica* Aschers and Graebn to the Pacific coast of North America. *Canadian Journal of Fish and Aquatic Science* 39: 1642–1648.

Hay, C. H. 1990. The dispersal of sporophytes of *Undaria pinnatifida* by coastal shipping in New Zealand, and implications for further dispersal of *Undaria* in France. *British Phycological Journal* 25: 301–313.

Hay, M. E. and S. D. Gaines. 1984. Geographic differences in herbivore impact: do Pacific herbivores prevent Caribbean seaweeds from colonizing via the Panama Canal? *Biotropica* 16: 24–30.

Hillson, C. J. 1976. *Codium* invades Virginia waters. *Bulletin of the Torrey Botanical Club* 103: 266–267.

International Council for the Exploration of the Seas (ICES). 1981. *Report of the Working Group on Introductions and Transfers of Marine Organisms.* Sète, France, June 5–1, 1991. CM. 1981/F:46.

International Council for the Exploration of the Seas (ICES). 1987. *Report of the Working Group on Introductions and Transfers of Marine Organisms.* Brest, France, June 10–12, 1987. CM. 1987/F: 35.

International Council for the Exploration of the Seas (ICES). 1989. *Report of the Working Group on Introductions and Transfers of Marine Organisms.* Dublin, Ireland, May 23–26, 1989. CM. 1989/F: 16.

International Council for the Exploration of the Seas (ICES). 1991. *Report of the Working Group on Introductions and Transfers of Marine Organisms.* Helsinki, Finland, June 5–7, 1991. CM. 1990/F:44.

International Council for the Exploration of the Seas (ICES). 1997. *Report of the Working Group on Introductions and Transfers of Marine Organisms.* La Tremblade, France, April 22–25, 1997. CM. 1997/E+F: 6.

Invasive Spartina Project. 2002. *San Francisco Estuary. Invasive Spartina Project.* <http://www.spartina.org>.

Jousson, O., J. Pawlowski, L. Zaninetti, A. Meinesz, and C. F. Boudouresque. 1998. Molecular evidence for the aquarium origin of the green alga *Caulerpa taxifolia* introduced to the Mediterranean Sea. *Marine Ecology Progress Series* 172: 275–280.

Kaiser, J. 2000. California algae may be feared European species. *Science* 289: 222–223.

Knoepffler-Péguy, M., T. Belsher, C. F. Boudouresque, and M. Lauret. 1985. *Sargassum muticum* begins to invade the Mediterranean. *Aquatic Botany* 23: 291–295.

Koch, W. 1951. Historisches zum Vorkommen der Rotalge *Trailliella intricata* Batters bei. *Archiv für Mikrobiologie* 16: 78–79.

Laborel, J. 1992. *Origine probable de la contamination du nouveau port des Lecques (Var) par Caulerpa taxifolia. Rapport préliminaire.* Rapport du Laboratoire de Biologie Marine et d'Ecologie du Benthos, Faculté des Sciences de Luminy, France.

Langar, H., A. Djellouli, K. Ben Mustapha, and A. El Abed. 2000. Première signalisation de *Caulerpa taxifolia* (Vahl) C. Agardh en Tunisie. *Bulletin International des Sciences de la Mer* 27: 7–8.

Lawson, G. W. and D. M. John. 1982. The marine algae and coastal environment of tropical West Africa. *Nova Hedwigia* 70: 5–455.

Leppäkoski, E. 1994. Non-indigenous species in the Baltic Sea. In *Introduced species in European coastal waters*, C. F. Boudouresque, F. Briand, and C. Nolan, eds., pp. 67–75. Luxembourg: European Commission.

Loosanoff, V. L. 1975. Introduction of *Codium* in New Jersey waters. *Fisheries Bulletin* 73: 215–218.

MacDonald, E. M. and R. D. Davidson. 1998. The occurrence of harmful algae in ballast discharges to Scottish ports and the effects of midwater exchange in regional seas. In *Proceedings of the 8th International Conference on Harmful Algae, Vigo.* pp. 220–223.

Maggs, C. A. and H. Stegenga. 1999. Red algal exotics on North Sea coasts. *Helgoländer Meeresuntersuchungen* 52: 243–258.

Malinowski, K. C. and J. Ramus. 1973. Growth of the green alga *Codium fragile* in a Connecticut estuary. *Journal of Phycology* 9: 102–110.

McMinn, A., G. M. Hallegraeff, P. Thomson, A. V. Jenkinson, and H. Heijnis. 1997. Cyst and radionucleotide evidence for the recent introduction of the toxic dinoflagellate *Gymnodinium catenatum* into Tasmania waters. *Marine Ecology Progress Series* 161: 165–172.

Meinesz, A. 1992. Modes de dissémination de l'algue *Caulerpa taxifolia* introduite en Méditerranée. *Rapports et procès-verbaux des réunions Commission internationale pour l'exploration scientifique de la Mer Méditerranée* 33: 44.

Meinesz, A. and C. F. Boudouresque. 1996. Sur l'origine de *Caulerpa taxifolia* en Méditerranée. *Comptes Rendus de l'Academie des Sciences. Serie III, Sciences de la Vie (Paris)* 319: 603–613.

Meinesz, A., J. M. Cottalorda, D. Chiavérini, N. Cassar, and J. De Vaugelas, eds. 1998. *Suivie de l'invasion de l'algue tropicale* Caulerpa taxifolia *en Méditerranée: situation au 31 décembre 1997.* LEML-UNSA.

Meinesz, A. and B. Hesse. 1991. Introduction et invasion de l'algue tropicale *Caulerpa taxifolia* en Méditerranée nord-occidentale. *Oceanologia Acta* 14: 415–426.

Mollion, J. 1984. Seaweed cultivation of phycocolloid in the Mediterranean. *Hydrobiologia* 116–117: 288–291.

Munda, I. and K. Lüning. 1977. Growth performance of *Alaria esculenta* off Helgoland. *Helgoländer Wissenschaftliche Meeresuntersuchungen* 29: 311–314.

Munro, A. L. S., S. D. Utting, and I. Wallentinus. 1999. *Status of introductions of non-indigenous marine species to North Atlantic waters 1981–1991.* ICES Cooperative Research Report No. 231.

Noda, M. 1987. *Marine algae of the Japan Sea.* Tokyo: Kazama Shobo.

Novaczek, I. and J. McLachlan. 1989. Investigations of the marine algae of Nova Scotia XVII. Vertical and geographic distribution of marine algae on rocky shores of the Maritime Provinces. *Proceeding of Nova Scotia Institute of Science* 38: 91–143.

Perez, R. 1997. *Ces algues qui nous entourent.* IFREMER Editions.

Perez, R., R. Kass, and O. Barbaroux. 1984. Culture expérimentale de l'algue *Undaria pinnatifida* sur les côtes de France. *Science Pêche* 343: 1–15.

Pollard, D. A. and P. A. Hutchings. 1990. A review of exotic marine organisms introduced to the Australian region. II. Invertebrates and algae. *Asian Fisheries Science* 3: 223–250.

Por, F. D. 1978. *Lessepsian migrations. The influx of Red Sea biota into the Mediterranean by way of the Suez Canal.* Berlin: Springer.

Por, F. D. 1989. *The legacy of Tethys. An aquatic biogeography of the Levant.* Kluwer, Dordrecht.

Por, F. D. 1990. *Lessepsian* migrations. An appraisal and new data. *Bulletin Institute Océanographique, Monaco* 7: 1–10.

Posey, M. H. 1988. Community changes associated with the spread of an introduced seagrass, *Zostera japonica. Ecology* 69: 974–983.

Reise, K., S. Gollasch, and W. J. Wolff. 1999. Introduced marine species of the North Sea coasts. *Helgoländer Meeresuntersuchungen* 52: 219–234.

Ribera, M. A. 1994. Les macrophytes marins introduits en Méditerranée: biogéographie. In *Introduced species in European coastal waters,* C. F. Boudouresque, F. Briand, and C. Nolan, eds., pp. 37–43. Luxembourg: European Commission.

Ribera, M. A. and C. F. Boudouresque. 1995. Introduced marine plants, with special reference to macroalgae: mechanisms and impact. In *Progress in phycological research,* Vol. 11, F. E. Round and D. J. Chapman, eds. Biopress Ltd., pp. 187–268.

Russell, D. J. 1992. The ecological invasion of the Hawaiian reefs by two marine red algae, *Acanthophora spicifera* (Vahl) Boerg. and *Hypnea musciformis* (Wulfen) J.Ag. and their association with two native species, *Laurencia nidifica* J.Ag. and *Hypnea cervicornis* J.Ag. *ICES Marine Science Symposium* 194: 1008–1013.

Sancholle, M. 1988. Présence du *Fucus spiralis* (Phaeophyceae) en Méditerranée occidentale. *Cryptogamie Algologie* 9: 157–161.

Sant, N., O. Delgado, C. Rodríguez-Prieto, and E. Ballesteros. 1996. The spreading of the introduced seaweed *Caulerpa taxifolia* (Vahl) C. Agardh in the Mediterranean Sea: testing the boat transportation hypothesis. *Botanica Marina* 39: 427–430.

Santelices, B. 1989. *Algas marinas de Chile. Distribución, ecología, utilización, diversidad.* Santiago: Ediciones Universidad Católica de Chile.

Scagel, R. F., P. W. Gabrielson, D. J. Garbary, L. Golden, M. W. Hawkers, S. C. Lindström, J. C. Oliveira, and T. B. Widdowson. 1989. *A synopsis of the marine algae of British Columbia, southeast Alaska, Washington and Oregon.* Phycological Contribution Number 3. Department of Botany, University of British Columbia.

Smith, T. E. and S. R. Kerr. 1992. Introductions of species transported in ships' ballast waters: the risk to Canada's marine resources. *Canadian Technical Report on Fisheries and Aquatic Sciences 1867.*

South, G. R. and I. Tittley. 1986. *A checklist and distributional index of the benthic marine algae of the North Atlantic Ocean.* St. Andrews: Huntsman Marine Laboratory and British Museum (Natural History).

Subba Rao, D. V., W. G. Sprules, A. Locke, and J. T. Carlton. 1994. Exotic phytoplankton species from ships' ballast waters: risk of potential spread to mariculture on Canada's east coast. *Canadian Data Report on Fisheries and Aquatic Sciences 937.*

Taylor, W. R. 1957. *Marine algae of the northeastern coasts of North America.* Ann Arbor: University of Michigan Press.

Taylor, W. R. 1960. *Marine algae of the eastern tropical and subtropical coasts of the Americas.* Ann Arbor: University of Michigan Press.

van den Hoeck, C., A. M. Breeman, and W. T. Stam. 1990. The geographic distribution of seaweed species in relation to temperature: present and past. In *Expected effects of climatic change on marine coastal ecosystems,* J. J. Beukema, W. J. Wolff, and J. J. W. M. Brouns, eds., Kluwer, Dordrecht, pp. 55–67.

Verlaque, M. 1989. Contribution à la flore des algues marines de Méditerranée: Espèces rares ou nouvelles pour les côtes françaises. *Botanica Marina* 32: 101–113.

Verlaque, M. 1994. Inventaire des plantes introduites en Méditerranée: origines et répercussions sur l'environnement et les activités humaines. *Oceanologica Acta* 17: 1–23.

Verlaque, M. 2001. Checklist of the macroalgae of Thau Lagoon (Hérault, France), a hot spot of marine species introduction in Europe. *Oceanologica Acta* 24: 29-49.

Verlaque, M. and C. F. Boudouresque. 1991. *Stypopodium schimperi* (Buchinger *ex* Kützing) Verlaque et moudouresque comb. nov. (Dictyotales, Fucophyceae), algue de Mer Rouge récemment apparue en Méditerranée. *Cryptogamie, Algologie* 12: 195–211.

Verlaque, M. and R. Riouall. 1989. Introduction de *Polysiphonia nigrescens* et d'*Antithamnion nipponicum* (Rhodophyta, Ceramiales) sur le littoral méditerranéen français. *Cryptogamie, Algologie* 10: 313–323.

Wallentinus, I. 1994. Concerns and activities of the ICES Working Group on introductions and transfers of marine organisms. In *Introduced species in European coastal waters,* C. F. Boudouresque, F. Briand, and C. Nolan, eds., pp. 76–84. Luxembourg: European Commission.

Welcomme, R. L. 1992. A history of international introductions of inland aquatic species. *ICES Marine Science Symposia* 194: 3–14.

Womersley, H. B. S. 1984. *The marine benthic flora of southern Australia.* Part I. Adelaide: Government Printer.

Womersley, H. B. S. 1987. *The marine benthic flora of southern Australia.* Part II. Adelaide: Government Printer.

Womersley, H. B. S. 1994. *The marine benthic flora of southern Australia.* Part IIIA. Canberra: ABRS.

Zibrowius, H. 1991. Ongoing modification of the Mediterranean marine fauna and flora by the establishment of exotic species. *Mésogée. Bulletin du Museum d'Histoire Naturelle de Marseille* 51: 83–107.

APPENDIX 8.1.

List of the reported nonnative species of marine flora by taxonomic group, pathway of introduction, and geographic region of occurrence. **Taxonomic Group:** **B**=Bacillariophyceae, **C**=Chlorophyceae, **Ch**=Chrysophyceae, **D**=Dinophyceae, **HP**=Higher plant, **P**=Phaeophyceae, **R**=Rhodophyceae, **Rh**=Rhaphydophyceae; **Pathway:** **AQ**=aquaria, **AQUA**=aquaculture, **ASSO**=associated species with shellfish transfer, **BA**=ballast, **FIS**=fishing, **FO**=fouling, **RES**=research, **REST**=marsh restoration, **SUEZ**=Suez Canal, **???**=uncertain; **Geographical Region:** **AUS**=Australia, **AZ**=the Azores, **CAN**=Canary Islands, **CHI**=China, **Eu A**=European Atlantic coast, **Gu G**=Guinea-Gulf, **HAW**=Hawaii, **MED**=Mediterranean Sea, **NA A**=North America Atlantic coast, **NA P**=North America Pacific coast, **N ZEA**=New Zealand.

SPECIES	GROUP	PATHWAY	GEOGRAPHICAL DISTRIBUTION						
			Eu A	MED	NA P	NA A	AUS	N ZEA	HAW
Acanthophora spicifera (Vahl) Boergesen	R	FO							HAW
Acrothamnion preissii (Sond.) E. M. Woll.	R	FO		MED					
Acrothrix gracilis Kylin	P	ASSO		MED					
Agardhiella subulata (C. Agardh) Kraft *et* M. J. Wynne	R	ASSO	Eu A	MED					
Aglaothamnion feldmanniae Halos	R	FO		MED					
Aglaothamnion tenuissimum (Bonnem.) Feldm.-Maz.	R	???			NA P				
Ahnfeltiopsis flabelliformis (Harv.) Masuda	P	ASSO		MED					
Alexandrium catenella (Whedon *et* Kofoid) Balech	D	BA	Eu A						
Alexandrium leeii Balech	D	BA	Eu A						
Alexandrium minutum Halim	D	BA	Eu A						
Alexandrium tamarensises (Lebour) Balech	D	BA	Eu A						
Anacyomene stellata (Wulfen) C. Agardh	C	???					AUS		
Anotrichium furcellatum (J. Agardh) Baldock	R	???	Eu A						
Antithamnion amphigeneum A. Millar	R	FO		MED					
Antithamnion densum (Suhr) Howe	R	???	Eu A						
Antithamnion pectinatum (Mont.) Brauner ex Athanas. *et* Tittley	R	ASSO		MED					
Antithamnion pectinatum (Mont.) Brauner ex Athanas. *et* Tittley	R	FO				NA A			
Antithamnionella spirographidis (Schiffn.) E. M. Woll.	R	FO	Eu A	MED					
Antithamnionella ternifolia (Hook. *et* Harv.) Lyle	R	FO	Eu A	MED				N ZEA	
Arthrocladia villosa (Huds.) Duby	P	FO					AUS		
Asparagopsis armata Harv.	R	FO	Eu A	MED					
Asparagopsis armata Harv.	R	ASSO	Eu A	MED					
Asperoccocus bullosus Lamour.	R	FO						N ZEA	
Asperoccocus compressus Griff. *ex* Hook.	P	???					AUS		
Audouinella sargassicola (Boergesen) Garbary	R	SUEZ		MED					
Audouinella spathoglossi (Boergesen) Garbary	R	SUEZ		MED					
Audouinella subseriata (Boergesen) Garbary	R	SUEZ		MED					
Avrainvillea amadelpha (Mont.) A. Gepp *et* E. Gepp	C	???							HAW
Bonnemaisonia hamifera Har.	R	FO	Eu A	MED	NA P	NA A			
Bonnemaisonia hamifera Har.	R	RES	Eu A						

220

Species		Code	Eu A	MED	NA P	NA A	AUS	N ZEA	Gu G
Bonnemaisonia hamifera Har.	R	**ASSO**	Eu A						
Caulerpa mexicana Sond. *ex* Kütz.	C	**SUEZ**		MED					
Caulerpa racemosa (Forssk.) J. Agardh	C	**???**		MED					
Caulerpa racemosa (Forssk.) J. Agardh	C	**SUEZ**		MED					
Caulerpa scalpelliformis (R. Brown *ex* Turner) C. Agardh	C	**SUEZ**		MED					
Caulerpa taxifolia (Vahl) C. Agardh	C	**???**			NA P		AUS		
Caulerpa taxifolia (Vahl) C. Agardh	C	**AQ**		MED					
Chaetomorpha melagonium (Weber *et* Mohr) Kütz.	C	**???**					AUS		
Champia affinis (Hook. *et* Harv.) Harv.	R	**FO**						N ZEA	
Chara connivens Salzm. *ex* A. Brown	C	**BA**	Eu A						
Chnoospora minima (Hering) Papenf.	P	**FO**						N ZEA	
Chondria arcuata Hollenb.	R	**???**					AUS		
Chondria coerulescens (J. Agardh) Falkenb.	R	**ASSO**		MED					
Chondria harveyana (J. Agardh) De Toni	R	**FO**						N ZEA	
Chondria pygmaea Garbary *et* Vandermeulen	R	**SUEZ**		MED					
Chondrus crispus Stackh.	R	**AQUA**		MED					
Chondrus giganteus Yendo f. *flabellatus* Mikami	R	**ASSO**		MED					
Chorda phyllum (L.) Stackh.	R	**ASSO**		MED					
Chrysymenia wrightii (Harv.) Yamada	R	**ASSO**		MED					
Cladophora dalmatica Kütz.	C	**???**					AUS		
Cladophora laetevirens (Dillw.) Kütz.	C	**???**					AUS		
Cladophora lehmanniana (Lindenb.) Kütz.	C	**???**					AUS		
Cladophora patentiramea (Mont.) Kütz.	C	**FO**		MED					
Cladophoropsis zollingeri (Kütz.) Reinbold	C	**SUEZ**		MED					
Cladosiphon zosterae (J. Agardh) Kylin	P	**ASSO**		MED					
Codium fragile ssp *atlanticum* (A. Cotton) P. C. Silva	C	**ASSO**	Eu A						
Codium fragile ssp *scandinavicum* P. C. Silva	C	**???**	Eu A						
Codium fragile ssp *tomentosoides* (Goor) P. C. Silva	C	**???**						N ZEA	
Codium fragile ssp *tomentosoides* (Goor) P. C. Silva	C	**FIS**				NA A			
Codium fragile ssp *tomentosoides* (Goor) P. C. Silva	C	**FO**			NA P	NA A			
Codium fragile ssp *tomentosoides* (Goor) P. C. Silva	C	**ASSO**	Eu A	MED		NA A			
Colpomenia durvillae (Bory) Ramirez	P	**FO**						N ZEA	
Colpomenia peregrina (Sauv.) Hamel	P	**???**		MED		NA A			
Colpomenia peregrina (Sauv.) Hamel	P	**ASSO**	Eu A						
Corynomorpha prismatica (J. Agardh) J. Agardh	R	**???**							Gu G

SPECIES	GROUP	PATHWAY	Eu A	MED	NA P	NA A	AUS	N ZEA	HAW
Corynophloea umbellata (C. Agardh) Kütz.	P	ASSO	Eu A						
Coscinodiscus wailesii Gran et Angst	B	BA	Eu A		NA P	NA A			
Cottoniella fusiformis Boergesen	R	???					AUS		
Cryptonemia hibernica Guiry et L. M. Irvine	R	???	Eu A						
Cryptonemia seminervis (C. Agardh) J. Agardh	R	???	Eu A						
Cutleria multifida (Sm.) Grev.	P	FO						N ZEA	
Dasya baillouviana (S. G. Gmel.) Mont.	R	FO	Eu A						
Dasya baillouviana (S. G. Gmel.) Mont.	R	ASSO	Eu A						
Derbesia rhizophora Yamada	C	ASSO		MED					
Derbesia tenuissima (De Not.) P. Crouan *et* H. Crouan	C	???					AUS		
Desmarestia viridis O. F. Müll.	P	???		MED					
Desmarestia viridis O. F. Müll.	P	ASSO		MED					
Dipterosiphonia dendritica (C. Agardh) F. Schmitz	R	???		MED					
Discosporangium mesarthrocarpum (Meneg.) Hauck	P	???					AUS		
Elachista orbicularis (Ohta) Skinner	P	???					AUS		
Elodea canadensis Michx.	HP	BA	Eu A						
Eucheuma denticulatum (Burman) Collins *et* Herv.	R	AQUA							HAW
Eucheuma isiforme (C. Agardh) J. Agardh	R	AQUA							HAW
Fibrocapsa japonica Toriumi *et* Takano	Rh	BA	Eu A						
Fucus evanescens C. Agardh	P	???	Eu A						
Fucus serratus L.	P	BA				NA A			
Fucus spiralis L.	P	FIS		MED					
Furcellaria lumbricalis (Huds.) Lamour.	R	BA				NA A			
Gelidium vagum Okamura	R	???			NA P				
Gracilaria armata (C. Agardh) Grev.	R	SUEZ		MED					
Gracilaria disticha (J. Agardh) J. Agardh	R	SUEZ		MED					
Gracilaria epihippisora M. D. Hoyle	R	AQUA							HAW
Gracilaria eucheumatoides Harv.	R	AQUA							HAW
Gracilaria salicornia (C. Agardh) Dawson	R	AQUA							HAW
Gracilaria tikvahiae McLachlan	R	AQUA							HAW
Grateloupia cf. *turuturu* Yamada	R	ASSO		MED					
Grateloupia doryphora (Mont.) Howe	R	ASSO	Eu A	MED					
Grateloupia doryphora (Mont.) Howe	R	FO				NA A			
Grateloupia filicina var. *luxurians* A. Gepp *et* E. Gepp	R	ASSO	Eu A	MED					

Species	Status	Method	Eu A	MED	NA P	NA A	AUS	N ZEA	CHI	HAW
Griffithsia corallinoides (L.) Trevisan	R	ASSO		MED						
Griffithsia crassiuscula C. Agardh	R	FO						N ZEA		
Gymnodinium catenatum Graham	D	BA	Eu A				AUS			
Gyrodinium aureolum Hulburt	D	BA	Eu A			NA A				
Halophila stipulacea (Kütz.) Reinke	HP	SUEZ		MED						
Halothrix lumbricalis (Kütz.) Reinke	P	ASSO		MED						
Halymenia actinophysa Howe	R	???			NA P					
Herposiphonia parca Setchell	R	ASSO		MED						
Heterosigma akashiwo (Hada) Hada ex Sournia	B	BA	Eu A							
Hydroclathrus clathratus (Bory ex C. Agardh) Howe	P	FO						N ZEA		
Hypnea cervicornis J. Agardh	R	???		MED						
Hypnea cornuta (Kütz.) J. Agardh	R	SUEZ		MED						
Hypnea esperi auctorum	R	SUEZ		MED						
Hypnea musciformis (Wulfen) Lamour	R	???			NA P					HAW
Hypnea nidifica J. Agardh	R	SUEZ		MED						
Hypnea spicifera (Suhr) Harv.	R	SUEZ		MED						
Hypnea valentiae (Turner) Mont.	R	FO		MED						
Hypnea valentiae (Turner) Mont.	R	ASSO		MED						
Hypnea valentiae (Turner) Mont.	R	SUEZ		MED						
Ishige isiforme Yendo	P	BA			NA P					
Kappaphycus alvarezii (Doty) Doty ex P. C. Silva	R	AQUA								HAW
Kappaphycus striatum (F. Schmitz) Doty ex P. C. Silva	R	AQUA								HAW
Laminaria japonica Aresch.	P	AQUA		MED					CHI	
Laminaria japonica Aresch.	P	ASSO		MED						
Laurencia brongiartii J. Agardh	R	ASSO	Eu A							
Laurencia japonica Yamada	R	ASSO		MED						
Laurencia microcladia Kütz.	R	???		MED						
Laurencia okamurae Yamada	R	ASSO		MED						
Leathesia difformis (L.) Areschoug	P	ASSO		MED						
Lithophyllum yessoense Foslie	R	ASSO		MED						
Lomentaria clavellosa (Turner) Gaillon	R	???				NA A				
Lomentaria hakodatensis Yendo	R	ASSO	Eu A	MED	NA P					
Lomentaria orcadensis (Harv.) Collins *et* W. R. Taylor	R	???				NA A				
Lophocladia lallemandii (Mont.) F. Schmitz	R	SUEZ		MED						
Macrocystis integrifolia Bory	P	AQUA			NA P				CHI	

SPECIES	GROUP	PATHWAY	Eu A	MED	NA	AUS	N ZEA	CHI	HAW	CAN
Macrocystis pyrifera (L.) C. Agardh	P	AQUA						CHI	HAW	
Mastocarpus stellatus (Stackh.) Guiry	R	RES	Eu A							
Monostroma obscurum (Kütz.) J. Agardh	C	ASSO		MED						
Myriophyllum sibiricum Komarov	HP	BA	Eu A						HAW	
Nemacystus decipiens (Suringar) Kuck.	P	???								
Odontella sinensis (Grev.) Grunov	B	BA	Eu A		NA A					
Olisthadiscus luteus Carter	Ch	BA	Eu A							
Padina boergesenii Alleder et Kraft	P	SUEZ		MED						
Padina boryana Thivy	P	SUEZ		MED	NA A					
Palmaria palmata (L.) Kuntze	P	AQUA	Eu A							
Pikea californica Harv.	R	FO	Eu A				N ZEA			
Pilayella littoralis (L.) Kjellm.	P	FO								
Pilayella littoralis (L.) Kjellm.	P	ASSO		MED						
Pilinella californica Hollenb.	C	???							HAW	
Platysiphonia caribaea D. L. Ballant. *et* M. J. Wynne	R	???								CAN
Pleonosporium caribaeum (Boergesen) R. E. Norris	R	FO	Eu A	MED						
Pleurosigma simonsenii Hasle	B	BA	Eu A							
Plocamium secundatum (Kütz.) Kütz.	R	???		MED						
Polysiphonia breviarticulata (C. Agardh) Zanardini	R	???			NA A					
Polysiphonia brodiaei (Dillwyn) Spreg.	R	FO				AUS	N ZEA			
Polysiphonia constricta Womersley	B	BA					N ZEA			
Polysiphonia denudata (Dillwyn) Grev. ex Harv.	R	???			NA P					
Polysiphonia fucoides (Huds.) Grev.	R	FIS		MED						
Polysiphonia harveyi Bailey	R	ASSO	Eu A							
Polysiphonia harveyi Bailey	R	???		MED						
Polysiphonia morrowii Harv.	R	ASSO		MED						
Polysiphonia paniculata Mont.	R	ASSO		MED						
Polysiphonia senticulosa Harv.	R	???	Eu A							
Polysiphonia senticulosa Harv.	R	FO				AUS	N ZEA			
Polysiphonia sertularioides (Gratel.) J. Agardh	R	FO					N ZEA			
Polysiphonia subtilissima Mont.	R	FO					N ZEA			
Polysiphonia subtilissima Mont.	R	ASSO					N ZEA			
Porphyra tenera Kjellm.	R	AQUA			NA P					
Porphyra yezoensis Ueda	R	ASSO		MED						
Predaea huismanii Kraft	R	???								CAN
Prionitis patens Okamura	R	ASSO		MED						

224

Species		BA	Eu A	MED	NA P	NA A	AUS	N ZEA	AZO
Prorocentrum minimun (Pavill.) J. Schiller	B								
Punctaria latifolia Grev.	P	**FO**	Eu A					N ZEA	
Rhodophysema georgii Batters	R	**ASSO**		MED					
Rhodothamniella cf. *codicola* (Boergesen) Bidoux *et* F. Magne	R	**ASSO**		MED					
Rhodymenia erythraea Zanardini	R	**FO**		MED					
Rosenvingiella polyrhiza (Rosenv.) P. C. Silva	C	**???**					AUS		
Sarconema filiforme (Sond.) Kylin	R	**SUEZ**		MED					
Sarconema scinaioides Boergesen	R	**SUEZ**		MED					
Sargassum muticum (Yendo) Fensholt	P	**FIS**			NA P				
Sargassum muticum (Yendo) Fensholt	P	**ASSO**	Eu A	MED	NA P				
Sargassum verruculosum C. Agardh	P	**FO**						N ZEA	
Schottera nicaeensis (Lamour. *ex* Duby) Guiry *et* Hollenb.	R	**???**					AUS		
Scytosiphon dotyi M. J. Wynne	P	**ASSO**	Eu A	MED					
Solieria chordalis (C. Agardh) J. Agardh	R	**FO**	Eu A						
Solieria dura (Zanardini) F. Schmitz	R	**SUEZ**		MED					
Solieria tenera (J. Agardh) M. J. Wynne *et* R. W. Taylor	R	**???**					AUS		
Sorocarpus micromorus (Bory) P. C. Silva	P	**FO**					AUS		
Spartina alterniflora Loisel.	HP	**REST**							
Spartina anglica C. E. Hubb.	HP	**BA**	Eu A						
Spartina anglica C. E. Hubb.	HP	**REST**			NA P				
Spartina densiflora Brongn.	HP	**REST**			NA P				
Spartina patens Muhl.	HP	**???**			NA P				
Spartina townsendii H. *et* J. Groves	HP	**???**					AUS		
Spatoglossum variabile (Fig.) De Not.	P	**SUEZ**		MED					
Spermatochnus paradoxus (Roth) Kütz.	P	**???**					AUS		
Sphacelaria mirabilis (Reinke *ex* Batters) Prud'homme	P	**???**				NA A			
Sphacella subtilissima Reinke	P	**???**					AUS		
Sphaerotrichia divaricata (C. Agardh) Kylin	P	**ASSO**		MED					
Stichyosiphon soriferus (Reinke) Rosenvinge	P	**???**				NA A	AUS		
Striaria attenuata (Grev.) Grev.	P	**???**					AUS		
Striaria attenuata (Grev.) Grev.	P	**FO**						N ZEA	
Stypopodium schimperi (Kütz.) Verlaque *et* Boudouresque	P	**SUEZ**		MED					
Symplocladia marchandioides (Harv.) Falkenberg	R	**???**							AZO
Thalassiosira angstii (Gran) I. V. Makarova	B	**BA**	Eu A						
Thalassiosira punctigera (Castrac.) Hasle	B	**BA**	Eu A						

SPECIES	GROUP	PATHWAY	GEOGRAPHICAL DISTRIBUTION						
Thalassiosira tealata Takano	B		Eu A						
Ulva pertusa Kjellm.	C	ASSO		MED					
Undaria pinnatifida (Harv.) Suringar	P	BA			AUS				
Undaria pinnatifida (Harv.) Suringar	P	AQUA	Eu A	MED					
Undaria pinnatifida (Harv.) Suringar	P	FO				N ZEA			
Undaria pinnatifida (Harv.) Suringar	P	ASSO	Eu A	MED					
Womersleyella setacea (Hollenb.) R. E. Norris	R	FIS		MED					
Womersleyella setacea (Hollenb.) R. E. Norris	R	FO		MED					
Wrangelia bicuspidata Boergesen	R	???					HAW		
Zostera japonica Asch. et Graebn.	HP	BA						NA P	NA A
Zostera japonica Asch. et Graebn.	HP	FIS						NA P	NA A
Zostera japonica Asch. et Graebn.	HP	ASSO							NA A

Chapter 9

Spatial and Temporal Analysis of Transoceanic Shipping Vectors to the Great Lakes

Robert I. Colautti, Arthur J. Niimi, Colin D. A. van Overdijk,
Edward L. Mills, Kristen Holeck, and Hugh J. MacIsaac

Anthropogenic introductions of nonindigenous species (NIS) are predicted to impact biodiversity of lakes more than any other major ecosystem type over the coming century (Sala et al. 2000). Freshwater ecosystems are highly vulnerable to invasions by NIS because of their close association with human activity, including exploitative uses for municipal and industrial water supplies, natural resource development (e.g., fishing, aquaculture), and commercial navigation and recreation. These varied uses provide countless invasion opportunities for NIS throughout the world. Consequences of these invasions have become well characterized, as many of the world's large lakes have been colonized by infamous nuisance invaders such as Nile perch (*Lates niloticus*), zebra mussels (*Dreissena polymorpha*), water hyacinth (*Eichhornia crassipes*), and hydrilla macrophytes (*Hydrilla verticillata*). Profound changes to the physical, chemical, and biological properties of lakes have followed invasions by these and other species of invertebrate and vertebrate animals, and micro- and macroscopic plants (e.g., Zaret and Paine 1973, Oliver 1993, Spencer et al. 1999, Ketelaars et al. 1999, MacIsaac 1999, Vander Zanden et al. 1999, Hall and Mills 2000, Lodge et al. 2000, Donald et al. 2001, Dick and Platvoet 2000, Schindler et al. 2001, Vanderploeg et al. 2002).

NIS are introduced to lakes through both intentional and inadvertent vectors. Government-sponsored stocking programs are the leading intentional vectors. Historically, many species of fishes were added to lakes around the world in an attempt to create fisheries where they previously did not exist (e.g., in fishless alpine lakes), to enhance preexisting fish stocks (e.g., through introductions of *Micopterus* spp. and *Lates nilotica*), or as a biological control agent of insects, snails, or nuisance plants (e.g., introductions of *Gambusia*, *Mylopharyngodon*, or *Cyprinus*) (see further discussion by Fuller in this volume, and references therein). Stocking of predatory Nile perch (*Lates niloticus*) into Lake Victoria represents one of the greatest evolutionary and ecological disasters precipitated by mankind, as up to 200 species of vulnerable endemic cichlid fishes were subsequently driven to extinction (Kaufman 1992).

Invertebrates have also been widely stocked to lakes throughout the world, typically with the intention of enhancing food supplies available to fishes. Crustaceans, such as amphipods, mysids, and crayfish, have been stocked most commonly, sometimes with catastrophic consequences. For example, mysids were stocked into lakes in Scandinavia, Kootenay Lake in British Columbia, and Flathead Lake in Montana. Amphipods and mysids were also introduced to many lakes in the former Soviet Union between 1940 and 1960, although the practice appears to have waned in recent years (see Grigorovich et al. 2002). The Baikal amphipod *Gmelinoides fasciatus* was first stocked in the Volga River system during the early 1960s, and later to many other lakes throughout western and northern Russia. It established in western Russia in Lake Ladoga in the early 1980s, and is now abundant in that system (Panov 1996). Rather than augmenting the food supply available to fishes in these systems, mysids can compete for zooplankton prey with young-of-year planktivorous fishes, often causing a collapse of the very fish populations they were intended to enhance (see Spencer et al. 1999). Stocking or aquaculture programs may indirectly facilitate introduction of other, nontarget species that parasitize, infect, or are similar in appearance to target species (Grigorovich et al. 2002).

Shipping activities constitute a very important vector for the inadvertent introduction of NIS to coastal marine habitats and some freshwater systems, such as the Laurentian Great Lakes of North America (e.g., Carlton and Geller 1993, Ruiz et al. 2000a, 2000b, Ricciardi 2001, Leppäkoski et al. 2002). Ships traveling between the world's ports have long employed ballast for stability and trim when traveling without cargo. Initially, solid materials were loaded as ballast (e.g., sand, soil, rock), which resulted in dispersal of seeds of many terrestrial and wetland plants (see Mills et al. 1993). Water replaced solid materials as the dominant ballast medium around 1900, resulting in the transport and release of many waterborne aquatic taxa (see Mills et al. 1993). Ballast water was the dominant vector of NIS to the Great Lakes between 1960 and 2001, a trend that has been accompanied by a dramatic increase in the number of new invaders (Mills et al. 1993, Ricciardi 2001, Grigorovich et al. 2003a).

Shipping interacts with the creation of dams and canals, which alter hydrology to provide access to new watersheds and thereby facilitate dispersal of NIS. For example, the Caspian Sea was invaded in 1999 (or earlier) by the ctenophore *Mnemiopsis leidyi*, which was likely introduced from the Black Sea or the Sea of Azov by a ship utilizing the Volga–Don Canal. This canal, which was opened in 1952, connects the Black and Azov Seas with the Caspian Sea (Ivanov et al. 2000). Similarly, invertebrate species have dispersed to the lower Rhine River and the Baltic Sea via a series of connecting rivers and canals within Europe (reviewed in bij de Vaate et al. 2002). In the Great Lakes basin, creation of canal systems along the St. Lawrence and Niagara Rivers has likewise facilitated dispersal of NIS from lower to upper lakes (Mills et al. 1993).

Certainly many other vectors contribute to the global transfer of species, the relative importance of which varies from system to system and over space and time. For example, release of sport baitfish and other organisms resident in bait water may result in establishment of NIS in systems utilized by anglers (Litvak and Mandrak 2000). Sport fisheries and pleasure boating may also result in inadvertent invasions if macrophytes and attached invertebrate fauna are stranded on boat trailers moved between systems (see Johnson et al. 2001). Releases of unwanted aquarium pets or live fishes intended for human consumption can also result in invasions (e.g., Fuller, this volume).

Efforts to prevent new invasions require an understanding of the invasion process, particularly the sources and mechanisms of propagule supply. In this chapter, we review the present state of knowledge of NIS transfer and invasion patterns in the Great Lakes, focusing especially on ship-mediated transfer.

Vectors to the Great Lakes

The Great Lakes are an excellent model system with which to analyze invasion vectors (MacIsaac et al. 2001). The system is well defined and studied, allowing for identification of invasion vectors and pathways, and is similar to many estuarine ecosystems and large inland water bodies where shipping dominates vector supplies of NIS. The lakes also serve as a gateway to invasion of adjacent inland lakes through a host of other vectors associated with human activities (e.g., see Johnson et al. 2001, Borbely 2001).

At least 162 NIS are established in the Great Lakes proper (Ricciardi 2001, Grigorovich et al. 2003a). All of these species are characteristic of lentic ecosystems, including a broad spectrum of organisms from phytoplankton to fish. The reported establishment rate of new NIS increased linearly between 1959 and 1989 at an annual rate of 0.621 new species; however, this rate accelerated to 1.880 new species per year (95% confidence interval: 1.543–2.216) between 1989 and 2001 (Grigorovich et al. 2003a). Many of the NIS that invaded the Great Lakes in recent years originated from Europe, notably the Baltic Sea and lower Rhine River areas (Ricciardi and MacIsaac 2000, bij de Vaate et al. 2002, Grigorovich et al. 2003a). For example, allozyme and mitochondrial DNA analyses have pinpointed the origins of Great Lakes populations of *Bythotrephes longimanus* and *Cercopagis pengoi* water fleas to the Baltic Sea region, and of *Echinogammarus ischnus* amphipods to the lower Rhine River (Berg et al. 2001, Cristescu et al. 2001, M. Cristescu, unpublished data). Internal waterways in Europe have facilitated the dispersal of species from the Black and Azov Seas to the Baltic Sea and lower Rhine (see bij de Vaate et al. 2002). Once established in major ports in western and northern Europe, Ponto-Caspian and other Eurasian species invade the Great Lakes in secondary invasions mediated by ships' ballast water (Ricciardi and MacIsaac 2000, bij de Vaate et al. 2002). A spate of ballast-mediated invasions by Ponto-Caspian species has transformed Great Lakes species communities in recent years (Ricciardi and MacIsaac 2000).

The Great Lakes currently receive NIS propagules from ballast tanks in two distinctive forms. First, they receive large volumes of water from each of a relatively small number of ships that enter the lakes loaded with saline ballast water (ballast-on-board or BOB ships). The introduction of biota in ships' ballast is not surprising, given the vast amount of water imported in this way. For example, in 1995 the Great Lakes received an estimated 5×10^6 m^3 of ballast water per year from oceangoing ships (Aquatic Sciences 1996). Canada implemented voluntary ballast water exchange regulations covering the Great Lakes in 1989. These regulations were made mandatory, effectively covering the entire Great Lakes basin, by legislation implemented by the U.S. Coast Guard in 1993 (U.S. Coast Guard 1993). Regulations require that vessels entering the lakes from foreign locations treat low-salinity ballast water before discharging into the Great Lakes. The only treatment identified to date consists of ballast water exchange, whereby ships flush their tanks while traversing open ocean to purge most organisms from their tanks, and to kill remaining freshwater organisms by osmotic stress (Locke et al. 1991, 1993, U.S. Coast Guard 1993). While this treatment reduces populations of freshwater organisms resident in ballast tanks, its efficacy is not complete (Locke et al. 1993, MacIsaac et al. 2002).

A large percentage of ships enter the lakes loaded with cargo (no-ballast-on-board or NOBOB ships) and carry only residual ballast water (i.e., 50–60 tons) and sediment. These vessels fill their ballast tanks with Great Lakes water when they discharge cargo in port. Great Lakes water loaded by these ships mixes with the "residuals." The mixed slurry is then discharged at a subsequent port, often on the Great Lakes. This ballasting activity is entirely legal, though it may predispose the Great Lakes to invasion by taxa living in ballast residuals or by their viable resting stages (Bailey et al. 2003; C. van Overdijk, unpublished data).

It is not clear why the reported invasion rate of the Great Lakes accelerated during the 1990s. Possibilities include greater attention and more researchers studying invasion phenomena, lag periods between introduction of NIS and their first reported discoveries in the system (Grigorovich et al. 2003a), or changes in vector supply (Carlton 1996).

Here we provide an overview of NOBOB ships as a potential vector to the Great Lakes. Our objectives are to (1) assess temporal variation in intensity of transoceanic ship traffic entering the lakes; (2) assess the relative frequency of transoceanic ships entering the entire system loaded with saline ballast water (BOB ships) or loaded with cargo and only residual water and sediments in ballast tanks (NOBOB ships); (3) determine the relative frequency of BOB and NOBOB vessels to each of the lakes, and compare these patterns with the establishment sites of recent invaders to the system; (4) characterize the regions of origin for NOBOB vessels and contrast this pattern with the recent invasion history of the Great Lakes. Overall, our intent is to gain a rough understanding of potential supply of NIS to the lakes by ships, using ballast activities as a proxy.

GREAT LAKES SHIPPING PROFILE

We compiled information on commercial ships originating from foreign ports and inbound to the Great Lakes from the period 1994–2000, using compiled reports on annual ship arrivals (Eakins 1995, 1996, 1997, 1998, 1999, 2000, 2001). We supplemented these data with shipping information collected by the St. Lawrence Seaway Management Corporation for 1986 through 1998 (C. Major, pers. comm.) to determine global ports of origin for ships visiting all ports on the Great Lakes. Both sources of data were used to build a comprehensive database of ship activity for all inbound ships during 1997, including port and country of origin, ports visited on

FIGURE 9.1. Location of major Great Lakes ports visited by foreign, transoceanic ships between 1986 and 2000. Ports on the St. Lawrence River were excluded from the study.

the Great Lakes (Fig. 9.1), and their ballasting/deballasting activities while operating on the Great Lakes; additionally, we determined whether ships entered under BOB or NOBOB status.

We selected 1997 to provide detailed analysis of vector traffic to the Great Lakes, although there appeared to be minimal variation in patterns between 1994 and 2000. We determined the last port of call (i.e., the last port that a vessel visited prior to entering the St. Lawrence Seaway) for all ships that entered the Great Lakes during 1997 (Table 9.1).

We classified vessels based on their ballast water status when entering the Great Lakes. In most cases, oceangoing ships that entered the lakes in NOBOB status deliver cargo and also load cargo before leaving the Great Lakes. Although such vessels load ballast water while in transit on the lakes, we continued to classify these vessels as NOBOB ships to distinguish them from those that entered the lakes with saline ballast water.

We made the following assumptions regarding ships' ballast water management activities:

• All vessels that deballast at their first port of call (BOB ships by definition) discharge only ballast water of oceanic origin, in compliance with existing regulations (U.S. Coast Guard 1993).
• All vessels that discharge cargo at their first port of call are in NOBOB

TABLE 9.1. Last country of origin and final port of call for NOBOB ships entering the Great Lakes during 1997.

NOBOB Vessels' Country of Origin	Lake				
	Superior	*Huron*	*Michigan*	*Erie*	*Ontario*
Belgium	33	1	4	2	1
Netherlands	8	2	6	4	4
Baltic Sea (German and Swedish Baltic ports)	58	2	5	4	3
Germany (North Sea ports)	8	0	0	1	0
Sweden/Norway (North Sea ports)	2	1	1	0	0
Mediterranean/Atlantic Europe	35	0	7	0	2
U.K.	8	1	5	0	0
Latin America	20	1	0	2	2
Brazil	12	0	0	3	0
Japan/China	5	0	0	1	1
Australia	16	0	0	2	0
South Africa	9	0	0	0	1
Ukraine	4	0	0	0	0
Romania	2	0	0	0	0
Indonesia	1	0	0	0	0
U.S.A.	0	2	0	0	0
Canada	0	0	0	1	1
Other European	0	1	0	0	0
Unidentified	7	1	0	1	0
Ballasted ships	15	3	0	6	1

Note: All vessels are assumed to have loaded and subsequently discharged Great Lakes water as ballast in one of the Great Lakes. Nine ships' records were discarded owing to lack of information pertaining to the Great Lake in which ballast water was discharged. Thirty-six ships visited two European ports, and three visited three European ports, prior to arriving in the Great Lakes; each port and country visited by these ships was tabulated separately because the order in which the ports were visited could not be ascertained. Ballasted ships carried (saline) water to the Great Lakes. Vessels arriving from German and Swedish ports were subdivided into those from the Baltic Sea and from the North Sea.

condition (i.e., they had no exchangeable ballast upon entering the Great Lakes).

- All vessels load some freshwater ballast at each Great Lake port where they discharge cargo, and discharge ballast water at subsequent ports on the Great Lakes (if any) where outbound cargo is loaded. Where ships loaded cargo at two consecutive ports, ballast values of one-half were given for each port. No ships were observed that loaded cargo at more than two ports in a single trip. NOBOB vessels are assumed to load Great Lakes water as ballast because cabotage legislation prevents foreign vessels from loading new cargo for transfer within the system to ports in the same country. We found no records of interlake movement of North American cargo by foreign NOBOB vessels.

- NOBOB vessels that leave the lakes without loading cargo for their outbound trip do not discharge ballast water loaded while operating on the Great Lakes.
- Ballast water discharges associated with particular ports are ascribed to the downstream lake basin. For example, discharges at Sault Sainte Marie were ascribed to Lake Huron, whereas those in the St. Clair and Detroit Rivers were considered to occur in Lake Erie. Ballast water discharged at Port Huron, Michigan, and Sarnia, Ontario, were deemed to occur in Lake Erie, and those in the Welland Canal (near Port Weller) were ascribed to Lake Ontario (Fig. 9.1).
- A uniform volume of ballast water is released by each ship operating within the Great Lakes, regardless of ship size, cargo, or other factors. This is similar to the approach of dividing ballast discharge evenly among ports loading cargo (above).

Our estimate of "propagule pressure" is admittedly coarse. It assumes that ballast water discharge patterns can be determined by the cargo loading patterns and that the density of organisms contained in ballast water of different ships is invariant. Ships arriving from different source regions may load different densities of live organisms in ballast water. Carlton (1985) and Ruiz et al. (2000a) reported that survival of organisms in ballast tanks is strongly time-dependent. Thus, variability in the number of viable propagules could be quite high depending on the duration of the trip, the source region, and the efficacy of ballast water exchange prior to entering the Great Lakes.

We have not included any information on ports on the St. Lawrence River that were visited by inbound ships, nor have we tracked the destination of Great Lakes ballast water loaded by NOBOB vessels that leave the lakes without discharging water. Many of these vessels visit ports on the St. Lawrence River on their outbound journey, and likely discharge water at these sites (R. Colautti, unpublished data).

SHIPPING AND BALLAST WATER DISCHARGE PATTERNS

The volume of inbound traffic to the Great Lakes by foreign vessels has varied tremendously over the past twenty-two years (Fig. 9.2). Traffic has

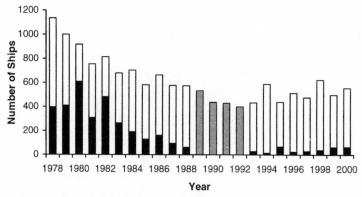

FIGURE 9.2. Total number of foreign, transoceanic vessels entering the Great Lakes through the St. Lawrence River system between 1978 and 2000. Ships carrying ballast water into the lakes (black bars) have declined both in absolute number and relative to those entering with cargo (NOBOB ships; white bars). Years for which no distinction was made between vessel types are shown in gray bars.

declined since the late 1970s, and has remained more or less stable during the past fifteen years, with some variability likely correlated to global economic activity. The fraction of inbound ships loaded with ballast water has strongly diminished in recent years, corresponding with enhanced economic efficiency of shipping companies during the late 1980s and the 1990s. Consequently, both the absolute number and the proportion of foreign ships entering the Great Lakes carrying ballast water (BOB ships) diminished sharply over the past twenty-five years, though both appear to have leveled off in recent years.

Inbound traffic to the Great Lakes between 1986 and 1998 was dominated by ships arriving from European ports, notably those in the lower Rhine River region (i.e., Belgium, the Netherlands), other localities on the North Sea (i.e., Germany, Norway, Denmark), and the Baltic Sea (i.e., Latvia, Lithuania, Poland, Estonia, Germany, Sweden, Russia, Finland) (Fig. 9.3). However, it is difficult to interpret these data since many of the vessels, particularly those from the late 1980s onward, arrived to the Great Lakes under NOBOB status. For these vessels, the last port of call was more likely to be a ballast water recipient than a ballast water donor. For example, Antwerp, Belgium, is one of the leading ports serving the Great Lakes, but most vessels originating at this site loaded cargo and potentially discharged ballast before departure. Collectively, the top ten vessel source regions represented an average of 88 percent of all inbound traffic to the Great Lakes.

We analyzed data on the movement of cargo to and from ships operat-

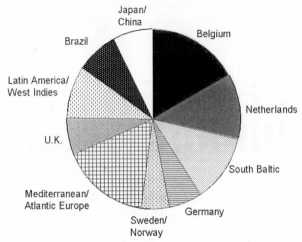

FIGURE 9.3. Average percent contribution of commercial ships entering the Great Lakes between 1986 and 1998, for the ten leading countries or regions, based upon last port of call.

ing at their first ports of call on the Great Lakes to infer their status as BOB or NOBOB. Between 1994 and 2000 inclusive, an average of 8.1 percent of foreign vessels bound to the Great Lakes declared BOB status (Table 9.2). A large fraction of these vessels (55.4%) proceeded directly through the lower lakes and discharged (saline) ballast water in Lake Superior. Lakes Ontario (17.4%) and Erie (17.9%) averaged fewer direct discharges of ballast water by inbound BOB ships than did Lake Superior, despite being the first lakes in the system (Table 9.2).

The first port of call for most NOBOB vessels entering the Great Lakes between 1994 and 2000 was located on Lake Ontario (40.6%) and Lake Erie (43.1%). Lake Superior was the initial port of call for only a very small fraction of inbound NOBOB vessels (0.6%). By contrast, the pattern of ballast water discharge by BOB and NOBOB vessels was focused on Lake Superior (Fig. 9.4). The majority of NOBOB vessels deballasted in Lake Superior (74.5% of total), irrespective of whether these ports represented the second, third, or final port of call. In 2000, for example, a total of 397 ships visited a second port on the Great Lakes after having off-loaded cargo at their first ports of call. Of these, approximately 37 percent proceeded to ports on Lake Superior where they loaded cargo (and presumably discharged ballast water) for their outbound voyage. An additional 13 percent loaded cargo at other Great Lakes ports, and 199 ships off-loaded cargo at their second port of call and continued on. Of the 199 ships that off-loaded cargo at their second port, 29 left the Great Lakes without loading cargo for their return trip, and 170 continued to a third port on the Great Lakes.

TABLE 9.2. Distribution of ships entering the Great Lakes that either discharge ballast water or discharge cargo at their first port of call.

Ship Entry Type	Ships per Year Entering the Great Lakes						
	1994	1995	1996	1997	1998	1999	2000
Ballast	15.0	67.0	24.0	28.5	39.0	62.0	64.0
Erie (%)	16.7	17.9	37.5	24.6	20.5	12.9	10.9
Huron (%)	0.0	4.5	0.0	0.0	5.1	0.0	10.2
Michigan (%)	16.7	7.5	0.0	0.0	5.1	4.8	6.3
Ontario (%)	20.0	22.4	29.2	19.3	23.1	8.1	11.7
Superior (%)	46.7	47.8	33.3	56.1	46.2	74.2	60.9
NOBOB	572.0	372.0	489.0	447.5	583.0	435.0	490.0
stayed (%)	67.3	69.9	68.9	74.0	74.7	77.7	81.0
departed (%)	32.7	30.1	31.1	26.0	25.3	22.3	19.0
TOTAL SHIPS	587.0	439.0	513.0	476.0	622.0	497.0	554.0

Note: All ballast water discharged at the first port of call is considered saline, in compliance with extant regulations (U.S. Coast Guard 1993). All ships that discharge cargo at the first port of call are considered NOBOB. NOBOB ships were classified into those that stayed within the Great Lakes (see Fig. 9.4) and those that departed the system, without deballasting at any port, following off-loading of cargo. Ships arriving with ballast water were categorized by the lake that ultimately received discharged water. Percentages are rounded off. (Source: Eakins 1995, 1996, 1997, 1998, 1999, 2000, 2001.)

Almost 50 percent of these 170 ships loaded cargo in Lake Superior, and an additional 69 ships off-loaded cargo and continued operating on the Great Lakes. Again, a small number (13) of ships left the Great Lakes without loading cargo after their third port of call, and the remaining (56) ships continued to a fourth, or (rarely) fifth or greater, port of call. In each case, Lake Superior was the primary recipient of NOBOB ships that loaded cargo for their outbound voyage from their final port of call (Fig. 9.4).

Although some interannual variation was observed, general patterns emerged. First, between 68 and 82 percent of NOBOB vessels at their first port of call remained as NOBOB vessels at their second one (i.e., they dropped cargo and loaded ballast water at both ports; Fig. 9.4). This value dropped to between 26 and 75 percent at the third port visited. Most of the NOBOB vessels that discharged water at the second or third ports of call did so in Lake Superior. Lake Superior received more discharges of Great Lakes ballast water than all of the other lakes combined, and this pattern was consistent across years (Fig. 9.4). Thus, a disproportionate number of BOB and NOBOB vessels discharge ballast water into Lake Superior, even though Lakes Ontario and Erie are the initial ports of call of many NOBOB vessels.

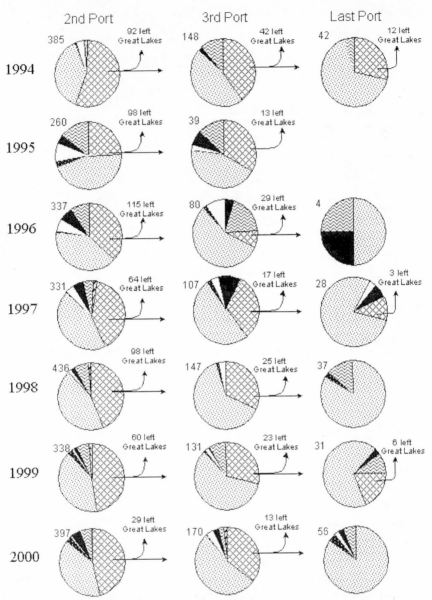

FIGURE 9.4. Spatial and temporal analysis of activity patterns of NOBOB ships entering the Great Lakes that (a) visited additional ports in the lakes before departure and (b) off-loaded cargo at their first port of call (see Table 9.2). All vessels are considered to have loaded ballast water during discharge of cargo in the first port of call. Each pie diagram illustrates the percentage of total ships (number above pie diagrams) that discharged additional cargo in that port of call (cross-hatched), or discharged Great Lakes ballast water in Lake Superior (stippled), Lake Michigan (dark stippled), Lake Huron (white), Lake Erie (black), Lake Ontario (wave), or at an unknown destination (diagonal). Many NOBOB ships left the Great Lakes for ports on the St. Lawrence River or other destinations without discharging Great Lakes ballast water into the Great Lakes; the number of these vessels is provided between pie diagrams. Activity of ships that discharged ballast at a fourth or later port is combined under "Last Port."

PROPAGULE PRESSURE: THE NULL HYPOTHESIS

In general, the propagule pressure model predicts that invasion success should be positively associated with the number and quality of inocula delivered to a recipient system. This propagule pressure model has been proposed as a possible explanation of NIS invasions in marine and other ecosystems (Carlton 1987, 1996, Carlton and Geller 1993, Ruiz et al. 1997, Kolar and Lodge 2001). Likewise, Ricciardi and MacIsaac (2000) reported that the pattern of NIS invasions of the Great Lakes by Ponto-Caspian species was consistent with the propagule pressure concept. So far, no effort has been made to quantify the relationship between NIS in the Great Lakes and propagule supply from donor regions. This analysis would require comprehensive information on the number of ships arriving to each of the lakes, the density and quality of organisms surviving transit in each of the ballast tanks, and the volume of ballast water discharged from each tank (see Carlton 1985). Ballast tanks in individual ships vary in location, size, accessibility, and biotic composition (Locke et al. 1991, 1993, Hamer et al. 2000, Bailey et al. 2003). Thus, comprehensive characterization of biological communities is a complex and difficult undertaking.

Ecologists have utilized both theoretical and empirical approaches to study determinants of invasion success, although most of these efforts have been directed at terrestrial ecosystems. For example, characteristics of the recipient community, notably its native biodiversity or natural or human-induced disturbance, are thought to affect invasion success (see Elton 2000). This area has received considerable examination in recent years (e.g., Levine and D'Antonio 1999, Lonsdale 1999, Shurin 2000, Levine 2000, Kolar and Lodge 2001). Availability of spatial or nutrient resources has also been related to invasion success (e.g., Burke and Grime 1996, Levine and D'Antonio 1999, Sher and Hyatt 1999, Stohlgren et al. 1999). In addition, invasion success may be influenced by biological characteristics and ecological interactions of potential colonists, including the number, size, and dispersing distance of individuals or resting stages from a population, or the order in which species invade communities (Drake 1993, Lodge 1993, Williamson 1996, Rejmánek 1996, Rejmánek and Richardson 1996, Grevstad 1999, Lonsdale 1999, Levine 2000, Shurin 2000, Kolar and Lodge 2002). It is likely that a combination of factors including an adequate and timely arrival of competent propagules, tolerance of physical and chemical conditions, and availability of spatial or nutrient resources are required for successful colonization by NIS.

We argue that the importance of propagule pressure is perhaps least understood, because of the difficulty inherent in quantifying the number of potential colonists involved in most natural invasions, as well as ethical and practical difficulties involved in experimentally manipulating NIS propagule pressure in most ecosystems (but see Grevstad 1999). Some propagule pressure models have been tested using inland lake systems. For example, Bossenbroek et al. (2001) developed mathematical models to predict invasions of zebra mussels (*Dreissena polymorpha*) based upon vector movement between invaded and noninvaded inland lakes, while Borbely (2001) did so for spiny water fleas (*Bythotrephes longimanus*) invading inland lakes in Ontario. These models have illustrated the importance of human vectors (trailered boats and contaminated fishing line, respectively) in the rapid dispersal of these Eurasian species in North America (e.g., see Johnson et al. 2001). Evidence has also accrued in terrestrial systems regarding the importance of propagule pressure. Lonsdale (1999), for example, reported that the number of nonindigenous plant species established in nature reserves was strongly related to the number of human visitors. It is important to note that transfer of propagules by vectors is but the first component of the invasion process, and that some ecosystems subjected to intense propagule pressure may, nevertheless, support few invaders if physical or chemical conditions are unfavorable (e.g., Chesapeake Bay; Smith et al. 1999). Nevertheless, the differential introduction of propagules is a key factor that must be accounted for in studies of invasion dynamics (Lonsdale 1999).

Lake Superior: An Invasion Haven?

Our study suggests that far more foreign BOB and NOBOB ships operating on the Great Lakes deballast in Lake Superior than on any of the other lakes. Although this lake has been the initial site of some NIS reports, most recently of ruffe (Pratt et al. 1992), the lower lakes dominate reports of initial NIS sightings (see Grigorovich et al. 2003a). Assuming that the frequency of vessel deballasting is a robust proxy of volume of ballast water discharged, more invasions of Lake Superior may have been expected. This discrepancy raises an interesting question: Is there something unique to Lake Superior that prevents establishment of NIS despite its relatively high inoculation rate, or have ecologists engaged in unintentionally biased reporting of NIS in the Great Lakes?

It is possible that Lake Superior is relatively inhospitable to NIS. Lake Superior is far less productive than the lower Great Lakes, and has a much greater ratio of limnetic to littoral habitat. Its thermal regime also exhibits much less seasonal variability than the lower lakes. Smith et al. (1999) reported that the upper Chesapeake Bay, despite receiving a large inocula of exotic species in ballast water, supports relatively few ballast-mediated NIS owing to adverse environmental conditions at the release sites. For Lake Superior, the relative lack of disturbance or invasions may also play a role. Disturbance of lower lakes or their watersheds, or presence of NIS that are "ecosystem engineers" (i.e., species that alter the physical/chemical properties of their environment) only in the lower lakes (e.g., zebra mussels), may have disproportionately facilitated invasions in these systems relative to Lake Superior (Simberloff and von Holle 1999, Ricciardi 2001).

Alternatively, Lake Superior may be more invaded than has been recognized, since many established NIS may remain undetected due to the large surface area of the lake and low sampling effort relative to the lower lakes. If this hypothesis is correct, intensive surveys should reveal heretofore unidentified NIS in the lake, particularly in regions where ballast water is discharged most commonly. A comprehensive survey of Lake Superior to test this hypothesis revealed a number of NIS range extensions from the lower Great Lakes, but no invaders new to the basin (Grigorovich et al. 2003b).

Although our intent was to provide a first approximation for ballast operations and associated propagule supply, it is likely that some of our assumptions, particularly those involving ballast water volume and content, are very coarse. For example, our assumption that vessels deballast only at the terminal port in the Great Lakes where they load cargo for the outbound journey may not be robust. BOB or NOBOB ships that discharge ballast en route to the terminal port could cause invasions in some of the lower lakes. Indeed, it has recently been reported that the sites of first discovery of NIS were concentrated around shallow, connecting channels in the Great Lakes, consistent with deballasting procedures that increase trim and improve maneuverability (Grigorovich et al. 2003a).

Mandatory ballast water exchange legislation covering the Great Lakes was implemented in 1993. This policy requires that all ships arriving from outside the EEZ (Exclusive Economic Zone) with freshwater exchange that water (or conduct an equally effective treatment) while on the open ocean in water not less than 2,000 meters deep and at least 320 kilometers from the nearest coastline (U.S. Coast Guard 1993). We assume that most freshwater organisms in the tanks would be purged, and the remaining ones killed when immersed in saline water. This procedure likely provides strong, but not absolute, protection of the Great Lakes from ballast-borne, freshwater invaders (Locke et al. 1993, MacIsaac et al. 2002). MacIsaac (1999) proposed that implementation of this policy should alter the pattern of invasions to the Great Lakes, with greater emphasis placed on invasions mediated by resting stages in ships' sediments, and less on ballast water itself. Resting stages are less likely to be purged with ballast water owing to their location in the bottom of the tanks, and less likely to be killed by saline ballast when the tanks are refilled. These resting stages could be expelled with ballast water in the Great Lakes, or later hatch when the tanks were filled with freshwater ballast. However, the relative importance of live organisms in ballast water, and of viable resting stages in ballast sediments, is only now being explored (Bailey et al. 2003). Even without considering resting stages in residual sediments, NOBOB vessels collectively appear to pose a greater risk of new invasions than BOB ships that comply with extant ballast water regulations (MacIsaac et al. 2002). Clearly, greater attention must be devoted to quantifying the volume and biological composition of ballast water delivered to each of the Great Lakes in order to provide a more rigorous test of the propagule pressure hypothesis.

Acknowledgments

H. J. M. is grateful to Drs. Greg Ruiz and Jim Carlton for the invitation to participate in the GISP pathways workshop. Claude Major, Al Ballert, Chris Wiley, and the U.S. Coast Guard kindly assisted with data acquisition and provided helpful comments. Comments from Greg Ruiz greatly improved the manuscript. We are grateful for financial support from the Great Lakes Fishery Commission to E. L. M., K. H., and H. J. M., the Natural Sciences and Engineering Research Council to H. J. M., and the Department of Fisheries and Oceans to A. J. N.

References

Aquatic Sciences, Inc. 1996. *Examination of aquatic nuisance species introductions to the Great Lakes through commercial shipping ballast water and assessment of control options. Phase I and Phase II.* ASI project E9225/E9285. St. Catharines, Ontario.

Bailey, S. A., I. C. Duggan, C. D. A. van Overdijk, P. T. Jenkins, and H. J. MacIsaac. 2003. Viability of invertebrate diapausing eggs collected from residual ballast sediment. *Limnology and Oceanography.* 48: 1701–1710.

Berg, D. J., D. W. Garton, H. J. MacIsaac, V. E. Panov, and I. V. Telesh. 2001. Changes in genetic structure of North American *Bythotrephes* populations following invasion from Lake Ladoga, Russia. *Freshwater Biology* 47: 275–282.

bij de Vaate, A., K. Jazdzewski, H. A. M. Ketelaars, S. Gollasch, and G. van der Velde. 2002. Geographical patterns in range extension of Ponto-Caspian macroinvertebrate species in Europe. *Canadian Journal of Fisheries and Aquatic Science* 59: 1159–1174.

Borbely, J. V. M. 2001. *Modelling the spread of the spiny waterflea* (Bythotrephes longimanus) *in inland lakes in Ontario using gravity models and GIS*. M. Sc. Thesis, University of Windsor, Windsor, Canada.

Bossenbroek, J. M., C. E. Kraft, and J. C. Nekola. 2001. Prediction of long-distance dispersal using gravity models: zebra mussel invasion of inland lakes. *Ecological Application* 11: 1778–1788.

Burke, M. J. W. and J. P. Grime. 1996. An experimental study of plant community invasibility. *Ecology* 77: 776–790.

Carlton, J. T. 1985. Transoceanic and interoceanic dispersal of coastal marine organisms: the biology of ballast water. *Oceanography and Marine Biology Annual Review* 23: 313–371.

Carlton, J. T. 1987. Patterns of transoceanic marine biological invasions in the Pacific Ocean. *Bulletin of Marine Science* 41: 452–465.

Carlton, J. T. 1996. Pattern, process, and prediction in marine invasion ecology. *Biological Conservation* 78: 97–106.

Carlton, J. T. and J. B. Geller. 1993. Ecological roulette: biological invasions and the global transport of nonindigenous marine organisms. *Science* 261: 72–82.

Cristescu, M. E. A., P. D. N. Hebert, J. D. S. Witt, H. J. MacIsaac, and I. A. Grigorovich. 2001. An invasion history of *Cercopagis pengoi* based on mitochondrial gene sequences. *Limnology and Oceanography* 46: 224–229.

Dick, J. T. A. and D. Platvoet. 2000. Invading predatory crustacean *Dikerogammarus villosus* eliminates both native and exotic species. *Proceedings of the Royal Society of London, Series B* 267: 977–983.

Donald, D. B., R. D. Vinebrooke, R. S. Anderson, J. Syrgiannis, and M. D. Graham. 2001. Recovery of zooplankton assemblages in mountain lakes from the effects of introduced sport fish. *Canadian Journal of Fisheries and Aquatic Science* 58: 1822–1830.

Drake, J. A. 1993. Community-assembly mechanics and the structure of an experimental species ensemble. *American Naturalist* 137: 1–26.

Eakins, N. 1995. *Ships on the Great Lakes in 1994*. Canadian Coast Guard.

Eakins, N. 1996. *Seaway lakers and salties 1995*. Canadian Coast Guard.

Eakins, N. 1997. *Seaway lakers and salties 1996*. Canadian Coast Guard.

Eakins, N. 1998. *Lakers and salties 1997–1998*. Canadian Coast Guard.

Eakins, N. 1999. *Lakers and salties 1998–1999*. Canadian Coast Guard.

Eakins, N. 2000. *Lakers and salties 1999–2000*. Canadian Coast Guard.

Eakins, N. 2001. *Salties 2000–2001*. Canadian Coast Guard.

Elton, C. S. 2000. *The ecology of invasions by animals and plants*. University of Chicago Press.

Grevstad, F. S. 1999. Experimental invasions using biological control introductions: the influence of release size on the chance of population establishment. *Biological Invasions* 1: 313–323.

Grigorovich, I. A., R. I. Colautti, E. L. Mills, K. H. Holeck, and H. J. MacIsaac. 2003a. Ballast-mediated animal introductions in the Laurentian Great Lakes: retrospective and prospective analyses. *Canadian Journal of Fisheries and Aquatic Science*. In review.

Grigorovich, I. A., A. V. Korniushin, D. K. Gray, I. C. Duggan, R. I. Colautti, and

H. J. MacIssac. 2003b. Lake Superior: an invasion coldspot? *Hydrobiologia*. In press.

Grigorovich, I. A., H. J. MacIssac, N. V. Shadrin, and E. L. Mills. 2002. Patterns and mechanisms of aquatic invertebrate introductions in the Ponto-Caspian region. *Canadian Journal of Fisheries and Aquatic Sciences* 60: 740–756.

Hall, S. R. and E. L. Mills. 2000. Exotic species in large lakes of the world. *Aquatic Ecosystem Health and Management* 3: 105–135.

Hamer, J. P., T. A. McCollin, and I. A. N. Lucas. 2000. Dinoflagellate cysts in ballast tank sediments: between tank variability. *Marine Pollution Bulletin* 40: 731–733.

Ivanov, V. P., A. M. Kamakin, V. B. Ushivtzev, T. Shiganova, O. Zhukova, N. Aladin, S. I. Wilson, G. R. Harbison, and H. J. Dumont. 2000. Invasion of the Caspian Sea by the comb jellyfish *Mnemiopsis leidyi* (Ctenophora). *Biological Invasions* 2: 259–264.

Johnson, L. E., A. Ricciardi, and J. T. Carlton. 2001. Overland dispersal of aquatic invasive species: a risk assessment of transient recreational boating. *Ecological Applications* 11: 1789–1799.

Kaufman, L. S. 1992. Catastrophic change in species-rich freshwater ecosystems: the lessons of Lake Victoria. *BioScience* 42: 846–857.

Ketelaars, H. A. M., F. E. Lambregts-van de Clundert, C. J. Carpentier, A. J. Wagenvoort, and W. Hoogenboezem. 1999. Ecological effects of the mass occurrence of the Ponto-Caspian invader, *Hemimysis anomala* G. O. Sars, 1907 (Crustacea: Mysidacea), in a freshwater storage reservoir in the Netherlands, with notes on its autecology and new records. *Hydrobiologia* 394: 233–248.

Kolar, C. S. and D. M. Lodge. 2001. Progress in invasion biology: predicting invaders. *Trends in Ecology and Evolution* 16: 199–204.

Kolar, C. S. and D. M. Lodge. 2002. Ecological predictions and risk assessment for alien fishes in North America. *Science* 298: 1233–1236.

Leppäkoski, E., S. Gollasch, P. Gruszka, H. Ojaveer, S. Olenin, and V. Panov. 2002. The Baltic—a sea of invaders. *Canadian Journal of Fisheries and Aquatic Science* 59: 1175–1188.

Levine, J. M. 2000. Species diversity and biological invasions: relating local process to community pattern. *Science* 288: 852–854.

Levine, J. M. and C. M. D'Antonio. 1999. Elton revisited: a review of evidence linking diversity and invasibility. *Oikos* 87: 15–26.

Litvak, M. K. and N. E. Mandrak. 2000. Baitfish trade as a vector of aquatic introductions. In *Nonindigenous freshwater organisms in North America*. R. Claudi and J. Leach, eds., Boca Raton: CRC Press LLC. pp. 163–180.

Locke, A., D. M. Reid, W. G. Sprules, J. T. Carlton, and H. van Leeuwen. 1991. Effectiveness of mid-ocean exchange in controlling freshwater and coastal zooplankton in ballast water. *Canadian Technical Report on Fisheries and Aquatic Science 1822.*

Locke, A., D. M. Reid, H. C. van Leeuwen, W. G. Sprules, and J. T. Carlton. 1993. Ballast water exchange as a means of controlling dispersal of freshwater organisms by ships. *Canadian Journal of Fisheries and Aquatic Science* 50: 2086–2093.

Lodge, D. M. 1993. Biological invasions: lessons for ecology. *Trends in Ecology and Evolution* 8: 133–137.

Lodge, D. M., C. A. Taylor, D. M. Holdich, and J. Skurdal. 2000. Nonindigenous crayfishes threaten North American freshwater biodiversity: lessons from Europe. *Fisheries* 25: 7–20.

Lonsdale, W. M. 1999. Global patterns of plant invasions and the concept of invasibility. *Ecology* 80: 1522–1536.

MacIsaac, H. J. 1999. Biological invasions in Lake Erie: past, present and future. In *State of Lake Erie: past, present and future,* M. Munawar and T. Edsall, eds., pp. 305–322. Leiden: Backhuys.

MacIsaac, H. J., I. A. Grigorovich, J. A. Hoyle, N. D. Yan, and V. E. Panov. 1999. Invasion of Lake Ontario by the Ponto-Caspian predatory cladoceran *Cercopagis pengoi. Canadian Journal of Fisheries and Aquatic Science* 56: 1–5.

MacIsaac, H. J., I. A. Grigorovich, and A. Ricciardi. 2001. Reassessment of species invasions concepts: the Great Lakes basin as a model. *Biological Invasions* 3: 405–416.

Mills, E. L., J. H. Leach, J. T. Carlton, and C. L. Secor. 1993. Exotic species in the Great Lakes: a history of biotic crises and anthropogenic introductions. *Journal of Great Lakes Research* 19: 1–54.

Oliver, J. D. 1993. A review of the biology of giant salvinia (*Salvinia molesta* Mitchell). *Journal of Aquatic Plant Management* 31: 227–231.

Panov, V. E. 1996. Establishment of the Baikalian endemic amphipod *Gmelinoides fasciatus* Stebb. in Lake Ladoga. *Hydrobiologia* 322: 187–192.

Pratt, D. M., W. H. Blust, and J. H. Selgeby. 1992. Ruffe, *Gymnocephalus cernuus:* newly introduced in North America. *Canadian Journal of Fisheries and Aquatic Science* 8: 1616–1618.

Rejmánek, M. 1996. A theory of seed plant invasiveness: the first sketch. *Biological Conservation* 78: 171–181.

Rejmánek, M. and D. M. Richardson. 1996. What attributes make some plant species more invasive? *Ecology* 77: 1655–1661.

Ricciardi, A. 2001. Facilitative interactions among aquatic invaders: is an invasional meltdown occurring in the Great Lakes? *Canadian Journal of Fisheries and Aquatic Science* 58: 2513–2525.

Ricciardi, A. and H. J. MacIsaac. 2000. Recent mass invasion of the North American Great Lakes by Ponto-Caspian species. *Trends in Ecology and Evolution* 15: 62–65.

Ruiz, G. M., J. T. Carlton, E. D. Grosholz, and A. H. Hines. 1997. Global invasions of marine and estuarine habitats by non-indigenous species: mechanisms, extent, and consequences. *American Zoologist* 37: 621–632.

Ruiz, G. M., P. W. Fofonoff, J. T. Carlton, M. J. Wonham, and A. H. Hines. 2000a. Invasion of coastal marine communities in North America: apparent patterns, processes, and biases. *Annual Review of Ecological Systematics* 31: 481–531.

Ruiz, G. M., T. K. Rollins, F. C. Dobbs, L. A. Drake, T. Mullady, A. Huq, and R. R. Colwell. 2000b. Global spread of microorganisms by ships. *Nature* 408: 49–50.

Sala, O. E., F. S. Chapin III, J. Armesto, E. Berlow, J. Bloomfield, R. Dirzo, E. Huber-Sanwald, L. F. Huenneke, R. B. Jackson, A. Kinzig, R. Leemans, D. M. Lodge, H. A. Mooney, M. Oesterheld, N. L. Poff, M. T. Sykes, B. H. Walker, M.

Walker, and D. H. Wall. 2000. Global biodiversity scenarios for the year 2100. *Science* 287: 1770–1774.

Schindler, D. E., R. A. Knapp, and P. R. Leavitt. 2001. Alteration of nutrient cycles and algal production resulting from fish introductions into mountain lakes. *Ecosystems* 4: 308–321.

Sher, A. A. and L. A. Hyatt. 1999. The disturbed resource-flux invasion matrix: a new framework for patterns of plan invasion. *Biological Invasions* 1: 107–114.

Shurin, J. B. 2000. Dispersal limitation, invasion resistance, and the structure of pond zooplankton communities. *Ecology* 81: 3074–3086.

Simberloff, D. and B. von Holle. 1999. Positive interactions of nonindigenous species: invasional meltdown? *Biological Invasions* 1: 21–32.

Smith, L. D., M. J. Wonham, L. D. McCann, G. M. Ruiz, A. H. Hines, and J. T. Carlton. 1999. Invasion pressure to a ballast-flooded estuary and an assessment of inoculant survival. *Biological Invasions* 1: 67–87.

Spencer, C. N., D. S. Potter, R. T. Bukantis, and J. A. Stanford. 1999. Impact of predation by *Mysis relicta* on zooplankton in Flathead Lake, Montana, USA. *Journal of Plankton Research* 21: 51–64.

Stohlgren, T. J., D. Binkley, G. W. Chong, M. A. Kalkhan, L. D. Schell, K. A. Bull, Y. Otsuki, G. Newman, M. Bashkin, and Y. Son. 1999. Exotic plant species invade hot spots of native plant diversity. *Ecological Monograph* 69: 25–46.

United States Coast Guard. 1993. *Ballast water management for vessels entering the Great Lakes*. Code of Federal Regulations 33-CFR Part 151.1510.

Vanderploeg, H. A., T. F. Nalepa, D. J. Jude, E. L. Mills, K. Holeck, J. R. Liebig, I. A. Grigorovich, and H. Ojaveer. 2002. Dispersal and emerging ecological impacts of Ponto-Caspian species in the Laurentian Great Lakes. *Canadian Journal of Fisheries and Aquatic Science* 59: 1209–1228.

Vander Zanden, M. J., J. M. Casselman, and J. B. Rasmussen. 1999. Stable isotope evidence for food web shifts following species invasions of lakes. *Nature* 401: 464–467.

Williamson, M. 1996. *Biological Invasions*. London: Chapman and Hall.

Zaret, T. M. and R. T. Paine. 1973. Species introductions in a tropical lake. *Science* 182: 449–455.

Part II

INVASION MANAGEMENT AND POLICY

Chapter 10

An Australian Perspective on the Management of Pathways for Invasive Species

Paul Pheloung

Australia has a unique native flora and fauna dominated by species of plants and animals not naturally present elsewhere in the world, including the majority of marsupials and the eucalyptus trees. Australian ecosystems can be richly diverse, but many are fragile and often subject to harsh conditions. The climate is hot and dry over most of the country, and soils are nutritionally poor (Reid 1990).

As a country entirely bound by water, Australia has been largely isolated from genetic exchanges by natural means. Following the arrival of Europeans in the eighteenth century, however, a huge number of exotic organisms were introduced deliberately or unintentionally. While some of these have been beneficial and contribute to the quality of Australian life, many have naturalized and become pests of production and/or the natural environment. Most of temperate Australia, where rainfall is sufficient to support some form of European-style primary production, is dominated by exotic plants and animals; only a small proportion of remnant areas remain as nature reserves and national parks. Approximately 20,000 species of plants are native to Australia (Hnatiuk 1990). More than 2,000 exotic plant species have been recorded as naturalized (Groves 1998).

Nevertheless, Australia remains free of many potentially invasive species, but strict quarantine protocols are necessary to preserve this status (Nairn et al. 1996).

This chapter discusses current policy and procedures to manage these risks from the perspective of the Department of Agriculture, Fisheries, and Forestry—Australia (AFFA).

LEGISLATIVE FRAMEWORK

The Quarantine Act of 1908 and the Quarantine Proclamation of 1998 (Anonymous 2002) are the principle legislative tools for preventing the introduction, establishment, or spread of diseases or pests affecting people, animals, or plants. These instruments allow quarantine officers to (1) deal with quarantine matters; (2) set out the legal basis for controlling imports of goods, animals, and plants, including the things a director of quarantine must take into account when deciding whether to grant a permit for importation into Australia; and (3) determine the penalties for noncompliance with the act.

The scope of quarantine should include measures for the examination, exclusion, detention, observation, segregation, isolation, protection, treatment, and regulation of vessels, installations, human beings, animals, plants, or other goods to prevent or control the introduction, establishment, or spread of diseases or pests that could cause significant damage.

The Environment Protection and Biodiversity Conservation (Wildlife Protection) Act of 2001 (Environment Australia 2001) provides legislative mechanisms to protect and conserve Australian flora and fauna, including preventing the introduction into Australia of potentially invasive species. The agencies administering these acts (AFFA and Environment Australia [EA], respectively) collaborate to ensure that pathways for the entry of pests and diseases, including invasive species, are managed appropriately.

THE ROLE OF AFFA

Within AFFA, three groups have specific responsibility for managing sanitary and phytosanitary (SPS) risks, including those presented by invasive species: the Australian Quarantine and Inspection Service (AQIS), Biosecurity Australia (BA), and the Offices of the Chief Plant Protection Officer and the Chief Veterinary Officer (OCPPO and OCVO). AQIS is responsible for quarantine operations, and BA is responsible for quarantine policy. The OCPPO and OCVO are responsible for post-border response and preparedness, including non-SPS aspects of invasive species.

AFFA aims to achieve two outcomes of relevance to this discussion through the services it provides: (1) protection of Australia's animal, plant, and human health and the environment through the application of SPS measures; and (2) improved market access opportunities for Australian food and other agricultural products. Although AFFA has traditionally focused on protecting the health of economically important animals and plants, the first outcome clearly identifies protection of the environment as part of the overall scope.

TERMINOLOGY AND STANDARDS

I adopt here the following definitions contained within the International Standards for Phytosanitary Measures (ISPM) published by the Food and Agriculture Organization of the United Nations (FAO 1999a):

- *Pest.* Any species, strain, or biotype of plant, animal, or pathogenic agent injurious to *animals, animal products,* plants, or plant products. (Note: The ISPMs were developed for plant health standards, but I have broadened them here to cover animal health as well.)
- *Quarantine pest.* A pest of potential economic importance to the area endangered thereby and not yet present there, or present but not widely distributed and being officially controlled.
- *Pathway.* Any means that allows the entry or spread of a pest.
- *Ecosystem.* A complex of organisms and their environment, interacting as a defined ecological unit (natural or modified by human activity, e.g., agro-ecosystem), irrespective of political boundaries.

The Convention on Biological Diversity (CBD) Subsidiary Body on Scientific, Technical, and Technological Advice (SBSTTA) has identified the need for but has yet to develop standard terminology on alien species (CBD 2002). Under guiding principles for the prevention, introduction, and mitigation of impacts of alien species that threaten ecosystems, habitats, or species, SBSTTA has adopted interim definitions:

- *Alien species.* A species, subspecies, or lower taxon introduced outside its normal past or present normal distribution; includes any part (e.g., gametes, seeds, eggs, or propagules) of such species that might survive and subsequently reproduce.
- *Invasive alien species.* An alien species whose establishment and spread threaten ecosystems, habitats, or species with economic or environmental harm.

The definition of an invasive alien species is closely allied to, but not the same as, that of a quarantine pest. The exceptions are the species that have become well established in the country and for whom continued quarantine would be of little value. Because quarantine pests are the par-

ticular focus of AFFA activities, the term quarantine pest is used throughout this chapter.

The current International Plant Protection Convention (IPPC) Pest
Risk Analysis standard (ISPM 11, FAO 2001) clearly identifies environmental damage as an impact to be mitigated. In joint consultations
between IPPC and CBD, there was agreement that the scope of the IPPC
applies to the protection of wild flora and thus makes an important contribution to the conservation of biological diversity. An IPPC expert working
group on risk analysis for environmental impacts of plant pests was established to further document principles and methodology as an annex to
ISPM 11 (FAO 2002a).

Pathways

For the purpose of this chapter, pathway can be considered the combination of point of origin, vector (such as a commodity), and mode of entry.
Table 10.1 is a matrix of potential vectors of quarantine pests and modes
of entry. Although not intended to identify all pathways, the table indicates vector/entry mode combinations that represent a significant risk and
for which quarantine measures have been established.

Point of Origin

The origin of a trade activity is an important consideration that can influence the required quarantine risk management because of the presence or
absence of pests at the origin. For example, imports of oranges from the
United States are currently allowed from California, Arizona, and Texas
(subject to conditions) but not from Florida (AQIS 2002b). This is principally because of the presence of citrus canker in Florida and the absence of
an agreed mechanism to manage the risk of introducing this disease with
oranges imported from Florida.

Mode of Entry

This refers to how a pest crosses the international border into Australia—through regulated channels at designated ports of entry (seaports
and airports) and international mail. These modes of entry are the focal
point for border inspections and associated risk management actions
such as fumigation or post-entry quarantine. Along with the Australian
Customs Service, the Department of Immigration and Multicultural
Affairs, and Australia Post, AQIS maintains a presence at international

TABLE 1O.1. Likely pathways for the entry of invasive species to Australia ("Y" denotes a likely pathway).

Vector (risk carrier)	Mail	Seaports	Airports	Natural Movement (wind or water assisted)	Illegal Activities (Fishing, smuggling, and illegal immigrants)	Extraordinary Activities (East Timor, Sydney Olympic Games)
			Mode of Entry			
Horticultural products						
Fresh fruit and vegetables	Y	Y	Y		Y	
Processed plant products	Y	Y	Y		Y	
Reproductive plant material*						
Grain	Y	Y	Y			
Seed		Y	Y	Y	Y	Y
Nursery stock (bulbs, corms, tubers, stem and root cuttings)		Y	Y	Y		
Pathogen spores				Y		
Forest products						
Bulk timber		Y				
Wood packing	Y	Y	Y		Y	Y
Wooden articles (e.g., furniture and carvings)	Y	Y	Y		Y	Y
Live animals*						
Horses		Y	Y			Y
Pets		Y	Y		Y	

continues

TABLE 1O.1. Continued

Vector (risk carrier)	Mode of Entry					
	Mail	Seaports	Airports	Natural Movement (wind or water assisted)	Illegal Activities (Fishing, smuggling, and illegal immigrants)	Extraordinary Activities (East Timor, Sydney Olympic Games)
Protected wildlife (e.g., smuggled reptiles and birds)	Y	Y			Y	
Feral animals		Y			Y	
Migrating birds				Y		
Flying insects		Y		Y		
Animal products						
Fresh meat and fish		Y	Y		Y	
Cooked and processed products	Y	Y	Y		Y	
Other (Hitchhikers)						
Ballast water, hull fouling		Y				
Machinery		Y	Y			Y
Shipping containers		Y				
Cabins, holds, and infrastructure		Y	Y			Y
Passengers and crew		Y	Y			Y

* These may be hosts for pests (e.g., disease-carrying birds), pests in their own right (e.g., weeds and rodents), or both (e.g., exotic bees that carry parasitic mites).

airports and seaports. AQIS and customs officers are also present at mail exchanges.

Quarantine pests can also be introduced as a consequence of activities or events occurring outside of designated ports of entry, such as illegal fishing, smuggling, asylum seekers, and natural means (e.g., windborne spores). Surveillance is the primary tool for managing such risks. The Northern Australia Quarantine Strategy (NAQS), discussed later in this chapter, is an AQIS program established for this purpose.

Vectors

The vectors for quarantine pests are diverse and do not necessarily constitute living or organic items. The pest may be present as an infection, infestation, or contaminant of a commodity (the commodity may be a quarantine pest in its own right as a weed or feral animal) or have a nonspecific association with trade carriers such as shipping containers or vessels. This last category, which includes ants, bees, and termites on shipping containers, Asian gypsy moth egg masses on ship structures, and marine organisms in ballast water, is arguably the greatest and most difficult of all risks to manage.

INTERVENTION MECHANISMS

Nairn et al. (1996) comprehensively reviewed the provision of quarantine services in Australia. The following themes were highlighted by the review, and endorsed by the Australian government:

- *Managed risk based on science.* Accepting that zero risk is an unachievable quarantine objective, scientifically sound steps should be taken to reduce quarantine risks to an acceptably low level. Determining what constitutes an acceptable or "Appropriate Level of Protection" for Australia is a difficult and contentious problem (Gascoine 2001).
- *A partnership approach.* Consultation among governments, industry, and the general public.
- *Increased consideration of the effects of the introduction of new species on Australia's unique natural environment.*
- *The continuum of quarantine.* Consisting of pre-border, border, and post-border components.

These themes are reflected in the discussion contained in the section "Risk

Identification and Management." The last theme, as it relates to pathways for invasive species, warrants some further discussion here.

Continuum of Quarantine

The continuum of quarantine refers to the following objectives:

1. Implementing measures outside of Australia (pre-border) to reduce the threat of entry of quarantine pests.
2. Using well-targeted border controls.
3. Having the capacity for early detection of pest incursions.
4. Having emergency responses (incursion management plans) to contain, control, and eradicate pest incursions.

The Australian government endorsed these objectives and provided additional resources specifically for this purpose (AFFA 1997a).

Border controls, the central element of the continuum, are managed by AQIS and include visual inspection, X-ray technology, and trained dogs to detect quarantine material. The recent outbreak of foot-and-mouth disease (FMD) in the United Kingdom led to the destruction of 10 million livestock. This, and the perceived human health concerns associated with mad cow disease (bovine spongiform encephalitis, or BSE) outbreaks in the European Union, had enormous trade implications and caused a significant reduction in demand for meat products in many parts of the world. In part as a result of these events, Australia has substantially increased resource allocations for quarantine border operations to raise the level of intervention on all types of traffic: travelers, air and sea cargo, and mail (AFFA 2002b). In the case of international mail, 100 percent intervention is now the practice, rather than sampling a portion of the total traffic based on risk profiles. This means that every item of mail is subject to visual inspection, brought within the detection range of a trained sniffer dog, or passed through X-ray machinery. AQIS is now seeking to ensure, through technical refinements and training, that these tools are effective.

Quarantine requirements to manage the risk of invasive species entering Australia are comprehensive and well documented (AQIS 2002a). The Import Conditions Database, called ICON (AQIS 2002b), is a searchable, online manual of these requirements. For example, when a commodity or country–commodity combination is specified, a complete description is given of the requirements necessary before the commodity will be allowed entry into Australia.

AQIS and Biosecurity Australia contribute to pre-border risk management through arrangements put in place by exporting countries to manage quarantine pests as part of the sanitary or phytosanitary requirements attached to imports of particular commodities. These include disinfestation treatments and pest management in the production systems or systematic surveillance to confirm that pests of concern are not present in the production area. Several international standards provide a basis for Australia's offshore management of quarantine risks (FAO 1996, 1997, 1999b, 2002b).

The Office of the Chief Plant Protection Officer (OCPPO) and Office of the Chief Veterinary Officer (OCVO) have key roles in the pre-border and post-border components of the quarantine continuum. The OCPPO was established to coordinate national responses to incursions of plant pests and to develop linkages and networks for information exchange both nationally and internationally (AFFA 2000b). The OCVO, established earlier than the OCPPO, deals with the threat of animal diseases.

The Northern Australia Quarantine Strategy (discussed further in the next section) is an AQIS program that typifies the theme of the continuum of quarantine. The program has a pre-border component that focuses on the identification and management of quarantine risks from Australia's nearest neighbors, Indonesia and Papua New Guinea, in collaboration with those countries, and a post-border component to provide early warning of incursions and raise public awareness of the risks (AFFA 1998).

Risk Identification and Management

The approaches taken by AFFA to manage the range of risks presented by movements of goods and people can be illustrated by the following examples. These examples reflect my familiarity with those areas, but I emphasize that many other programs manage a wide range of other complex areas of risk.

1. Adopting a prescriptive process to analyze and manage the risks associated with imports of commodities such as fruits and vegetables.
2. Managing the risk of pest introductions to northern Australia.
3. Managing risks associated with the importation of viable plant material.
4. Managing risks associated with the importation of wood products.

Key elements of risk management for examples 2, 3, and 4 are listed in Table 10.2.

TABLE 1O.2. Overview of effectiveness of the management of three classes of pathways for quarantine pests.

	Northern Australia Quarantine Strategy	Introduction of Potential New Weeds	Forest Product Imports
Intervention Mechanism	Surveillance and monitoring in NAQS region; Inspection of travelers and vessels	Weed Risk Assessment of applications to import new plants; Targeted sampling of mail traffic based on risk profiles	Inspection-based intervention system in place with defined clearance and failure criteria; Mandatory treatment on material from high-risk sources; Target arthropod pests
Intervention Data	NAQS Database—Detailed record of surveillance and monitoring	Evolving list of approved/permitted plants, and biological information upon which decisions are based; Postal interceptions and seizures (seeds)	Pest and Disease Information (PDI) Database—recording interceptions by quarantine officers (mainly arthropod interceptions)
Management Limitations	Very large, remote, and sparsely populated areas to be monitored; Natural pathways for pest introduction cannot be managed	Reliability of desktop prediction of weed potential; Community understanding of potential risk and quarantine requirements; Volume of postal traffic; Seeds are easily ordered, particularly over Internet, and difficult to detect if undeclared	Very large trade volume; Wooden packing materials are associated with a large proportion of all trade; Direct inspections on less than 5% of trade; Reliability of documented certification of phytosanitary status of product

Management Strengths	Clearly defined strategy and focus; Experienced specialist scientific staff; Comprehensive public awareness program	Well-defined system in place to screen all new plant introductions; Resources dedicated to optimizing mail inspection; Public awareness initiatives to encourage voluntary declaration of plant introductions	Range of approved treatments documented; Experienced inspectors; Rigorous requirements where a risk has been identified
Management Efficacy	Public awareness program is effective, quarantine in northern Australian is well understood and supported; Early warning provided for a number of recent incursions	AQIS has opportunity to assess risk of most proposed new plant introductions; Increasing proportion of referred mail items seized— i.e., targeting of high-risk categories of mail is effective	Interception data indicates a substantial risk is currently borne; Offshore risk management protocols implemented by exporters evaluated by inspection; Formal IRAs commenced to review current protocols

Import Risk Analysis (IRA)

Biosecurity Australia (BA) has developed and implemented an Import Risk Analysis (IRA) process (AFFA 1997b, 2001a, 2002c) that is highly consultative, involving all those affected by a type of trade, and consistent with the Agreement on Sanitary and Phytosanitary Measures (SPS Agreement) ratified by the World Trade Organization (WTO). An IRA includes the technical pest risk analysis, undertaken within a prescribed process that ensures that the principles of the SPS Agreement are adhered to while also ensuring that Australia's conservative approach to the management of quarantine risk continues.

The SPS Agreement identifies international standard-setting organizations (the Codex Alimentarius Commission, the International Office of Epizootics, and the International Plant Protection Convention) that provide the technical framework under which the identification and management of quarantine risk are assessed.

Import Risk Analyses are generally initiated by Biosecurity Australia in response to a request to allow or modify current requirements for a type of trade, such as bulk feed maize from the United States or salmon products from Canada.

The IRA on bulk maize importation from the United States was initiated by the Australian feedlot industry because of a desire to ensure security of supply when domestic stocks become limited, for example, by drought. This IRA has been finalized and the recommendation is that the pest and disease risks are too great to allow entry of U.S. maize into Australia that has not been treated to fully devitalize the grain and any associated organisms (AFFA 2003).

The salmon IRA is an example of a request by Canada for access to Australian markets for its salmon products in accordance with WTO/SPS guidelines. The IRA established quarantine conditions under which Canadian fresh salmon products (and fresh salmon products from other countries) could be safely imported to Australia (AFFA 1999).

AQIS may also initiate an IRA to review existing requirements for trade if there are indications that the existing requirements are inadequate or inappropriate. Bulk imports of coniferous timber from Canada, the United States, and New Zealand are a current example that I discuss in more detail later.

The IRA process includes identification of quarantine pests associated with each pathway, and the SPS measures required, if any, to manage the risk of introduction of these pests. An assessment of the environmental impact is also done, in consultation with the Australian Commonwealth agency responsible for protection of the natural environment, Environment Australia (EA 2002).

The Market Access and Biosecurity Web pages are within the AFFA Website (AFFA 2002a). IRA documents are available on this site.

The Northern Australia Quarantine Strategy (NAQS)

The Northern Australia Quarantine Strategy (NAQS) was established in 1989 to address the quarantine risks from Australia's nearest neighbors, Indonesia and Papua New Guinea. The program typifies the theme of the continuum of quarantine: collaborative surveillance and monitoring activities in Indonesia and Papua New Guinea, monitoring and inspection of cross-border activities along the northern coastline, and surveillance and monitoring within northern Australia.

NAQS pathways include natural movement by wind or water, illegal fishing vessels and refugees, traditional trading between indigenous communities in the Torres Strait and Papua New Guinea, smuggling, and extraordinary activities such as the large cargo shipments arising from Australian military and aid activity in East Timor (Table 10.1). Figure 10.1 shows the NAQS region and surveillance activity within this region.

To provide some focus for the surveillance and monitoring program, target lists were developed for invertebrate pests, pathogens, and weeds. These lists are for pests likely to enter Australia from its northern neighbors and are periodically reviewed. These target lists can be viewed in the AQIS/NAQS pages of the AFFA Web site.

NAQS scientists have provided early warning and contributed to follow-up activities to ensure that detected pests do not become established and spread. They have been responsible for the early detection of exotic fruit flies, Asian honeybees (as carriers of parasitic mites), Japanese encephalitis, and Siam weeds. Surveys in Timor, West Papua, and Papua New Guinea have provided knowledge of the spread and proximity of quarantine pests. These species include animal pathogens such as surra and classical swine fever, plant pathogens such as the causal agent of Panama disease in bananas, exotic fruit flies such as papaya fruit fly, and weeds such as witchweeds.

Because of the social unrest associated with the move to independence in East Timor, Australia took a lead role in the provision of a peacekeeping presence in that country. As a consequence there were substantial movements of military and aid personnel, vehicles, and machinery between Australia and East Timor. Timor contains many pests of serious quarantine concern to Australia (e.g., Siam weed, papaya fruit fly, classical swine fever,

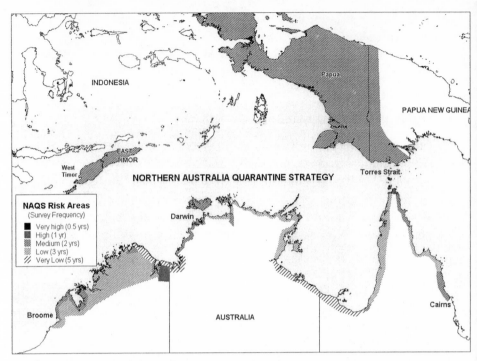

FIGURE 10.1. NAQS monitoring and surveillance in Australia, Indonesia, Papua New Guinea, and East Timor. Surveys are conducted at least annually in the Torres Strait Islands at the tip of Cape York.

and Japanese encephalitis), necessitating an evaluation of the risks and implementation of emergency measures to manage these risks. These measures include provision of quarantine staff in East Timor and additional staff in Darwin, and training of military personnel on the risks and inspection procedures. Risk management, again, embodies the principle of the continuum of quarantine—inspection and cleaning of all clothing and equipment before their departure from East Timor, follow-up inspection and treatment on arrival in Australia, and ongoing surveillance within Australia at destination points for military machinery to provide early warning of any quarantine pests that do escape detection during border operations (AFFA 2000a). East Timor also faces risks from the movement of goods and people from Australia. Following the establishment of East Timor as a sovereign nation, AQIS has assisted the country in establishing a quarantine service and legislation to manage the risk of pest and disease introductions (AQIS, pers. comm. 2003).

Screening and Monitoring the Importation of Plant Material

Nonindigenous plant species have been brought into Australia continuously, beginning with the first European colonization (Groves 1998). The demand to import new plants continues; in a global environment of free trade and easier access to novel plants, this demand is likely to continue to grow. Many of these plants have the potential to become weeds in Australia (Nairn et al. 1996, Groves 1998).

The government's traditional approach was to maintain a list of prohibited exotic plants that are potential weeds in Australia. This approach was limited by the difficulty of identifying all potential weeds proactively. Australia has recently adopted the alternative approach of implementing a list of permitted plant species that are not quarantine pests. This list is necessarily very large (currently >5,000 taxa) to accommodate native plants as well as established exotic plants. The list is dynamic, and additions are made as new plants proposed for introduction to Australia are assessed as low weed risks. AQIS uses a three-tier system to administer this list and screen new plant introductions (Fig. 10.2) (Pheloung 2001, Walton 2001). Central to this system is a weed risk assessment system (WRA); a set of predominantly yes/no questions on the history of weediness, climate preferences, undesirable attributes, and biology of a plant. The system generates a score that indicates the potential weed risk.

Figure 10.3 shows a breakdown of WRA outcomes processed during 1998–99. Since 1997, 1,136 applications to import new plant species were assessed. The proportion of species rejected has remained close to 20 percent, but the proportion accepted decreased from about 60 percent in 1997 to 40 percent in 2001, with a corresponding increase in the proportion that required more information from the applicant or further evaluation before a final decision could be made (David Porritt, employee of BA, pers. comm. 2002). This change reflects the growing proportion of applications to import taxa that are poorly described or documented. Further, evaluation is required when a clear decision could not be made based on answers to the questions. In those cases, the species are prohibited unless the importer is prepared to meet the cost of a more detailed three-tier assessment involving experimentation.

New species may also enter the country undeclared. Seeds entering Australia through the postal system are of particular concern. Mail-order catalogs of seeds, readily found on the Internet, contain lists of many species that are potential weeds in Australia. Figure 10.4 shows mail items seized because of quarantine breaches since 1998. Between 5 and 10 per-

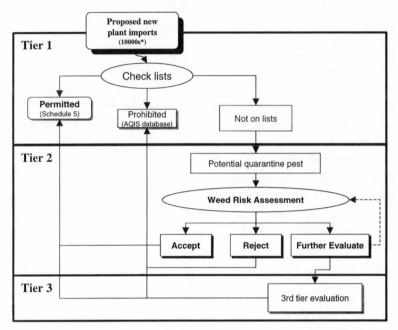

FIGURE 10.2. Three-tiered system for screening proposed plant imports for weediness. The "Not on lists" criteria for proceeding to Tier 2 includes determining that the species is not present in Australia. If the species is present in Australia, and is not the subject of an official program of eradication or containment, it is added to the permitted list.

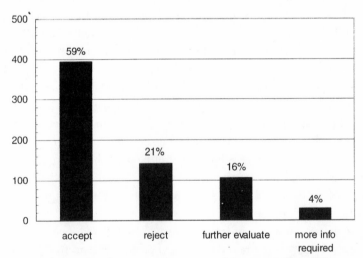

FIGURE 10.3. Weed Risk Assessments processed during 1998–99 (Biosecurity Australia, pers. comm. 2001).

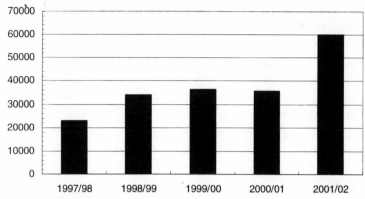

FIGURE 10.4. Mail items seized by quarantine inspectors each financial year since 1997 (AQIS, pers. comm. 2002).

cent of these seizures are seeds. During the 2001–02 financial year the move toward 100 percent intervention resulted in 60,000 seizures, an increase over the previous year of more than 24,000. More than 6,000 of these seizures were seeds. The volume of mail prevents opening and inspecting all traffic and emphasizes the importance of increased public awareness and voluntary compliance. That message, coupled with the deterrence of heavy penalties, is relatively easy to convey at ports of entry. Print and electronic media are needed to reach people ordering seed by mail. AQIS is attempting to use the Internet to convey this message, but has yet to devise an effective method of ensuring that anyone attempting to place such an order is made aware of the legal requirements (1) to ensure the plant is a permitted import, and (2) that the package is properly labeled so that it can be identified and inspected by quarantine officers to confirm the product is as described and free of contaminants.

Contamination of clothing, machinery, and commodities are other areas of risk that are targeted by inspection at airports, mail centers, and seaports.

Managing the Importation of Bulk Timber and Wood Packaging Associated with Trade

Australia imports forest products in many forms and has done so for many years. The types of products are diverse but can be broadly categorized into bulk timber in the form of sawn timber and logs (which must be milled to 200 mm or less in at least one cross section dimension and pass an inspection before release from quarantine), products such as furniture and particle boards that contain processed timber, and wood packing materials associated with most types of trade (AFFA 2001b).

Bulk timber and wood packing are the major concerns, evident from the large number of pest interceptions recorded by quarantine officers inspecting these shipments on arrival. A total of 836 interceptions of insect pests were recorded on sawn coniferous timber from New Zealand, Canada, and the United States from July 1999 to June 2001 on an interceptions database maintained by Biosecurity Australia (John Caling, employee of BA, pers. comm. 2001). AQIS has established a protocol to manage these risks that is primarily based on inspection and treatment where inspection indicates the need. Timber must be free of bark, and logs must remain on quarantine-approved premises and be regularly inspected until they are milled so that at least one dimension does not exceed 200 mm. The protocol is fully described on the AQIS ICON database <www.aqis.gov.au/icon>.

The Quarantine Review Committee (Nairn et al. 1996) noted that the risks might not be adequately managed under the existing arrangements. The concerns included the following:

• The large number of potential quarantine pests that are intercepted on bulk timber and wood packing.
• 100 percent inspection could not be achieved; for bulk timber, 5 percent or less of the product (external surfaces of bundles, for example) were inspected.
• Inspection and treatment regimes were directed to and effective on only invertebrate pests; pathogen risks were not being addressed.

The approach taken by AFFA to address these concerns has both operational and policy components.

The operational components are intended to improve effectiveness under current protocols. Pesticide timber treatments, by overseas suppliers, and the protocols for applying these treatments in a manner acceptable to AQIS are being reviewed. AQIS is also encouraging overseas suppliers to develop quality assurance arrangements in exchange for quicker and less costly clearance from quarantine on arrival in Australia. These approaches seek to have the risks dealt with offshore before the product arrives.

Biosecurity Australia has commenced an Import Risk Analysis to examine in detail the phytosanitary risks associated with sawn coniferous timber from the United States and New Zealand (AFFA 2001b). The IRA was prompted because of the number of pest interceptions on these categories of trade and the increasing concern shown by other countries, including New Zealand, Canada, and the United States, about the phytosanitary risks associated with imported timber and wood packing. The IRA will examine pathogen risks in detail. The finalized IRA will determine the need for changes to the quarantine measures required by Australia.

CONCLUSIONS AND FUTURE DIRECTIONS

Australia is conscious of its unique status and relative freedom from pests. Growth in trade and tourism activity has been complemented by increased vigilance against the introduction of undesirable exotic organisms, a vigilance that is scientifically based on sound and careful examination of the risks.

Revised legislation and the substantial resources allocated to border intervention, coupled with rigorous import risk analysis of pathways, should go a long way toward preventing the introduction of new invasive species.

Nevertheless, no amount of activity will provide absolute protection against such invasions. The post-border phase of the continuum of quarantine will continue to benefit from improvements on the systems currently in place:

- Preparedness (pest target lists, incursion management plans)
- Early warning (surveillance and monitoring, public awareness)
- Response capability (streamlined resource allocation and mobilization for containment and eradication of pest incursions, diagnostic capability)

AQIS's NAQS program has gone a long way toward providing an early warning system in northern Australia. Extending this early warning capability to the rest of Australia and improving response capability and effectiveness are the tasks ahead.

ACKNOWLEDGMENTS

My sincere thanks to all those in AQIS, Biosecurity Australia, OCPPO, and other parts of AFFA who provided information and comment on drafts of the manuscript.

REFERENCES

Agriculture, Fisheries, and Forestry—Australia (AFFA). 1997a. *Australian Quarantine—A Shared Responsibility: The Government Response.* Canberra:

Commonwealth of Australia. Available from <www.affa.gov.au> via Quarantine and Export Services > About AQIS > Quarantine Reports.

AFFA. 1997b. *The AQIS Import Risk Analysis Process Handbook.* Australian Quarantine and Inspection Service. Australia: Agriculture, Fisheries, and Forestry. Available from <www.affa.gov.au/content/publications.cfm> under Biosecurity Australia.

AFFA. 1998. *QEAC report: Review of the Northern Australian Quarantine Strategy.* Australian Quarantine and Inspection Service. Available from <www.affa.gov.au/content/publications.cfm>.

AFFA. 1999. *Finalised Import Risk Analysis: Salmon.* Biosecurity Australia. Available from <www.affa.gov.au/content/publications.cfm>.

AFFA. 2000a. *Evaluation of the Quarantine Risks Associated with Military and Humanitarian Movements between East Timor and Australia.* Australian Quarantine and Inspection Service. Available from <www.affa.gov.au/content/publications.cfm>.

AFFA. 2000b. *OCPPO Functional Review 2000.* Animal and Plant Health. Available from <www.affa.gov.au/content/publications.cfm>.

AFFA. 2001a. *Draft Guidelines for Import Risk Analysis.* Biosecurity Australia. Available from <www.affa.gov.au/docs/market_access/biosecurity/index.html>.

AFFA. 2001b. Sawn coniferous timber from Canada, NZ and the USA - Issues paper. Biosecurity Australia. Available from <www.affa.gov.au/content/publications.cfm>.

AFFA. 2002a. Agriculture, Fisheries, and Forestry—Australia. Available from <www.affa.gov.au>.

AFFA. 2002b. *AQIS Report to Clients.* Australian Quarantine and Inspection Service. Available from <www.affa.gov.au/content/publications.cfm>.

AFFA. 2002c. *Draft Administrative Framework for Import Risk Analysis—A Handbook.* Biosecurity Australia. Available from <www.affa.gov.au/docs/market_access/biosecurity/index.html>.

AFFA. 2003. *Finalized Import Risk Analysis—Maize from the USA.* Biosecurity Australia. Available from <www.affa.gov.au/docs/market_access/biosecurity/index.html>.

Anonymous. 2002. *Quarantine Act of 1908.* Available from <http://scaletext.law.gov.au/html/pasteact/0/71/top.htm>.

Australian Quarantine and Inspection Service (AQIS). 2002a. <www.aqis.gov.au>.

AQIS. 2002b. *Import Conditions Database, ICON.* Available from <www.aqis.gov.au/icon/asp/ex_querycontent.asp>.

Convention on Biological Diversity (CBD). 2002. Report of the Sixth Meeting of the Subsidiary Body on Scientific, Technical and Technological Advice. Conference of the Parties to the Convention on Biological Diversity. Sixth meeting, The Hague. Available from <www.biodiv.org/convention/sbstta.asp>.

Environment Australia (EA). 2001. *Environment Protection and Biodiversity Conservation Act.* Available from <http://www.ea.gov.au/epbc/index.html>.

Environment Australia (EA). 2002. <www.ea.gov.au>.

Food and Agriculture Organization of the United Nations (FAO). 1996. *Require-*

ments for the Establishment of Pest-free Areas. ISPM Publication No. 4. Available from <www.ippc.int/IPP/En/ispm.htm>.

FAO. 1997. *Guidelines for surveillance. ISPM Publication No. 6.* Available from <www.ippc.int/IPP/En/ispm.htm.>

FAO. 1999a. *Glossary of Phytosanitary Terms, International Standards for Phytosanitary Measures. ISPM Publication No. 5.* Available from <www.ippc.int/IPP/En/ispm.htm>.

FAO. 1999b. *Requirements for the Establishment of Pest-free Places of Production and Pest-free Production Sites. ISPM Publication No. 10.* Available from <www.ippc.int/IPP/En/ispm.htm>.

FAO. 2001. *Pest Risk Analysis for Quarantine Pests. ISPM Publication No. 11.* Available from <www.ippc.int/IPP/En/ispm.htm>.

FAO. 2002a. *Analysis of Environmental Risks. Supplement to ISPM Pub. No. 11.* Contact ippc@fao.org.

FAO. 2002b. *The Use of Integrated Measures in a Systems Approach for Pest Risk Management. ISPM Publication No. 14.* Available from <www.ippc.int/IPP/En/ispm.htm>.

Gascoine, D. 2001. Appropriate level of protection from an Australian perspective. In *The Economics of Quarantine and the SPS Agreement*, K. Anderson, C. McRae, and D. Wilson, eds., pp. 132–140. Adelaide: Centre for International Economic Studies, and Canberra: AFFA Biosecurity Australia.

Groves, R. H. 1998. *Recent Incursions of Weeds to Australia 1971–1995. CRC for Weed Management Systems.* Technical series No. 3. January 1998. Adelaide: CRC for Need Management Systems.

Hnatiuk, R. J. 1990. *Census of Australian Vascular Plants. Australian Flora and Fauna Series No. 11.* p. 650. Canberra: AGPS.

Nairn, M. E., P. G. Allen, A. R. Inglis, and C. Tanner. 1996. *Australian Quarantine. A Shared Responsibility.* Canberra: Department of Primary Industries and Energy: Commonwealth of Australia. Available from <www.affa.gov.au> via Quarantine and Export Services > About AQIS > Quarantine Reports.

Pheloung, P. C. 2001. Implementation of a permitted list approach to plant introductions to Australia. In *Weed Risk Assessment*, R. H. Grove, F. D. Panetta, and J. G. Virtue, eds., pp. 83–92. Collingwood: CSIRO Publishing.

Reid, R. L. 1990. *The Manual of Australian Agriculture.* Sydney: Butterworths.

Walton, C. S. 2001. Weed risk assessment for plant introductions to Australia. In *Weed Risk Assessment*, R. H. Grove, F. D. Panetta, and J. G. Virtue, eds., pp. 93–99. Collingwood: CSIRO Publishing.

Chapter 11

Invasive Species Management in New Zealand

Barbara J. Hayden and Carolyn F. Whyte

Because New Zealand's economy relies heavily on meat, wool, seafood, dairy, and forestry products, the introduction of exotic organisms is regarded with great concern. This reliance on primary industries makes the country particularly vulnerable to exotic invasions, whether aquatic, terrestrial, or marine organisms. New Zealand is a relatively small, sparsely populated country. As a result of the low population density (approximately 30 people per square mile) and the absence of significant heavy manufacturing, New Zealand still has thousands of kilometers of unpolluted coastline. Maintenance of high water quality and natural biodiversity is of major concern to environmental organizations and the seafood and tourism industries. New Zealand's image as a clean and green country is fiercely defended, but it is far from unsullied by introduced species. New Zealand has many introduced species, some of which have become major pests.

Several independent agencies are involved in the management of introduced species in New Zealand, and there is a strong historical bias on prevention and control of agricultural pests. The Biosecurity Act of 1993 has provided an integrated biosecurity management framework, with enough flexibility to manage biosecurity risks from specific taxa (terrestrial, aquatic, marine) or from vectors, or both. In this chapter, we outline the components of New Zealand's biosecurity management system and then provide examples of its use to manage a vector (imported used cars) and a specific organism (a bacterial pathogen of beehives). Finally, we briefly discuss New Zealand's ballast water management strategy, which highlights not only the flexibility with which the Biosecurity Act can be used but also some of the difficulties encountered in implementation of a management strategy.

The five components of New Zealand's integrated biosecurity system are border control, import health standards, post-entry quarantine, surveillance, and emergency disease and pest response (management once in New Zealand; e.g., pest management strategies).

BORDER CONTROL

Border control is a major line of defense against the introduction of exotic species and provides the statistics on which subsequent components of the management strategy are based. As an island nation, New Zealand enjoys natural quarantine barriers and hence is able to implement effective control systems. Inspection serves as border protection to stop the import of pests that are not already in New Zealand and is used at all entry pathways: passengers, vessels (both aircraft and ships), mail, personal effects, and commercial cargo.

Border control is most visibly seen at New Zealand's international airports, where passengers and luggage coming into the country are checked for products that have been banned or may be a transport medium for a pest. All seven international airports and the International Mail Centre in Auckland have Rapiscan X-ray machines, which have features for the detection of goods deemed to be of biological risk. In the year ending 30 June 2000, about 9,000 travelers per day arrived in New Zealand. From these passengers and crew, there were approximately 238 detections per day of goods that posed a high risk and required seizure, of which more than 64 were undeclared. Yet, the delay to each traveler who has his or her baggage x-rayed is, on average, just one minute (N. Hyde, pers. comm.). A Quarantine Detector Dog program supplements the X-ray inspections. Specially trained dog teams, which are highly proficient at detecting banned plant and animal products in baggage and mail, operate in all the international airports and at the Auckland Mail Centre.

Ministry of Agriculture and Forestry (MAF) border statistics (Whyte 2000) show that between 1993 and 1999, between 11.4 and 15.7 percent of passengers entering New Zealand by air had declared "risk goods" (not all high risk) and 1.5–2.3 percent were found with goods that they had not declared (Fig. 11.1). The increase in the percentage of passengers with declared goods, and the decrease in the percentage with undeclared goods, between 1995–96 and 1996–97 probably reflect greater passenger aware-

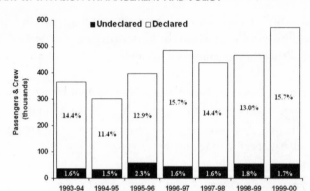

Figure 11.1. Passengers arriving by air with declared and undeclared risk goods.

ness of biosecurity issues, resulting from a much publicized Mediterranean fruit fly incursion, and the introduction of X-ray machines and detector dogs at international airports. The impact of the MAF Quarantine Service's public awareness strategy is further confirmed by the steady increase in the percentage of seizures of goods that were declared by passengers between 1995/96 and 1999/00 (Fig. 11.2). Although not presented here, these data are available for individual classes of risk goods (e.g., fruit fly host material, seeds, nursery stock, bee products, meat, poultry, and dairy products).

In the year ended 30 June 2000, approximately 51 million mail items (letters and parcels) arrived in New Zealand. Virtually all mail is cleared at the International Mail Centre in Auckland where more than 43 million of the items (84%) were x-rayed. Of these, more than 94,000 were opened, and 28, 969 (30.7%) contained risk goods.

The proportion of sea containers arriving in New Zealand that are inspected varies annually but is usually about 20 percent. In 1999–2000, 356,047 sea containers were landed in New Zealand, and 24.8 percent of

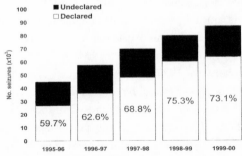

Figure 11.2. Percentage of seized risk goods declared by air passengers

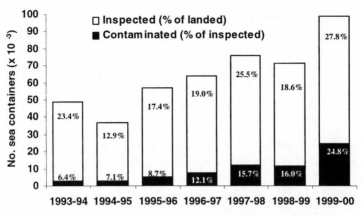

FIGURE 11.3. Sea containers inspected and contaminated with risk goods

those inspected were contaminated and required quarantine action for either plant or animal health reasons (Fig. 11.3). The increase in the proportion of inspected containers that were contaminated is cause for concern. The Quarantine Service also inspected more than 130,000 used cars in 1999–2000. This is a 100 percent inspection rate because imported used cars, most of which come from Japan, are known to be high-risk vectors. About 28 percent of these vehicles required some type of decontamination because of the presence of soil or other organic material. As with sea containers, there has been a steady increase in the proportion of used vehicles found to be contaminated, from 7.4 percent in 1993–94 to 27.8 percent in 1999–2000. Because of the high risk associated with these imported vehicles, approximately 45 percent of the total number are now inspected by MAF Quarantine staff in Japan prior to shipping. Of 31,441 consignments of personal effects imported in 1999–2000, 8,421 (27%) were inspected for risk goods.

Statistics such as these, which are collected and collated by the MAF Border Management Group (Whyte 2000), are an invaluable resource for estimating the quantities and types of biosecurity risk material being delivered to the border by a range of vectors.

IMPORT HEALTH STANDARDS

An Import Health Standard (IHS) must be in force before any risk goods can be imported into New Zealand. Risk goods are defined in the Biosecurity Act (1993) as

> any organism, organic material, or other thing or substance, that (by reason of its nature or origin) it is reasonable to sus-

pect to constitute, contain, or otherwise pose a risk that its presence in New Zealand will result in:

(a) exposure of organisms in New Zealand to damage, disease, loss, or harm; or
(b) interference with the diagnosis, management, or treatment, in New Zealand, of pests or unwanted organisms.

An IHS sets out the requirements that must be met for the effective management of risk goods before they may be imported, moved from the border, or given a biosecurity clearance. Its purpose is to provide explicit directions to potential importers on what measures must be met before the goods can be imported into New Zealand and to provide criteria for inspectors to assess whether the goods should be given biosecurity clearance.

When biosecurity clearance is given, the goods are released unconditionally into New Zealand. Therefore, before giving biosecurity clearance, inspectors must be satisfied that imported goods are not a biosecurity risk, or that the goods comply with the relevant IHS. Only the Director-General of Agriculture may issue an IHS, and then only on the recommendation of a Chief Technical Officer. Chief Technical Officers have been appointed in each of the four government departments with biosecurity responsibilities: the Ministries of Agriculture and Forestry, Fisheries, and Health, and the Department of Conservation (Fig. 11.4).

There is no obligation to have an IHS in force for all risk goods, and one will not be issued unless sufficient requirements can be imposed to provide for the effective management of risks associated with the importation of the goods. In this regard, the development of an IHS for ballast water was a particular challenge because there are few techniques available for effectively dealing with ballast water. Although New Zealand has had voluntary guidelines for the management of ballast water since 1992, an IHS for ballast did not come into effect until 1998 (Ministry of Fisheries 1998).

All IHSs must be based on risk assessments, the quality of which varies according to the availability of data on which to make the assessments. For example, there are more robust data on which to assess the risks of many of the terrestrial pests in New Zealand than there are on marine pests. Some IHSs are vector-focused (e.g., used cars, ballast water, and ships' hulls). As an example of the management options afforded by an IHS, for imported cars (Ministry of Agriculture and Forestry 1997 is listed in Appendix 11.1 (see <http://www,maf.govt.nz/biosecurity> for updates).

The key element of the standard for imported used cars is that all parts of each used vehicle must be free of contamination by animals, including

OPERATIONAL RESPONSIBILITY

 Director General
of MAF

Chief Technical Officers

Ministry of
Agriculture
& Forestry

Ministry of
Fisheries

Department of
Conservation

Ministry
of Health

Import Health Standards, Permits & Pest management strategies

FIGURE 11.4. Government departments involved in biosecurity management under the Biosecurity Act, 1993.

insects and other invertebrates, organic matter of animal origin, plants or plant products, soil, or water. Every used car is inspected because a prior risk analysis has identified them as high-risk vectors. When a used car arrives in New Zealand, an inspector checks the documentation, including any cleanliness or cleaning certificates, and then inspects all parts of the car, such as the undercarriage, backs of wheels, suspension, drive shaft, and exhaust. If any contamination is found, the importer is given the options of either having the vehicle decontaminated at an approved facility or reshipping it. In either case, the action is at the importer's expense. Decontamination is carried out under the direction of an inspector, and the vehicle is then reinspected. The importer must pay the initial inspection and reinspection charges. If the car meets the conditions of the IHS or has been decontaminated to the satisfaction of an inspector, biosecurity clearance is granted.

Import Permits

Before 1997, an Import Permit was required, in addition to compliance with an IHS, before risk goods could be brought into New Zealand. A 1997 amendment to the Biosecurity Act removed the requirement for an Import Permit in most cases so long as there is compliance with a relevant IHS. However, some situations still require a permit, such as before importation of live animals requiring quarantine. The purpose of the permit is to ensure that adequate arrangements are made prior to arrival of the animals. A permit cannot be issued unless an IHS is in force.

Environmental Risk Management Authority (ERMA)

An Import Permit must be obtained in order to intentionally import a new organism into New Zealand. The permit must be issued by the Environmental Risk Management Authority (ERMA), which operates under its own parliamentary legislation, the Hazardous Substances and New Organisms Act of 1996. ERMA's requirements are very stringent. It is, for example, very difficult to import a new species for mariculture because of the risk that it may harbor parasites or pathogens. Most permits issued by ERMA are for importation into containment such as zoos or research institutes, which have approved quarantine facilities. For example, biological control agents are imported so that host specificity can be tested before use. Some examples of applications to introduce (release) new species into New Zealand received in 1998–99 and ERMA's response are listed in Table 11.1.

TABLE 1 1 . 1 . Examples of applications to ERMA in 1998–99 to introduce (release) new species into New Zealand.

Organism	Decision
Camelus dromedarius (Arabian camel)	Refused because there was insufficient information about the relative harmful and beneficial effects of introducing camels.
Gonioctena olivacea (leaf-feeding beetle used for biological control of broom)	Refused because the beetle can also complete its life cycle on tree lucerne, and there was insufficient information about the relative harmful and beneficial effects of its introduction.
Osyptilus pilosellaea (moth used for control of hieracium)	Approved. The Chief Veterinary Officer noted that there was a risk of some harm but that all but one of the submitters believed that the potential benefits outweighed the risk to native plants; i.e., ERMA decisions are risk assessment based.

Post-Entry Quarantine

The Ministry of Agriculture and Forestry (MAF) classifies quarantine pests into three groups according to the quarantine measures required:

- pests requiring offshore treatment prior to export to New Zealand
- pests requiring a declaration in addition to the import certification
- pests for which import certification is adequate

Import Health Standards for live animals from specified countries define the conditions that must be met in the country of origin or of export, during transit, importation, and quarantine. These standards state whether an Import Permit or post-entry quarantine is required.

Entry of plant nursery stock, cut flowers, and cuttings depends on the country of origin. Some are allowed free entry; others are restricted or banned. In many cases, a certificate from the country of origin is all that is required. Restrictions on seeds vary depending on type and species. Some are permitted entry provided they are free of insect pests; others are prohibited or may require specific treatment. Bulbs are regarded as nursery stock that can carry a variety of pathogens; therefore, they can be brought into New Zealand only under permit for growing in quarantine.

Surveillance

Surveillance programs are used to monitor the health status of New Zealand's plant, animal, fish, and bee populations and export crops. They seek to achieve the following objectives:

- facilitate exports by being able to credibly certify New Zealand's true pest status and minimize the sanitary requirements that exports must meet
- fulfill international treaty obligations, including the prompt reporting of animal or plant health events to international organizations and trading partners
- develop and establish technically justifiable requirements for animals, animal products, plants, and plant products entering New Zealand
- enable the prompt notification of exotic diseases in New Zealand
- support the development of pest management strategies
- facilitate the formulation of public health policies for the control of animal pathogens and parasites that can affect human health

PEST MANAGEMENT STRATEGIES

The fifth component of New Zealand's biosecurity system is Pest and Disease Management Strategies (PDMSs), which are designed to manage or eradicate pests, including weeds, that are present in New Zealand. All PDMSs require approval from the Minister of Agriculture and Forestry to ensure that they are sound, feasible, and affordable. The four government departments shown in Figure 11.4 initiate most pest management strategies, although any interested organization can propose one. For instance, the National Beekeepers' Association was granted regulatory power under the Biosecurity Act to control American foulbrood (AFB) using a PDMS. This PDMS illustrates the use of the act to manage a specific organism.

American foulbrood, which is caused by a bacterium, is the most serious disease of honeybees in New Zealand. Therefore, the Beekeepers' Association has developed a PDMS in consultation with scientists and government regulatory authorities that has a strong emphasis on training and education of beekeepers but also places legal duties on beekeepers. These duties include reporting and destruction of beehives with clinical or visual signs of AFB and supplying annual data. The AFB PDMS allows appointment of authorized persons who have legal authority to enter private land, inspect beehives, and destroy beehives with clinical signs of AFB. While MAF has delegated responsibility for management of the PDMS to the National Beekeepers' Association, the beekeepers contract MAF staff to manage many of the provisions of the PDMS (e.g., maintenance of a database, audits of hive inspections, and training in diagnosis and elimination of AFB).

The use of antibiotics to control AFB disease is not allowed under the PDMS. Despite this, the management regimes in the AFB PDMS contained the disease to 2.8 percent of apiaries and 0.38 percent of hives in 1998–99. These levels are as low as, or lower than, those reported from countries that allow antibiotic feeding for the control of AFB, and they have been achieved without product contamination and bacterial resistance associated with the use of antibiotics. New Zealand is free from many of the world's most serious bee pests and diseases. In order to maintain that status, the government continues to provide funding for bee disease surveillance and test development, for border control and inspection, and for maintaining the ability to identify and respond to cases of exotic bee pests or diseases.

Coordination of Biosecurity Management: The Role of the Biosecurity Council

The Minister for Biosecurity is responsible for the administration and coordinated implementation of the Biosecurity Act. Coordination is achieved through a Biosecurity Council that comprises the chief executives of the Department of Conservation and the Ministries of Agriculture and Forestry, Health, Fisheries, Environment, and Research, Science, and Technology, the Environmental Risk Management Authority, a representative of regional councils, the Group Director of MAF Biosecurity Authority, and an independent chair.

The council advises the minister on the following:

- priorities for purchasing biosecurity services (including research)
- appropriate framework(s), methodologies, and procedures for risk assessment, risk management, and risk communications that ensure consistency in approach across departments
- coordination of biosecurity-related research (including PGSF and departmental operational research), and the need for a National Science Strategy for Biosecurity
- protocols for cross-agency cooperation (including funding bids)
- appropriate location and structure of biosecurity-related border inspection work (and the relationship to Customs)
- investigations related to biosecurity initiated by the minister
- responsibility for newly identified risks
- departmental responsibility for new pest incursions
- a strategic overview of an information and education strategy for biosecurity surveillance and awareness raising
- legislative or institutional barriers to biosecurity management and how these may be overcome
- departmental capacity and capability to respond to biosecurity risks and how these systems may be enhanced and adapted for new and emerging risks

Policies developed by the Biosecurity Council include the following:

- the process to be used by departments when developing national pest management strategies
- the interdepartmental consultation required in the conduct of risk analyses and the development and modification of Import Health Standards
- standard definitions of unwanted organisms under the Biosecurity Act

• the responsibilities of Chief Technical Officers, ERMA, Regional Councils, and the Director-General of MAF

Management of Ballast Water from Ships under the Biosecurity Act of 1993

New Zealand has had voluntary controls on ballast water discharges, based on the International Maritime Organization's Guidelines, since 1992. Here are the five main features of the New Zealand controls:

1. If possible, ballast water should not be discharged within New Zealand.
2. If ballast water has to be discharged, then it should be ballast that was exchanged or loaded in the open ocean. Details of the exchange, and the original source of the ballast water, must be provided to an inspector. Ballast water, which has been loaded within the territorial waters of another country, cannot be discharged without reporting it to an inspector prior to discharge.
3. Ballast water may be discharged if there is documented evidence to show that the ballast has been disinfected. As yet, no effective treatment options are available.
4. If none of the above three options can be fulfilled, then the master of the vessel has the option to discharge ballast in an approved area of New Zealand or to an onshore facility, to treat the ballast, or to have the ballast tested to show it is not a risk. Currently, there are no areas approved as ballast dumping areas nor any onshore discharge or ballast water treatment facilities.
5. No sediment or mud from the cleaning of the holds or ballast tanks, or anchor or chain lockers can be landed in New Zealand waters. This clause of the controls is mandatory because satisfactory alternatives to dumping sediment in the sea exist (e.g., by disposal in a landfill not immediately adjacent to the sea).

Although four ballast water management options are listed, the only practical option currently available to vessels having to discharge their ballast water in port is to exchange it with oceanic water before the ship arrives in New Zealand's territorial waters, a practice known to be only partially effective at minimizing the risk of exotic introductions. With the enactment of the Biosecurity Act in 1993, ballast water can be classified as a "risk good." Therefore, if an inspector "reasonably suspects" that ballast water arriving into New Zealand poses a risk to the flora and fauna already in New Zealand, the ballast water can be declared a risk good and the powers of the Biosecurity Act used. The Biosecurity Act has also allowed the Voluntary Guidelines to be replaced with an IHS (Ministry of Fisheries 1998), providing a legal framework for the controls (see Appendix 11.2).

Under the ballast water IHS, all ships are required to fill out a Vessel Ballast Report Form giving details of ballast water source, volumes discharged, and which of the management options were used. Providing incorrect information to an inspector is an offense under section 154(b) of the Biosecurity Act. It carries a penalty for individuals of up to twelve months' imprisonment and/or a fine not exceeding NZ$15,000, and for corporations a fine not exceeding NZ$75,000. Failure to obey the directions of an inspector is an offense under section 154(o). It carries a penalty for individuals of a fine not exceeding NZ$1,000, and for corporations a fine not exceeding NZ$15,000.

The data from the Vessel Ballast Report Forms allow New Zealand authorities to determine the volumes of ballast water being discharged and how much of it has been exchanged. However, because an effective method to determine whether ballast water has been exchanged has yet to be developed, there is currently no way to verify these data, which are based solely on the word of the vessel masters.

Thus, although there are adequate powers in the Biosecurity Act to deal with ballast water, there is only limited use of it as an enforcement mechanism because of the lack of proven, effective management strategies for ballast water. That situation is unlikely to change markedly until there are more effective, safe, practicable, economically sound, and environmentally acceptable options for dealing with ballast water. Currently, the act is used only in limited ways:

- to ensure that masters of vessels provide correct written information about their ballasting operations
- to prevent the discharge of sediment and tank cleaning residues in New Zealand waters
- to prevent the discharge of ballast water from Tasmania and Port Philip Bay, Melbourne, during the months when larvae of the invasive seastar *Asterias amurensis* may be in the water

Although the Biosecurity Act of 1993 and the Hazardous Substances and New Organisms (HSNO) Act of 1996 provide a comprehensive legal and administrative biosecurity framework, the ballast water issue highlights several problems related to implementation of management strategies under these acts. It is difficult to predict which organisms will become pests once outside their native range, and the current lack of adequate pest control methods and technologies hampers the effectiveness of controls and preventive measures in some cases.

CONCLUSION

The statistics collated by the MAF Border Management Group provide evidence of the risk goods arriving at the border daily and hint at the additional numbers that are probably entering the country undetected despite the best efforts of biosecurity agencies. Unintentional introductions are, however, only part of the picture. The species already established in New Zealand also threaten indigenous biodiversity, and many of these were intentionally introduced prior to the 1996 HSNO Act. Almost half of all vascular plant species growing wild in New Zealand, about 2,100 species, are introduced. At least another 17,000 introduced plant species are present in New Zealand private gardens and collections, or are being used in agriculture, horticulture, or forestry (Buddenhagen 1999). Many of these are likely to naturalize in the future. The Department of Conservation's weed database lists more than 240 naturalized land, wetland, freshwater, and marine plants as actual or potential invasive weeds. More than 70 percent of invasive weeds were deliberately introduced into New Zealand as ornamental plants. A further 12 percent were introduced for agriculture, horticulture, or forestry. Only 11 percent were introduced accidentally. Thus, the ability to control deliberate introductions under the HSNO Act is equally as important as the ability to control nondeliberate introductions under the Biosecurity Act.

ACKNOWLEDGMENTS

Thanks to Mike Alexander, National Advisor Border Inspection, MAF Biosecurity Authority, for assistance in locating data on New Zealand's biosecurity status, and to Maj De Poorter and an unknown reviewer for comments on the manuscript.

REFERENCES

Buddenhagen, C. 1999. Managing weed threats to the public conservation estate in New Zealand. *Aliens* 9: 18–19.

Ministry of Agriculture and Forestry. 1997. The Generic Import Health Standard (Biosecurity Act 1993) for Cars, Vans and Utility Vehicles (Commodity Class: Vehicles), 3 December 1997. New Zealand.

Ministry of Fisheries. 1998. Import Health Standard (Biosecurity Act 1993) for ships' ballast water from all countries. New Zealand.

New Zealand Government. 1993. Biosecurity Act 1993: An act to restate and reform the law relating to the exclusion, eradication, and effective management of pests and unwanted organisms.

Whyte, C. F. 2000. MAF Quarantine Service: Annual Statistics Report, 1993/94–1999/00.

Import Health Standard
(Biosecurity Act 1993) for Cars, Vans, and
Utility Vehicles from Any Country
3 December 1997

Note that this version of the IHS is included as an illustration only. Current versions of the IHSs can be obtained from the Ministry of Agriculture and Forestry website: http://www.maf.govt.nz/biosecurity.

1. GENERAL CONDITIONS

It is the responsibility of the importer to ensure that the used car, van or utility vehicle complies with the conditions in this standard. A used car, van or utility vehicle that does not comply with the conditions of this standard will be decontaminated prior to release from the port/airport area (or transitional area if inspected off the port or airport area) or may be reshipped or destroyed.

2. NEW VEHICLES

There are no conditions on the importation of new vehicles into New Zealand unless, upon arrival, an inspector considers that the vehicles have been contaminated in transit. In such a circumstance the vehicles will be treated as used vehicles.

3. USED VEHICLES

All used vehicles are prohibited entry into New Zealand unless they comply with the specifications given below:

(i) Each used vehicle has documentation available stating:

(a) identification;

(b) origin;

(c) shipment details;

(d) consignor; and

(e) consignee.

(ii) Every used vehicle must be free of contamination by any of the following:

    (a) animals, insects or other invertebrates (in any life cycle stage), eggs, egg casings or rafts, or any organic material of animal origin (including blood, bones, fiber, meat, secretions, excretions, etc);

    (b) plants or plant products (including fruit, seeds, leaves, twigs, roots, bark, saw dust, decayed or infested wood/timber or other organic material); or

    (c) soil or water.

(iii) All parts of the used vehicle must be free of contamination including the radiator, air-vents, chassis, tray, under-carriage, wheels, tyres, suspension, drive shaft, exhaust, the cab or passenger compartment, glove compartment, baggage compartment, spare wheel well, engine compartment, and any other interior or exterior surfaces.

(iv) In the case of vehicles with winches, wire, fiber ropes or cables shall have been removed from winches and treated by heating to 121°C for 15 minutes, or removed and not replaced, or replaced with new ropes or cables after the winches have been thoroughly cleaned and shall be accompanied by a certificate that they are new or that they have been treated according to the above specification.

4. INSPECTION ON ARRIVAL IN NEW ZEALAND

All vehicles (including wire ropes) will be inspected on arrival in New Zealand to the extent necessary to ensure they meet the conditions of 3. All vehicles shall be run over a ramp or an inspection pit constructed in a manner that allows a comprehensive inspection of the under surface.

5. USE OF CERTIFICATION

An importer may present documented evidence that the vehicle has undergone a cleaning process or otherwise meets the requirements of 3. An inspector may judge that a certificate of cleanliness (or a certificate that the vehicle has been cleaned, including a description of the cleaning process used) is sufficient to limit the degree of inspection to only what is necessary, in the opinion of that inspector, to confirm the veracity of the certificate. Importers who have proved the veracity of their certification may apply to the Ministry of Agriculture for a reduced inspection regime.

6. Inspection Overseas

In order to reduce delays at the port/airport of arrival in New Zealand importers may request the inspection of vehicles at the point of export.

7. Costs

The costs of inspection (including re-inspection), cleaning, fumigation, reship or destruction and any demurrage relating to delays in clearance due to any vehicle not meeting the requirements of this standard are the responsibility of the importer.

8. Enquiries

Unless indicated to the contrary all communications concerning this import health standard should be addressed to:

[The relevant Chief Technical Officer]

Chief Plants Officer
Ministry of Agriculture
P O Box 2526
Wellington
NEW ZEALAND
Fax: 64-4-474 4240

Requirements for the Supplier of the Inspection of Motor Vehicles

1. Inspection Area

Vehicles may only be landed at ports/airports that have decontamination/cleaning facilities that have been approved as transitional facilities under the Biosecurity Act 1993.

Vehicles shall be inspected on a sealed area at the port/airport area at which they have arrived.

Vehicles may not be taken off that area until they are inspected. Vehicles within a container may be taken directly to an approved transitional facility for inspection.

2. INSPECTION

Inspection shall be carried out under daylight conditions. Inspections inside or at night may only take place where artificial lighting provides the equivalent of daylight.

Every compartment and surface that is normally accessible without recourse to tools is to be inspected. If any compartments or structural areas cannot be readily opened by an inspector, because of accident damage or any other reason, the inspector will require those areas to be opened by the importer or his/her agent.

All parts of the used vehicle must be free of contamination including the radiator, air vents, chassis, tray, under-carriage, wheels, tyres, suspension, drive shaft, exhaust, the cab or passenger compartment, glove compartment, baggage compartment, spare wheel well, engine compartment, and the exterior surface.

All wood/timber associated with imports e.g. truck decking shall be inspected for fungal decay and/or insect attack.

3. CONTINGENCY

Contaminants are those listed in the Import Health Standard for Cars, Vans, and Utility Vehicles from any Country.

Should any contamination be found during inspection on arrival and the inspector is not able to conveniently deal with the contamination at the time and place of the inspection, the importer shall be given the options of either having the used car, van or utility vehicle decontaminated at a transitional facility approved for decontamination or reshipped. In either case the action will be at the importer's expense.

Transport to an approved transitional facility for decontamination shall be in a manner directed by an inspector that will minimize the biosecurity risk from the contamination.

The decontamination direction shall clearly describe the actions to be undertaken to decontaminate the vehicle. Decontamination must be carried out under the direction of an inspector.

Once decontamination is carried out, the used car, van or utility vehicle shall be re-inspected.

4. Charges

Initial inspection and any re-inspection charges shall be met by the importer. The charges will be in accordance with the Biosecurity (Costs) Regulations 1993. Any third party charges for activities such as transport to an approved transitional facility for decontamination or decontamination itself shall also be met by the importer. The charges for such services shall be at commercial rates and not subject to regulatory control. Importers shall be advised of the approved transitional facilities. The importer may choose which of the approved facilities will be used.

5. Biosecurity Clearance

Directions for decontamination shall be in writing for each vehicle.

A biosecurity clearance will be given when the vehicle meets both the general and specific conditions of this standard or it has been decontaminated to the satisfaction of an inspector. The biosecurity clearance shall be in writing and shall be traceable to the imported vehicle.

Import Health Standard for Ships' Ballast Water from All Countries (Biosecurity Act 1993)

---

1. SCOPE

This import health standard applies to ballast water loaded within the territorial waters of a country other than New Zealand and intended for discharge in New Zealand waters. Emergency discharge of ballast water is not covered by this standard.

2. GENERAL CONDITIONS

It is the responsibility of the Master to ensure that the ballast water, and any associated sediment, intended for discharge in New Zealand complies with the conditions in the standard. Ballast water that does not comply with the conditions must not be discharged in New Zealand waters.

Compliance with these controls must be consistent with the safety of the crew and the vessel. Nothing in these controls is to be read as relieving Masters of their responsibility for the safety of the vessel.

3. DEFINITIONS

Ballast water—water, including its associated constituents (biological or otherwise), placed in a ship to increase the draft, change the trim or regulate stability. It includes associated sediments, whether within the water column or settled out in tanks, sea chests, anchor lockers, plumbing, etc.

Inspector—an inspector under the Biosecurity Act 1993.

New Zealand waters—means:

a. the internal waters of New Zealand; and
b. the territorial sea of New Zealand.

Internal waters—means:

a. harbors, estuaries, and other areas of the sea that are on the landward side of the baseline of the territorial sea of New Zealand; and
b. rivers and other inland waters of New Zealand that are navigable by ships.

Territorial sea—the sea within 12 nautical miles of the seaward side of the baseline of the territorial sea of New Zealand. (See definition in section 3 of the Territorial Sea, Contiguous Zone and Exclusive Economic Zone Act 1977 for baseline.)

4. Requirements for Ballast Water

No ballast water may be discharged into New Zealand waters without the permission of an inspector. An inspector will only permit ballast water to be discharged if satisfied that the Master has met one of the criteria in section 5 below.

Part I of the Vessel Ballast Reporting Form approved by the Ministry of Fisheries must be completed for all vessels on arrival in New Zealand. Permission to discharge ballast water is granted when an inspector signs this part of the form.

Part II of the Vessel Ballast Reporting Form must be completed and submitted to an inspector at the last port of call in New Zealand.

Sediment that has settled in ballast tanks, ballasted cargo holds, sea chests, anchor lockers or other equipment must not be discharged into the sea, but must be taken to a landfill approved by an inspector.

5. Options for Satisfying an Inspector

Option 1

Demonstrating that either:

a. the ballast water has been exchanged en route to New Zealand in areas free from coastal influences, preferably on the high seas. (Accepted techniques are either emptying and refilling ballast tanks/holds or pumping through the tanks a water volume equal to at least three times the tank capacity.); or

b. the ballast water is fresh water (not more than 2.5 parts per thousand NaCl).

Option 2

Ballast water has been treated using an approved shipboard treatment system. Note—there are presently no approved shipboard treatment systems.

Option 3

Ballast is discharged in an approved area or onshore treatment facility. Note—there are presently no approved areas or onshore treatment facilities in New Zealand.

6. EXEMPTIONS

It is accepted that in some circumstances exchange may not be possible. An exemption will generally be granted by an inspector when it can be demonstrated that:

Exemption 1

a. The weather conditions on the voyage in combination with the construction of the vessel have precluded safe ballast water exchange; and

b. the ballast water was not loaded in any area listed in Annex 1.

Exemption 2

a. The construction of the vessel has precluded ballast water exchange; and

b. the ballast water was not loaded in any area listed in Annex 1.

7. COSTS

The costs of inspection, analysis, identification, delays, and any other costs associated with this standard are the responsibility of the owner and/or charter. These costs shall be actual, fair and reasonable.

Chapter 12

Vectors and Pathways of Biological Invasions in South Africa— Past, Present, and Future

David M. Richardson, James A. Cambray, R. Arthur Chapman,
W. Richard J. Dean, Charles L. Griffiths, David C. Le Maitre,
David J. Newton, and Terry J. Winstanley

Biological invasions—the spread of organisms following their introduction to areas outside their natural distribution as a result of human actions—are a by-product of the globalization of regional economies. Large parts of the world are currently dominated by human-modified ecosystems that often comprise a greater biomass of introduced than native organisms (Vitousek et al. 1997). The demand for an ever-increasing range of goods and services from organisms is growing worldwide, and these are, for many reasons, increasingly derived from nonnative organisms. People are designing, intentionally or by accident, ecosystems that are more productive and/or congenial to them. Reasons behind the selection of the species used in building these new ecosystems are complex and change over time. Besides the obvious need for food, shelter, and other essential requirements, there is a rapidly growing trade in ornamental species, pets, and other nonessential commodities. Such demands increase the vectors (physical agents that introduce aliens) and pathways (here taken as a more complex concept, embodying the economic/cultural forces that initiate and sustain the human-mediated movement of organisms between regions) for the introduction of nonnative species, and for their further dissemination within regions. These factors, acting in concert, are causing the homogenization of the world's biota.

The magnitude and net effects of biological invasions escalated rapidly over the twentieth century. During each decade, more species became invasive, more ecosystems were irreversibly altered, and an ever-increasing array of functions and processes was impacted by invasive alien organisms. Biological invasions are now one of the most serious threats to biodiversity worldwide, and coping with invasions is a daunting challenge. To manage biological invasions, we need detailed knowledge of the ecology of invading species and invaded systems. Our understanding of the various traits associated with invasiveness in different taxa, and the features that render certain systems more susceptible to invasion, has improved considerably (e.g., Rejmánek et al. 2004 for plants). However, there is little point in improving our understanding of the mechanisms of invasions without improving our understanding of the complicated dynamics that determine which alien organisms (and in what quantities, over what periods, in what circumstances, and in what combinations) are in demand by human societies. That these are the real "seeds" of biological invasions is frequently overlooked. Knowledge of the actual and potential pathways by which alien species arrive and disseminate within a region is lacking. Our ignorance in this arena is a serious obstacle to the effective management of biological invasions and biodiversity in the long term.

The formulation of systematic plans for dealing with alien species at national and regional scales demands that urgent attention be given to (1) identifying and considering the relative strengths of pathways by which different alien taxa arrived, are arriving, and are likely to arrive in the future; (2) identifying the pathways of rapid spread within the region; (3) evaluating assumptions (implicit or explicit) regarding the modes and importance of different pathways; and (4) identifying critical gaps in our knowledge about invasion pathways.

Urgent attention is also needed to identify and evaluate strategies for managing vectors and pathways. This entails, at least, (1) identifying management strategies to reduce future invasions that are currently in place; (2) examining methods for evaluating the effectiveness of management strategies; (3) identifying critical gaps in our knowledge and development of effective prevention strategies; and (4) highlighting key opportunities and difficulties for managing pathways in the context of global trade practices and policies.

South Africa is emblematic of the changing dynamics of vectors and pathways of biological invasions. Humans originated in this region, and were present here for many thousands of years before any alien organisms were introduced (Deacon and Deacon 1999). The first alien organisms arrived gradually, via human and livestock migrations from further north in Africa. As far as we know, none of these early aliens had any marked influence on natural systems. A tide of introductions followed the colonization of the region by Europeans in the seventeenth century. In the three and a half centuries since 1652 we can distinguish distinct phases of introductions driven by the needs and activities of humans; we can construct chronologies of introduction and invasion episodes. The last third of the twentieth century saw massive social transformations in South Africa, resulting in momentous changes in human demographics, macro- and microeconomic climates, and marked changes in the region's role in southern African, African, and global economies (Fox and Rowntree 2000). These factors have all influenced (or will influence in the future) the reliance of the region's human societies on nonnative organisms and hence the vectors and pathways for alien taxa.

The ravages of particular biological invaders are often seen in isolation. This chapter explores the history of human enterprises in this region to improve our understanding of the complex ways in which factors interact to drive the arrival and spread of nonnative organisms. It is necessary to revisit past events to clarify the links between human motives and pathways before describing the current situation and sketching likely scenarios for the near future (we, the authors, have taken the year 2020 as our planning horizon when discussing "the future").

Alien organisms from a range of taxonomic groups have invaded South African ecosystems (Macdonald et al. 1986). By far the most research has been on vascular plants and vertebrate animals in terrestrial and freshwater systems. In addressing the topic of pathways of biological invasions in terrestrial, freshwater, and marine systems of South Africa, we have selected groups of taxa (vascular plants, birds, mammals, freshwater fishes, and marine organisms) and sectors of human activities (aquaculture, forestry, and horticulture). We describe the "anatomies" of pathways that have driven alien introductions and invasions. In particular, we concentrate on more recent changes and discuss the implications of the pathways for the formulation of policies, legislation, and state-imposed controls that are clearly needed to mitigate the harmful effects of alien taxa already in South Africa and to reduce the magnitude of future impacts. In this chapter we do not deal with weeds of agriculture and forestry, and only touch briefly on commensal animals in human-dominated environments.

For each of the selected taxon groups and activity sectors, we explore (1) the driving forces affecting the introduction and dissemination of alien organisms (and how these forces have changed over time); (2) the management (intervention) mechanisms, including legal controls, currently in place to prevent or reduce invasions; (3) the rationale for existing management practices; and (4) the limitations (gaps and constraints) and/or merits surrounding these practices.

VASCULAR PLANTS

Invasive alien plants are prominent components of all main categories of freshwater and terrestrial ecosystems in South Africa. Richardson et al. (1997) provided a list of 84 important environmental weeds for southern Africa that includes taxa from many life-forms and from many parts of the world. Species from South America and Australia are overrepresented among those alien plant species that change the character, form, or nature of natural ecosystems over substantial areas ("transformer species" sensu Wells et al. 1986). Different taxa and life-forms have shown different levels of success in different biomes, partly because of their inherent habitat requirements, but also because of the history of introduction and dissemination in different parts of the region.

For our purposes in this chapter, the most useful database for shedding light on pathways of invasion for plants is the "Catalogue of Problem Plants in Southern Africa" (Wells et al. 1986). This source contains information on invasive status, reason for introduction (and/or current use), and many other aspects for 1,653 plant taxa, including 711 alien taxa. We analyzed data from this source to determine the main reasons for the introduction of alien taxa in various "pest plant" categories.

This database lists 272 alien taxa (38%) that have no obvious current use. We assigned these taxa to the "unspecified" category, which includes both accidental introductions (we suspect that most weeds of agriculture fall in this group) and taxa that were introduced intentionally, but which proved unsuitable for these purposes and are now not used for any specific purpose. For trees, examples are *Acacia decurrens* and *A. dealbata*, which were probably both introduced along with *A. mearnsii* by accident rather than design. The first two species were initially used along with *A. mearnsii* for tannin production, but neither proved suitable. For those taxa that are clearly used in one of the main "use categories" considered, ornamental plants (220 taxa) have caused most problems, followed by taxa used in agriculture (83), for fodder (49), in silviculture (24), as barrier plants (6), or as cover/binder plants (6) (Fig. 12.1A). Further analysis of the data shows that most of the introduced plants in the "unspecified" category have become "ruderal," "flora," "agrestal," or "pastoral" weeds (sensu Wells et al. 1986). Plants introduced for agriculture are also well represented in these categories of pest plants (Fig. 12.1B). Although silvicultural, fodder-producing, and cover/binding alien plants have contributed relatively few taxa to the total list of pest plants, the taxa from these groups that have invaded natural systems have caused by far the biggest impacts. More than 85 percent of the 84 most widespread and damaging "environmental weeds" in South Africa (Richardson et al. 1997) were introduced as ornamentals (32 taxa), for agriculture (11) or forestry (12), or fall in the "unspecified" category (17; see above). These data are for *numbers of taxa* and underrate the importance of forestry and horticulture as pathways responsible for the introduction and dis-

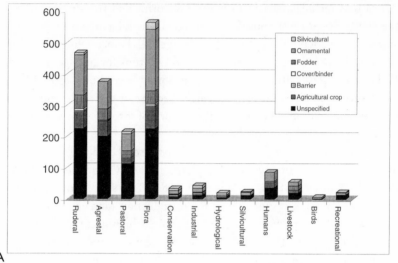

A

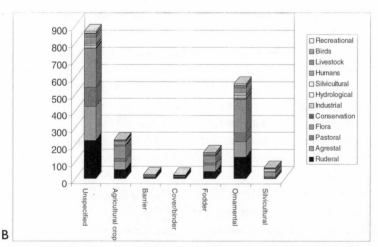

B

FIGURE 12.1. Classification of the 711 alien plant taxa cataloged by Wells et al. (1986) by (a) reasons for introduction and (b) the elements affected. Note that many taxa were introduced/cultivated for more than one reason; many taxa are also listed in more than one problem plant category.

semination of the most invasive and destructive invaders (most "trans-formers" sensu Richardson et al. [2000] arrived via these pathways). Taxa introduced for use in agriculture feature prominently in the pest flora as detailed above, but most taxa are problematic mainly in highly disturbed environments. Agriculture as a pathway of invasion is not discussed further in this chapter. We now consider the pathways of introduction facilitated by forestry and horticulture.

Forestry as a Pathway for the Introduction and Dissemination of Invasive Alien Plant Species

A national survey in 1996–97 showed that about 10 million hectares (ha) (8% of the area of South Africa) have been invaded to some degree by woody alien plants (Versfeld et al. 1998). A very large proportion of these species were introduced and disseminated within South Africa for purposes of "forestry." Forestry here means the introduction and planting of trees primarily for the production of wood and tanbark, with shelter, shade, and amenity value as secondary benefits. Trees planted primarily for fruit production are excluded. Although several important woody invaders are not used in "plantations" in the strict sense, they were propagated and promoted by government forestry organizations; these include *Acacia cyclops* and *A. saligna*, *Jacaranda mimosifolia*, *Melia azedarach*, and *Prosopis* species. Eleven species used in commercial plantations were mapped as invaders, and some have invaded more than a million hectares. Another 36 mapped species were actively promoted by government forestry, and many of these were introduced as potential plantation species, for driftsand reclamation, and for woodlots (Keet 1929, 1974, Poynton 1979a, 1979b, Shaughnessy 1986).

The most damaging group of invaders comprises the Australian *Acacia* species that have invaded more than 4.6 million ha (Versfeld et al. 1998). This area includes extensive invasions by *Acacia* species introduced for driftsand control in the dune fields and coastal plains of the Western and Eastern Cape provinces. *Acacia mearnsii* has invaded some 2.5 million ha, more than twenty times the area currently under cultivation (Versfeld et al. 1998). *Pinus pinaster* has invaded about 1.2 million ha, and all pines combined cover an area of about 3.0 million ha (see Richardson and Higgins 1998 for details on which species have invaded which habitats). Eucalypts have invaded about 2.4 million ha, particularly along rivers draining the eastern escarpment and major rivers in the Western Cape (Versfeld et al. 1998). The ecological and economic impacts of these invasions have been fairly well assessed (for a recent review, see van Wilgen et al. 2001).

Forestry before 1970

The pre-1970 history of alien tree species in South African forestry may be divided into a number of periods. Salient features of these periods and the main trends in forestry are summarized in Table 12.1. The period up to 1871 was characterized by mainly hit-and-miss introductions aimed primarily at timber and fuelwood production. The establishment of the Forestry Department in 1872 resulted in the systematic expansion of plantations using introduced species that had already displayed vigorous growth in South Africa and, later, new species selected for their potential to grow well based on similarities in climate and natural vegetation. These developments created pathways for invasions for scores of alien tree taxa (Fig. 12.2). Not only were species selected specifically for their ability to establish with minimal tending, but plantations were also established in many parts of the country, particularly in mountain catchments, and often over entire landscapes. These activities created the foci of the current widespread invasions (Rouget et al. 2002).

Forestry after 1970

A major thrust in commercial forestry since about 1970 has been to increase plantation productivity. This is partly because most of the best areas for forestry have already been planted (Cellier 1994) and partly because of the demand for better-quality saw timber and improved pulp yield. Constraints imposed by the Afforestation Permit System (see p. 305) have also played a role. Some new species have been introduced, but the greatest efforts have been concentrated on tree breeding (Cellier 1994). Research also identified the benefits of weed control resulting in improved tree establishment and growth (Denny and Schumann 1993).

·The area of commercial plantations of pines and eucalypts increased rapidly after 1960, with *Pinus patula*, *P. elliottii*, and *Eucalyptus grandis* (and its hybrids) accounting for most of the added area (van der Zel and Brink 1980, Forestry statistics). All three species are invasive (Versfeld et al. 1998), and the risk of invasions has grown because of increases in the size of source populations and the contact area between plantations (seed sources) and habitat potentially subject to invasion. Many decades of spread from plantations in some regions have masked the overwhelming importance of initial plantations as seed sources. Environmental variables, rather than distance from plantations, are now the primary determinants of distribution of self-sown stands in many areas (Rouget et al. 2002).

TABLE 12.1. A summary of important events in the development of forestry using alien tree species in South Africa up to 1970.

Dates	Period	Afforestation	Other Important Events	Main Reasons for Tree Introductions and Plantings	Key References
1652–1871	Experimental introductions	Dutch colonists at the Cape faced with shortage of wood for construction.	The Cape colony established and	Timber production for construction and shipbuilding,	Keet (1936, 1974) King (1943)
	Dutch and British colonial period, Cape government formed	Early introductions: *Quercus robur* (shipbuilding); *Populus nigra* (construction wood); *Pinus pinaster* (various purposes). Other species introduced >1830 that are now important invaders include *Acacia longifolia, A. melanoxylon, A. saligna, Allocasuarina verticellata, Hakea drupacea, H. gibbosa, Leptospermum laevigatum, Paraserianthes lophantha,* and *Pinus halepensis.* 1854: first plantation of *Acacia mearnsii* established in KwaZulu–Natal midlands for tanbark production. Forest Act passed in 1856 but not effective.	expanded northward, inland, and eastward. Farmsteads and newly established towns planted trees for shade, shelter, and fuelwood. More rapid expansion following the Great Trek (1834 onwards) and diamond discoveries in 1867. Attempts to control drift sands on the Cape Flats (1845–) and tree planting competitions.	plantings for shade, amenity value, shelter and windbreaks, and for fuelwood. Botanical collectors also introduced many species that later invaded; drift sand control and reclamation and tanbark production. Early drift sand control and reclamation and tanbark production. Colonial botanist introduces many new plant species.	Poynton (1979a, 1979b) Shaughnessy (1986) Wells et al. (1986) Cameron (1991)

continues

TABLE 12.1. Continued

Dates	Period	Afforestation	Other Important Events	Main Reasons for Tree Introductions and Plantings	Key References
1872–1914	Organized government involvement in forestry and unification of the Republic of South Africa A single department to conserve indigenous forest and promote tree planting	1872: First properly organized forestry department established in the Cape. Tree Planting Act (1876) promotes tree growing. First plantation of *Eucalyptus globulus* in 1876 and wattles and pines in Eastern Cape from 1877 onward to provide alternatives to indigenous species. The ability of some species, notably *Pinus pinaster* and *Acacia mearnsii*, to spread unaided noted. 1880s onward: Expansion of private *Acacia mearnsii*, *A. dealbata*, and *A. decurrens* plantations for bark production in KwaZulu–Natal and later in the Eastern Cape and Western Cape. Forest Act of 1888 continues systematic promotion of alien tree planting. *Prosopis* species introduced in about 1879–80, followed by more tree trials in the arid interior in 1898. The potential for introducing Mexican pines was recognized in 1903. Tree trials begun in Free State. Rapid expansion of eucalypt plantations for mining timber needs.	Gold discoveries (1886–87) resulted in gold rush. Sugar industry in KwaZulu–Natal expands significantly; imported labor brings in Asian tree species. Increasing economic activity introduces many alien species to the interior and promotes introductions of new species. 1875 onwards: Large-scale drift sand reclamation in the Western Cape, extending to the Eastern Cape by 1893. Rapid population and economic growth.	As above, plus lathes and poles for hut building (rural African population). Trees also planted in the belief that forests protect the soil, minimize surface runoff, and maximize the sustained yield of clear, high-quality water. Alien trees were promoted because they grew faster than any native trees. Systematic introduction of new species, especially Central American pines, for trials on the interior plateau and escarpment. Increasing emphasis on amenity plantings in gardens and urban open spaces.	Above references, plus: Anonymous (1902) Keet (1929) Lückhoff (1973) van der Zel and Brink (1980) Poynton (1990)

1914–1970	Expansion of state and private forestry			Above references, plus: Legat (1917)

Chronic and serious timber shortages during World War I stimulated direct, large-scale involvement of the state in afforestation after 1916, which was matched by the private sector.

Successes achieved by the state with plantations for production of sawlogs stimulated private investments in pine and eucalypt plantations for timber and later pulpwood production.

Expansion of wattle plantations continued rapidly; total area peaked at 363,650 ha in 1960. The collapse of the tanbark industry (after ±1961) led to a virtually complete abandonment of wattle plantations outside KwaZulu–Natal and reduced commercial plantations to about 130,000 ha at present.

Large *Pinus pinaster* plantations were established in the Western Cape. Other important invaders that were extensively planted during this period included *Eucalyptus grandis*, *Pinus elliottii*, *P. patula*, and *P. radiata*.

New species were being introduced (esp. more frost-tolerant eucalypts), but the main emphasis was on (a) improving silvicultural practices to maximize yields; (b) importing new seed stocks of tree species that were proving successful; and (c) tree selection to improve the quality and yields of existing growing stock using seed orchards and other techniques.

The South African economy expanded at an unprecedented pace, driven primarily by exports of minerals. The rapid expansion of towns and populations in the interior, especially in the Gauteng area, resulted in the introduction of numerous species new to these areas. These species have subsequently spread into the river systems that drain this key watershed area, namely the Vaal (and Orange), Limpopo, and Olifants.

Afforestation for sawlog and mining timber production continued to be important and there was a significant increase in plantings, specifically for pulp and paper production. By the early 1970s the annual yield of sawlogs, mining timber, and pulpwood was on a par.

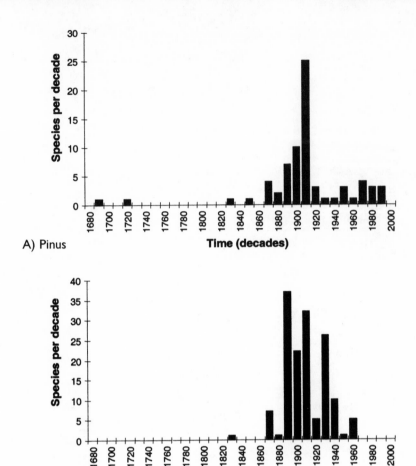

A) Pinus

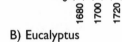

B) Eucalyptus

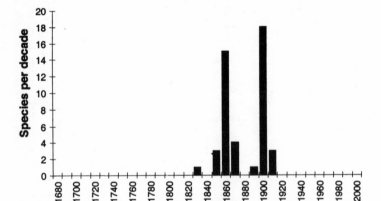

C) Acacia

FIGURE 12.2. First introductions of alien (A)*Pinus,* (B) *Eucalyptus,* and (C) *Acacia* species to South Africa by decade between 1680 and 2000, based on data from Poynton (1979a, 1979b; unpublished data for *Acacia*), Coetzee (1985), and Kietzka (1988).

A second important thrust was the revival of small-scale tree farm-ing—both small-grower schemes launched by the big companies, and community or social forestry and agroforestry driven mainly by govern-ment initiatives (Cellier 1993, DWAF 1996, 1997). Woodlots were initi-ated in the last decade of the nineteenth century to supply fuelwood, poles, and lathes, and were widely promoted until the 1960s. Most of these efforts failed, and interest in this form of forestry waned (Gandar 1984, DWAF 1997, Ham and Theron 1999). Small-grower schemes were promoted by the forestry industry from the late 1980s (Cellier 1991), and there are now about 10,000 growers involved (afforesting 14,000 ha), with about 2,500 new growers (4,000 ha) being added per year (Anony-mous 1999). The new forest policy has also seen the national forestry department switch its emphasis to promoting "rural development forestry" as a way of addressing pressing social and community issues in rural areas (DWAF 1996).

The emphasis on small-grower forestry and particularly rural development forestry is aimed largely at the production of fuelwood and fodder (agroforestry), especially in marginal sites. Little attention is being given to considerations of the potential weediness of some taxa; at best only mild warnings are given about potential invasions. Poynton (1984) recommended 35 species for fuelwood plantations, including 13 that, he notes, spread naturally; these include the top 5 invaders among the Australian *Acacia* species (Versfeld et al. 1998). Esterhuyse (1989) also noted the invasiveness of some recommended species; whereas Donald (1994) presented a weak case in arguing that *Paulwonia* species are unlikely to invade. *Paulwonia* spp. are already invasive in the United States, and climatic matching suggests ample suitable habitat on the eastern coastal seaboard of South Africa (R. Randall, unpublished data).

Important events in recent decades that have affected the afforestation pathway were the slowdown in economic growth in the 1980s and the dra-matic political reforms of the 1990s, culminating in the election of the first representative government in 1994. The policies of the new government, as seen in the Reconstruction and Development Programme (Simon and Ramutsindela 2000), have resulted in significant reforms affecting rural development and thus forestry (Cellier 1993). These initiatives include land reallocation, establishment of small-scale farming, and a renewed emphasis on rural development. The main difference between this period and those summarized in Table 12.1 is the emphasis on introducing new taxa with potential for fuelwood production and agroforestry. Many of these taxa are aggressive invaders and/or are closely related to known invaders, and/or have similar ecological traits to known invaders. These policy changes could have a significant impact on invasion patterns in the future, and they highlight the urgent need for a national strategy for dealing with invasions. Such a strategy must identify protocols for dealing with these conflicts of interest.

Existing Measures for Dealing with Pathways for the Spread of Alien Trees and Shrubs Introduced for Forestry

Developments described above have resulted in the current configuration (spatial arrangement of stands of different species in the landscape) of the South African forestry estate and have created an effective pathway for the spread of alien plants into many natural systems (Richardson et al. 1997). The challenge is to manage the existing pathways while giving due cognizance to new developments in the industry that are changing existing pathways and creating new ones.

ENVIRONMENTAL CERTIFICATION

During the late 1980s private forestry companies started investing more in environmental management, culminating in a set of in-house guidelines (Bigalke 1990, FIEC 1995). Much stricter standards are now being applied following the application of international standards as required by the Forest Stewardship Council (FSC), which specifically requires companies to control and. actively manage alien species to minimize their ecological impacts (FSC 1999). Foreign customers increasingly require South African forestry companies to demonstrate compliance with recognized environmental management standards; this has encouraged most companies to obtain FSC certification. As an additional measure, and to ensure that FSC certification is maintained, many companies have also initiated environmental management systems such as International Standards Organization 14001 (ISO 1996, Lawes et al. 1999, pp. 465–466). Commercial forestry companies also gave strong financial support to the alien-plant clearing projects of the national Working for Water Programme (<http://www-dwaf.pwv.gov.za/wfw/>; Working for Water Programme 1998). This, and the systematic clearing operations in the formerly state-owned plantations, has definitely reduced the importance of the commercial forestry pathway for invasive alien trees and shrubs. Huge damage, however, has already been done, and the existing configuration of plantations guarantees further invasions. Much work is required to mitigate past practices through the implementation of incentives and legislation, and to ensure the development of environmentally sound forestry. Lawes et al. (1999) review recent initiatives in this regard.

AFFORESTATION PERMIT SYSTEM

Another event that had a profound effect on afforestation in South Africa during this period was the formalization of the Afforestation Permit System as a result of the growing concerns about the known impacts of plantations on surface water production (van der Zel 1990, 1995). After 1972, a permit was required for any planting involving areas greater than 10 ha. The permit was issued on the basis of the predicted impact of the reduction in streamflow that would be caused by the planting. (This system has now been subsumed into a general License Assessment for forestry as required by the new National Water Act [36 of 1998]; see "Water Security" on p. 338 for further discussion.)

BIOLOGICAL CONTROL OF COMMERCIALLY IMPORTANT SPECIES

In the 1970s *Acacia mearnsii* growers succeeded in blocking the release of insects for the biological control of this species (Pieterse and Boucher 1997), one of the most important and widespread invaders in South Africa (Versfeld et al. 1998). This delay resulted in a huge cost to the South African taxpayer (De Wit et al. 2001); but the industry's position has now changed, and future releases of biocontrol agents are unlikely to be blocked. Forestry companies have, albeit cautiously, supported investigations aimed at assessing the feasibility of using introduced seed-attacking insects to reduce seed production of invasive, but commercially important, pines (van Wilgen et al. 2000; see also <http://www-dwaf.pwv.gov.za/Projects/wfw/Biocontrol.htm>).

The Future

The National Forests Act (84 of 1998) empowers the Minister of Water Affairs and Forestry to determine criteria to assess whether forests are being managed sustainably, and to develop indicators that may be used to measure this. The extent to which alien species are allowed to invade beyond forestry boundaries is currently not specified among the criteria or indicators. The list is not limited to those specified in the Act, and such a criterion or indicator could be included (e.g., see Lawes et al. 1999, p. 462, Box 2).

Horticulture as a Pathway for the Introduction of Invasive Alien Plant Species

Horticulture (the use of ornamental plants) is an important pathway for the introduction of alien plants to South Africa (Fig. 12.1). Exploring the dimensions of this pathway is hampered by there never having been a central body in South Africa that has orchestrated, managed, or chronicled the importation of plant species for horticulture. That many of the worst invaders of natural systems are showy plants is emphasized in the title of a recent guide to invasive plants of the former Cape Province: *Plant Invaders: Beautiful but Dangerous* (Stirton 1978). Horticulture is an effective and difficult-to-manage pathway for the introduction and dissemination of alien plants. For example, common wisdom regarding which species grow well, require minimal tending, are not susceptible to prevailing climatic extremes or local pests, attract birds (often potential dispersers), and so on, spreads fast among nurserymen. "Winners" are recognized quickly and are soon offered for sale at many nurseries, resulting in the establishment of numerous foci across a very wide range of conditions, thus creating an ideal scenario for initiating invasions.

Despite a recent upsurge in the popularity of "wild gardens" comprising mainly native plants (e.g., Botha and Botha 2000), alien plants conspicuously dominate most public and private gardens throughout South Africa. We know of no surveys done to determine the level of dependence of South African horticulture on nonnative taxa. To get some idea of this dependence, we compared the numbers of native and alien species in different categories of garden plants in the most widely used "handbook" for gardeners in South Africa. In choosing which book to use, we excluded those that promote indigenous plants—our aim was to get a snapshot of plant usage. Results of this analysis (Fig. 12.3) confirm the obvious: gar-

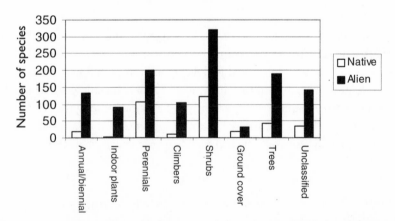

FIGURE 12.3. The number of indigenous and alien plant taxa in eight categories of plants used by gardeners in South Africa. (Unclassified = taxa not placed in any of the other categories, including, e.g., fruit trees, cycads.) Data are extracted from the index of *The A–Z of Gardening in South Africa* (Sheat 1982), one of the most widely used general-purpose gardening books in the country.

deners rely much more heavily on alien plants than natives in every plant category.

The deep-seated preference for non-native taxa has a cultural origin. Early European settlers wanted to reconstruct the gardens of Europe at the Cape. These introductions were, in many cases, assimilated into local culture, which perpetuated their further use. More recently, the ease of communication has exposed South African gardeners to popular garden subjects from other parts of the world. Many gardeners search diligently for "something different" (e.g., Mack 2001). The last decade has seen an explosion of large and small organizations specializing in the dissemination of plants (usually in the form of seeds or bulbs) all over the world. Despite existing phytosanitary regulations and other legislation relating to the introduction of alien organisms, a large proportion of introductions bypass official controls. Representative data on which taxa (or categories of taxa) are in greatest demand have proved impossible to assemble.

A recent survey of many commercial nurseries in the Western Cape showed that few nurseries still sell the very worst invasive species (Brown-Rossouw and Richardson 2000). These results demonstrate an important change in attitude, and probably the influence of a small yet influential group of nongovernmental organizations (NGOs) (notably, the Botanical Society of South Africa; Ivey and Heydenrych 1995) that have publicized the dangers of the most invasive alien plants. More recently, awareness campaigns run by the Working for Water Programme have been influential in this regard. There are still problems, however, as many nurseries sell several species that have only recently started to invade or that currently invade only in localized situations (e.g., *Metrosideros excelsa*). Many nurseries sell a wide range of taxa that are well-known invaders in other parts of the country or elsewhere in the world. A prominent example is *Schinus terebinthifolius*. Unlike the situation in commercial forestry reviewed earlier, where requirements of international buyers have been a major driver of changes in environmental policy, the market for horticultural products constitutes millions of individuals, with no central representative. Because of the vast number of small plantings of ornamentals, it would be difficult or impossible to trace the source of an invasive plant at a particular locality and thus to instigate any legal proceedings against any party. The proposed amendments to the Conservation of Agricultural Resources Act (see pp. 334–335) offer the best prospects of reducing the role of horticulture as a pathway for invasions.

Data are available for CITES-listed plants for which permits are required for (legal) introduction. Data for the past ten years show that the dominant families represented are Anacardiaceae, Cactaceae, Euphor-

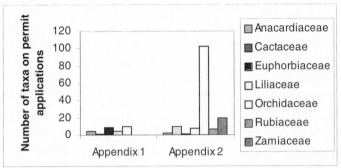

FIGURE 12.4. The number of taxa in the seven main plant families represented in applications for permits for introducing plant taxa listed in Appendixes 1 and 2 of the Convention on International Trade in Endangered Species of Wild Fauna and Flora (CITES) to South Africa between 1991 and early 1999. CITES Appendix 1 includes taxa that are in danger of extinction and are, or could be, affected by trade. Taxa listed in CITES Appendix 2 are not necessarily threatened with extinction, but may become threatened unless trade is strictly regulated. Data source: TRAFFIC East/Southern Africa: South Africa.

biaceae, Liliaceae, Orchidaceae, Rubiaceae, and Zamiaceae (Fig. 12.4). With the exception of Cactaceae, these families are generally underrepresented in weed floras worldwide. Although there are several striking examples of rare plants becoming invasive when moved to new areas (e.g., several South African Iridaceae in Australia), we suggest that introductions of CITES-listed plants are unlikely to contribute significantly to South Africa's invasive plant problem.

BIRDS AND MAMMALS IN TERRESTRIAL SYSTEMS

Reasons for Introducing Alien Vertebrates

Most alien terrestrial vertebrates in South Africa were introduced as novelties or ornamentals for private collections, for game viewing, or for hunting (Brooke et al. 1986, Boshoff et al. 1997). Trading of animals between landowners provides potential for invasion, or for genetic pollution of native species through hybridization. Very few vertebrates have been introduced for biological control of pests (cf. the situation with invertebrates; e.g., Olckers and Hill 1999), or for the production of fur or other fibers, or for commercial use other than hunting (Tables 12.2 and 12.3). Most alien bird taxa, apart from those deliberately introduced by C. J. Rhodes (1853–1902), and the house crow (*Corvus splendens*) that arrived in South Africa without human aid, appear to have been imported for

TABLE 12.2. Mammal species (excluding domestic livestock) that have been deliberately introduced to, or have invaded (as immigrants), or have escaped from captivity in South Africa (Brooke et al. 1986).

Common Name	Species	Date	Reason for Introduction to South Africa	Dispersal and Pathways for Spread within South Africa
rabbit**	Oryctolagus cuniculus	1654	Introduced to Robben Island as a source of meat	None, dispersed by human agency
grey squirrel**	Sciurus carolinensis	1890s	To improve amenities at the Cape	Colonization from place of release
nutria*	Myocastor coypus	1950s	Commercial fur production	Escaped from captivity
capybara*	Hydrochaeris hydrochaeris	??	Unknown, but deliberately introduced	??
house mouse***	Mus musculus	1600+	Immigrant	Feral
house rat***	Rattus rattus	< 800	Immigrant	Feral
brown rat**	Rattus norvegicus	1700+	Immigrant	Feral
North American mink*	Mustela vison	1950s	Commercial fur production	Escaped from captivity
cat***	Felis catus	??	Domestic pet	Feral, dispersed by human agency
dog***	Canis familiaris	??	Domestic pet	Feral, dispersed by human agency
donkey**	Equus asinus	??	Domestic livestock	Feral, dispersed by human agency
wild boar*	Sus scrofa	1920s	Biocontrol of pinetree emperor moth	Feral
llama*	Lama glama	??	Novelty and/or commercial fiber production	??
dromedary**	Camelus dromedarius	??	Transport in desert areas	Feral, dispersed by human agency
blackbuck	Antilope cervicapra	??	Novelty and/or hunting	??
fallow deer***	Cervus dama	1869	Novelty, kept in zoo at Newlands House, Cape Town	Dispersed by human agency
sika	Cervus nippon	??	Novelty and/or hunting	??
sambar	Cervus unicolor	??	Novelty and/or hunting	??

continues

TABLE 12.2. Continued

Common Name	Species	Date	Reason for Introduction to South Africa	Dispersal and Pathways for Spread within South Africa
red deer	Cervus elaphus	??	Novelty and/or hunting	??
hog deer	Cervus porcinus	??	Novelty and/or hunting	??
Pere David's deer	Elaphurus davidianus	??	Novelty and/or hunting	??
Himalayan tahr**	Hemitragus jemlahicus	1935	Novelty (kept in zoo)	Escaped from captivity
addax	Addax nasomaculatus	??	Novelty and/or hunting	??
scimitar-horned oryx	Oryx dammah	??	Novelty and/or hunting	??
mouflon	Ovis musimon	??	Novelty and/or hunting	??
Barbary sheep	Ammotragus lervia	??	Novelty and/or hunting	??
Indian water buffalo	Bubalus bubalis	??	Novelty and/or hunting	?, dispersed by human agency?
Nilgai	Boselaphus tragocamelus	??	Novelty and/or hunting	??

Note: * = extinct or probably extinct in South Africa; ** = population and geographic range in southern Africa stable or decreasing, species usually with restricted range; *** = population and range probably increasing

TABLE 12.3. Alien bird species that have been deliberately introduced or have invaded (as immigrants) southern Africa (Brooke et al. 1986).

Common Name	Species	Date	Reason for Introduction to South Africa	Dispersal and Pathways for Spread within South Africa
mute swan*	Cygnus olor	1918 and 1941	Novelty	Colonization from place of release (HA)
black swan	Cygnus atratus	1926	Novelty	Escaped from captivity
common shelduck	Tadorna tadorna	1974	Novelty	Escaped from captivity
mandarin**	Aix galericulata	1970	Novelty	Escaped from captivity
mallard***	Anas platyrhynchos	<1940	Novelty and/or hunting	Escaped from captivity (HA)
North American black duck*	Anas rubripes	1975	Novelty	Escaped from captivity
European pochard*	Aythya ferina	??	Novelty	Escaped from captivity
tufted duck*	Aythya fuligula	??	Novelty	Escaped from captivity
merlin*	Falco columbarius	??	Novelty?	Escaped from captivity
chukar**	Alectoris chukar	??	Hunting	Colonization from place of release
bobwhite*	Colinus virginianus	??	Hunting	Colonization from place of release
feral chicken**	Gallus gallus	??	Novelty	Escaped from captivity
silver pheasant*	Lophura nycthemera	??	Novelty and/or hunting	Colonization from place of release
ring-necked pheasant*	Phasianus colchicus	c. 1900–1950s	Hunting	Colonization from place of release
common peafowl	Pavo cristatus	1968?	Novelty	Escaped from captivity
Gough Island moorhen*	Gallinula comeri	1893	Novelty	Escaped from captivity
American coot*	Fulica americana	1891	Novelty	Escaped from captivity

continues

TABLE 12.3. Continued

Common Name	Species	Date	Reason for Introduction to South Africa	Dispersal and Pathways for Spread within South Africa
feral pigeon***	Columba livia	1652	Novelty	Escaped from captivity?
turtle dove**	Streptopelia turtur	??	Cage bird	Escaped from captivity
collared dove**	Streptopelia decaocto	??	Cage bird	Escaped from captivity
diamond dove**	Geopelia cuneata	??	Cage bird	Escaped from captivity
sulphur-crested cockatoo*	Cacatua sulphurea	??	Cage bird	Escaped from captivity
grey-headed lovebird	Agapornis cana	late 1890s	Novelty?	Deliberately introduced
cockatiel**	Nymphicus hollandicus	??	Cage bird	Escaped from captivity
ring-necked parakeet***	Psittacula krameri	1850s	Cage bird?	Escaped from captivity
budgerigar**	Melopsittacus undulatus	??	Cage bird	Escaped from captivity
plum-headed parakeet*	Psittacula cyanocephala	1970s	Cage bird	Escaped from captivity
jandaya conure*	Aratinga jandaya	??	Cage bird	Escaped from captivity
brown-throated conure*	Aratinga pertinax	??	Cage bird	Escaped from captivity
dusky-headed conure*	Aratinga weddellii	??	Cage bird	Escaped from captivity
nanday conure*	Nandaynus nenday	??	Cage bird	Escaped from captivity
green-rumped parrotlet	Forpus passerinus	??	Unknown	Deliberately introduced

Common name	Scientific name	Date			Method
house crow***	*Corvus splendens*	early 1970s	Self-introduced		Ship assisted
rook*	*Corvus frugilegus*	late 1890s		To improve the amenities of the Cape	Colonization from place of release
blackbird*	*Turdus merula*	late 1890s		To improve the amenities of the Cape	Colonization from place of release
song thrush*	*Turdus philomelas*	late 1890s		To improve the amenities of the Cape	Colonization from place of release
nightingale*	*Luscinia megarhynchos*	late 1890s		To improve the amenities of the Cape	Colonization from place of release
Eurasian starling***	*Sturnus vulgaris*	1899		To improve the amenities of the Cape	Colonization from place of release
common myna***	*Acridotheres tristris*	1888, 1902, 1930s	Cage bird		Probably escaped from captivity
emerald starling	*Coccycolius iris*	1993	Cage bird		Escaped from captivity
house sparrow*	*Passer domesticus*	1880–1890	Cage bird?		Probably escaped from captivity
Java sparrow*	*Padda oryzivora*	??	Cage bird		Escaped from captivity
strawberry finch*	*Estrilda amandava*	??	Cage bird		Escaped from captivity
red-crested cardinal*	*Paroaria coronata*	1958	Cage bird		Escaped from captivity
red-cowled cardinal*	*Paroaria dominicana*	c. 1960	Cage bird		Escaped from captivity
goldfinch*	*Carduelis carduelis*	1891, 1900	Cage bird, deliberately introduced?		Colonization from place of release
chaffinch**	*Fringilla coelebs*	1890s		To improve the amenities of the Cape	Colonization from place of release

Note: * = may be extinct; ** = population and geographic range in southern Africa stable or decreasing, species usually with restricted range; *** = population and range increasing; HA = spread facilitated by human agency aviaries. Introductions of alien birds to the Cape Peninsula were to "improve the amenities of the Cape by diversifying the bird fauna damned as poverty-stricken" (Shelley 1875, in Brooke et al. 1986).

Mammals

Many alien mammal species were introduced to South Africa (Table 12.2), but very few have become invaders (see Skinner and Smithers 1990). The history of many of the species that have become feral is obscure; species have escaped from captivity or from ships, and it may have been some time before they became invasive and obvious. Native Africans introduced several domestic mammals, including sheep, cattle, goats, and dogs, before A.D. 700 (Deacon 1986). The house rat (*Rattus rattus*) was probably introduced accidentally by Arab traders moving down the east coast, with the gene pool supplemented by later, undocumented arrivals (Brooke et al. 1986). The brown rat (*R. norvegicus*) was carried by ships from Europe, the house mouse (*Mus musculus*) probably by a similar pathway. Some thriving populations of alien mammals stem from escapes from zoos, as in the case of Himalayan tahr (*Hemitragus jemlahicus*) on Table Mountain, or escapes from enclosures on farmlands, as in the case of European fallow deer (*Cervus dama*). Fallow deer are usually kept as novelties or ornamentals, and often there is little control of their population growth.

Few alien mammals have been deliberately introduced into South Africa. The grey squirrel (*Sciurus carolinensis*) was introduced to the Cape Peninsula by C. J. Rhodes as part of his program to "improve" the amenities at the Cape (Brooke et al. 1986). This species has persisted and spread in areas where alien pines and oaks occur, but cannot colonize widely separated patches of these trees, except with assistance from humans (Smithers 1983). North American mink (*Mustela vison*) and the nutria (*Myocastor coypus*) were imported as stock for commercial fur production, and escapees subsequently founded feral populations in the Eastern Cape (Siegfried 1962), but are now extinct in that region. The other mammals in Table 12.2 were imported for a variety of reasons, including biocontrol (*Sus scrofa* to feed on the larvae of the pinetree emperor moth, *Imbrasia cytherea*, in the Western Cape). Other taxa, particularly alien antelope, sheep, and goats, were apparently imported as novelties, or for commercial fiber production (llama, *Lama glama*), or for transport in arid areas (dromedary, *Camelus dromedarius*). Among these species, probably only the mouflon (*Ovis musimon*) and Barbary sheep (*Ammotragus lervia*) are potentially invasive because of their ability to live in rugged and fairly arid habitats.

Birds

Forty-seven alien bird species were recorded in South Africa (Dean 2000). Some of these were intentional introductions, but others were escapees from captivity (Table 12.3) (Liversidge 1985, Brooke et al. 1986). Some species were successful colonists; that is, their distributions are sufficiently widespread to have been mapped in *The Atlas of Southern African Birds* (Harrison et al. 1997). Only seven species have established viable populations, and four of these species (feral pigeon, *Columba livia*; Eurasian starling, *Sturnus vulgaris*; common myna, *Acridotheres tristris*; and house sparrow, *Passer domesticus*) are widespread. All are commensals and are increasing in abundance and expanding in distribution (Brooke et al. 1986).

The house crow is the only alien species to be "self-introduced" (i.e., to have arrived in South Africa without human aid) in southern Africa and has markedly increased in numbers since the first records of its occurrence in the early 1970s (Berruti 1997). This species is known to use ships to "hitch lifts" around the Indian Ocean, and the increase in marine traffic down the east African coast during the closure of the Suez Canal from 1967 to 1980 probably facilitated the arrival of this species at Durban (Brooke et al. 1986).

With the exception of chukar (*Alectoris chukar*), bobwhite (*Colinus virginianus*), ring-necked pheasant (*Phasianus colchicus*) (apparently all introduced for hunting); silver pheasant (*Lophura nycthemera*) (probably introduced for hunting); rook (*Corvus frugilegus*), blackbird (*Turdus merula*), song thrush (*T. philomelas*), nightingale (*Luscinia megarhynchos*), Eurasian starling (*Sturnus vulgaris*), and chaffinch (*Fringilla coelebs*) (introduced to the Cape Town area by C. J. Rhodes), all other alien birds apparently escaped from captivity. Of the six bird species introduced, only the Eurasian starling and the chaffinch have survived—the starling to become abundant and invasive. Chaffinches initially spread on the Cape Peninsula, but their range has subsequently contracted and the species now occupies a tiny range (Maclean 1993).

What Have We Learned from the Past?

The irresponsible importation of alien vertebrates, coupled with careless supervision of caged animals and deliberate releases of captive animals, has contributed to the invasion of natural habitats by several species. Legislation has, on the whole, failed to regulate the importation of alien animals.

Once alien vertebrates are imported, there are virtually no controls in place to limit their spread within the country. Insufficient screening of potentially invasive species is an important factor, but only recently have attempts been made to identify characteristics of invasive alien animals (Williamson 1996, Green 1997).

What Factors Are Driving Introductions Now?

There appears to be little interest in importing large alien mammals into South Africa nowadays, but there is some interest in importing smaller mammals (notably, marmosets, monkeys, and lemurs) and considerable pressure from aviculturalists and other bird fanciers to import alien bird species (particularly parrots, parakeets, and waterbirds). Table 12.4 lists, for birds only, the numbers of species in each family that were recorded on CITES permits for importation into South Africa between 1991 and early 1999. Because of the difficulties of inspecting all such imports at points of entry, it is not known whether all the species listed were actually imported, and some of the species may have been exported or reexported to other countries. Table 12.4 does not include non-CITES species, so potentially invasive birds may be being brought into South Africa without any control. Again, legislation has largely failed to address this issue. Under current legislation, it is necessary to assess the environmental impact of such animals only if they are to be imported and confined in a structure for mass commercial production, imported in circumstances where they were declared an invasive alien species, or where they are to be released for the biological control of pests (Items 3, 4, and 5 on Schedule 1 of GNR 1182 in Government Gazette 18261 of 5 September 1997 as amended in GNR 670 on 10 May 2002). Aside from that, domestic (within South Africa) control of animals is restricted to those that are hunted or kept in captivity. The only direct control on the release of alien wild animals is contained in Nature Conservation Ordinances, such as section 31A of the Nature and Environmental Conservation Ordinance 19 of 1974 (former Cape Province), which prohibits a person from releasing an exotic wild animal without a permit. There are no published guidelines stipulating the circumstances in which such a permit might be issued.

Existing Measures for Dealing with Pathways for the Spread of Alien Birds and Mammals

Although current legislation prevents the importation of potentially invasive birds and mammals, there are apparently no measures for dealing

TABLE 12.4. Number of alien bird species in each family that are listed on CITES permits for importation into South Africa between 1991 and 1999 (TRAFFIC, East/Southern Africa: South Africa unpublished data). The potential to invade is a subjective assessment based on the characteristics of the family with reference to the natural experiment of introductions and invasions of birds worldwide.

Family	No. of Species	Potential to Invade
Accipitridae (hawks, eagles)	2	Low
Anatidae (ducks)	63	High
Bucerotidae (hornbills)	5	Low
Capitonidae (Rhamphastinae: toucans)	11	Low
Columbidae (pigeons and doves)	10	High
Corvidae (crows and magpies)	1	High
Cracidae (curassows, guans)	6	Low
Cracticidae (Australian butcher-birds)	1	High
Estrildidae (waxbills, firefinches, etc.)	40	Medium
Falconidae (falcons and kestrels)*	11	Medium
Fringillidae (canaries and buntings)	7	Low
Gruidae (cranes)	6	Low
Irenidae (leaf-birds)	3	Low
Musophagidae (turacos)	16	Low
Phasianidae (pheasants, quails, etc.)	23	Medium
Phoenicopteridae (flamingos)	2	Low
Ploceidae (sparrows, weavers, etc.)	7	High
Psittacidae (parrots)**	236	Low
Pycnonotidae (bulbuls)	2	High
Rallidae (crakes and rails)	1	Low
Spheniscidae (penguins)	1	Low
Strigidae (owls)	6	Low
Sturnidae (starlings and mynas)	9	High
Timaliidae (Asian babblers)	3	Low

Note: * = Most "species" in this group are hybrids between *Falco peregrinus* and other *Falco* species; ** = Some species in this group have been successful invaders of nonnative ecosystems, but most of the species in this family appear to lack invasive potential.

with pathways for the spread of alien birds and mammals within the country. The importation of an alien species is controlled by a permit system, whereby permits are granted if taxa are considered noninvasive or unlikely to be problematic in other ways. However, different provinces in South Africa apply different policies regarding which alien birds and mammals may be imported. A potentially invasive species that would be banned in one province may be imported into another province, and from there can be moved freely throughout the country. Aviculturalists can trade among themselves, dispersing potentially invasive bird species throughout the country; and alien mammals, particularly large herbivores, can be bought and sold at game sales without restriction.

What Vectors and Pathways of Introduction Are Likely to Exist in the Future?

If informal trade in alien mammals and birds within South Africa continues, there is considerable potential for fecund, aggressively competitive, preadapted species to become invasive. Legislation controlling the importation of alien animals and trading in animals in South Africa is directed principally at CITES-listed species, or indigenous animals (and plants), but is also aimed, to some extent, at the control of potentially invasive alien species.

However, among the families of birds listed in Table 12.4 are several groups that have the potential to invade, or to hybridize with indigenous taxa. Some species of Anatidae (ducks), in particular, have hybridized with indigenous southern African species, and have produced fertile F1 progeny. Similarly, the species of Falconidae (falcons) that are listed on CITES permits are mostly hybrids between *Falco peregrinus* and other large falcons. The Sturnidae (starlings) have a history of invasions in many parts of the world (Lever 1987). Informal trade between aviculturalists could move potentially invasive species into new bioclimates or habitats where they may become successful invaders.

New Measures Needed to Prevent Problems in the Future

It is critical that lists of bird and mammal taxa that are potentially invasive in South Africa, based on their invasive potential elsewhere in the world, be compiled to allow the formulation of legislation that would enable permit applications to be assessed objectively. Furthermore, personnel who are familiar with invasive alien animals elsewhere should assess permit applications. Applications should be assessed using several appropriate criteria. Currently, applications for the importation of species are evaluated by nature conservation authorities and by the Department of Health. A potentially invasive species that slips through the nature conservation screen could be allowed entry because it is not a known vector of diseases.

The present policy is to allow "more responsible" organizations (e.g., zoos, bird parks) more freedom to import alien birds and mammals, but this policy should be revised and imports more tightly controlled. Table 12.5 gives a simplified scheme of the changing importance of six pathways

TABLE 12.5. The relative importance of known pathways for the introduction of alien birds and mammals to South Africa: past, present, and future. Based on information reviewed in this chapter.

Pathway	Past	Present	Future
Agriculture	High	Medium	Medium
Biocontrol	Medium	Low	Low
Commercial fur trade	High	Low	Low
Hunting	Low	Medium	Medium
Pet trade (including aviculturalists)	High	High	High
Self-introductions	Medium	High	High

TABLE 12.6. Present tactics and recommendations to reduce future invasions in each of the identified pathways for alien birds and mammals in South Africa.

Pathway	Present Management Strategy	Recommended Management Strategy
Agriculture	Each case considered on its merits, and permits issued under nature conservation agencies' ordinances and health regulations. CITES species generally not allowed in.	The legislation governing the importation of any animal to be national, not provincial. Prescribed list of potentially invasive species to be compiled to supplement CITES list, with coordination and agreement between provinces. The issue of permits to be completely objective.
Biocontrol	Uncertain	First, a very strong case to be made for the necessity to import the animal. Extensive testing of the animal's potential to be invasive before importation permitted.
Commercial fur trade	Each case considered on its merits, and permits issued under nature conservation agencies' ordinances and health regulations. CITES species generally not allowed in.	The legislation governing the importation of any animal to be national, not provincial. Prescribed list of potentially invasive species to be compiled to supplement CITES list, with coordination and agreement between provinces. The issue of permits to be completely objective.
Hunting	Each case considered on its merits, and permits issued under nature conservation agencies' ordinances and health regulations. CITES species generally not allowed in.	The legislation governing the importation of any animal to be national, not provincial. Prescribed list of potentially invasive species to be compiled to supplement CITES list, with coordination and agreement between provinces. The issue of permits to be completely objective.

continues

TABLE 12.6. *Continued*

Pathway	*Present Management Strategy*	*Recommended Management Strategy*
Pet trade	Each case considered on its merits, and permits issued under nature conservation agencies' ordinances and health regulations. CITES species generally not allowed in.	The legislation governing the importation of any animal to be national, not provincial. Prescribed list of potentially invasive species to be compiled to supplement CITES list, with coordination and agreement between provinces. The issue of permits to be completely objective.
Self-introductions	Low-key, ineffectual attempts to destroy the relevant species.	Elimination of self-introduced alien mammals and birds should be given higher priority and should be coordinated by one authority.

by which alien birds and mammals are introduced into and disseminated within South Africa. Table 12.6 summarizes existing management strategies influencing (or potentially influencing) each of these pathways and the changes that we consider necessary to make them effective.

FRESHWATER FISH

Habitat destruction or degradation and introduced species have had the greatest impacts on native freshwater fishes in South Africa (Skelton 1987, De Moor and Bruton 1988, Davies and Day 1998). These and other threats usually occur together as a lethal "cocktail" threatening biodiversity (Cambray and Bianco 1998). This section describes the main vectors and pathways that have enabled alien fish to invade freshwater ecosystems in South Africa.

Main Vectors and Pathways for Invasions

Anglers—Game Fish

The introduction of game fish to South Africa began in the colonial era. Anglers often belong to well-organized clubs that focus on alien fishes, notably bass and trout species. Angling organizations form a network across South Africa and the world, creating an efficient pathway for dispersing alien taxa. South African anglers also move bass illegally between water bodies by keeping live fish in their boats. Current Nature Conservation Ordinances in each province prohibit the transport of live fish without a permit, but effective policing has been impossible (Impson et al. 1999). Punitive measures and public education are needed to prevent further distribution of alien predaceous fish. One of the fundamental problems is that the angling industry has grown around the introduction and spread of these species, especially trout. One nature conservation body still supplies trout for stocking in public waters, arguing that trout are part of the region's "cultural heritage."

Anglers—Food Fish

Anglers who fish for food, rather than recreation, are not as well organized as the recreational anglers. There are, nonetheless, several recent cases of such anglers moving invasive alien species (e.g., carp and sharp-toothed catfish) between rivers. Enforcement and public awareness campaigns are needed to prevent further intentional movement of alien fish species through this pathway.

Aquaculture—Food Fish

Aquaculture is an important pathway of invasion. Sharp-toothed catfish (*Clarias gariepinus*) are widely available, are easily transported, escape easily from dams, and are potentially highly invasive (De Moor and Bruton 1988, Cambray, in press). In the Kouga River, in the Eastern Cape, a farmer introduced catfish into his in-stream dam in 1993. The first rains after the introduction caused the dam to overflow, starting the invasion of the entire Gamtoos River system, one of the major river systems in the Eastern Cape (Cambray, in press). Tighter control in the issuing of permits is required to prevent potentially invasive aquaculture species from spreading. Once fish are in an area, they can easily be moved between farms. Again, education and awareness campaigns are needed, as is more effective policing.

Aquarium Trade

South Africa has a well-established trade in aquaria fish. Thus, there is the potential for escapes during transit, escapes from breeding ponds, public releases of "pets," and the like, with the additional problem of introductions of pathogens into wild populations of indigenous species. Fish imported from other countries should pass through customs. Some species are prohibited from entering the country, but the effectiveness of this prohibition is unknown. Once in the country, live aquaria fish are moved around quite freely. The pet trade should be closely monitored by nature conservation agencies. The effectiveness of customs procedures also needs to be evaluated. Awareness campaigns need to be aimed at the pet industry and consumers—for example, informing collectors of the implications of discarding alien fish.

Public Aquaria

Public aquaria are a potential source of invasive species. Aquarium managers may have permits to import species, but fish can be removed illegally from aquaria and then invade river systems. Managers of public aquaria currently require permits to move fish. Security must be ensured and staff properly educated on the consequences of illegal movements. Potentially invasive fish species should not be imported into aquaria.

Private Individuals (Garden Ponds)

Some ornamental ponds are located along river courses and are connected to rivers during floods, allowing fish to escape. As with aquarium fish, there is also the potential that owners of garden ponds will dump fish into nearby rivers. Garden pond fish are purchased through the pet trade, but there are also private pond-to-pond movements. Awareness campaigns are needed, and the pet trade industry should be closely monitored by nature conservation agencies to ensure that potentially invasive species are not stocked in garden ponds near river courses.

Interbasin Transfers

The tunnel connecting the Fish and Orange Rivers is one of the best known examples of an engineered pathway allowing invasive freshwater fish species to move between river systems in South Africa. This interbasin transfer (IBT) scheme was a major engineering feat (<http://www.dwaf. gov.za/orange/orange-f.htm>), but an ecological disaster (Cambray and Jubb 1977). Not only could Orange River fish cross into another catchment, but anglers could transport them to other rivers in the Eastern Cape (the sharptooth catfish, *Clarias gariepinus*, is a notable example), causing extensive ecological damage. No preventive action was taken before the opening of this tunnel in 1975. Since then, there have been other IBT projects that have led to the translocation of fish species. An example is the Tugela–Vaal pumped storage scheme pipeline, which allowed three fish species from the Free State (*Barbus aeneus*, *Labeo capensis*, and *L. umbratus*) to invade the extreme headwaters of the main Tugela River (Mulder and Engelbrecht 1998; Albany Museum, unpublished records). These species survived the 550-meter altitude difference and the turbines in the pipeline, and now pose threats, through hybridization and competition, to the endemic fish in the Tugela River system. Further IBT projects are planned, and funds are required to prevent the translocations of species by this pathway. The organization overseeing these projects must be held accountable for any translocation of species across natural barriers. This accountability would put a real cost to the possibility of introducing species, and funds could be allocated in the planning stages for preventive action.

Researchers

Researchers import fish for scientific experiments from across natural barriers (both within South Africa and from other parts of the world). Fish from these sources can escape as a result of poor security measures in ponds and aquaria. Researchers currently require permits to move fish, but stricter control is needed, with punitive measures if fish escape or are "loaned."

Internationally Shared Rivers

It is often noted that aquatic ecosystems are poorly understood, relatively unmanageable, shared by multiple users, and highly vulnerable to human interventions. The situation is even more complex where rivers cross national boundaries. Good legislation, policies, and enforcement in one country are useless if such measures do not exist or are not enforced in other countries through which the same river flows. International agreements are required to prevent invasive species spreading from one country to another.

The Changing Importance of Vectors and Pathways over Time

Pathways provided by nature conservation agencies have become less important, whereas all other categories of pathways—notably, those formed by anglers and interbasin transfer projects (Table 12.7)—have become more important. The pathways of nine alien (mostly game fish) and five translocated freshwater fishes in South Africa that are considered to have significant past, present, and possibly future impacts are given in Table 12.8.

Impacts of pathways provided by engineering projects (mainly interbasin transfers) are probably the most underestimated. With the increasing number of interbasin transfer projects, this pathway could be the main conduit of introductions in the future. Engineers and aquatic biologists need to cooperate to limit the impact of this pathway.

Current and recommended preventive measures are outlined in Table 12.9 for each of the eleven identified pathways. Overall, there is a need for

TABLE 12.7. The relative importance of known pathways for the introduction of alien freshwater fish into open waters of South Africa: past, present, and future. Based on information reviewed in this chapter

Pathway	Past	Present	Future
Nature conservation agencies	High	Medium	Low
Anglers (game fish; including translocated native taxa)	High	High	High
Anglers (food)	Medium	Medium	Medium
Aquaculture (game fish)	Medium	Medium	Medium
Aquaculture (food fish)	Medium	Medium	Medium
Aquarium trade	Medium	Medium	Medium
Public aquaria	Very low	Very low	Very low
Private individuals (garden ponds)	Low	Low	Low
Interbasin transfers	Medium	High	High
Researchers	Low (few*)	Very low	Very low
International rivers	Low	Low	Low

* Movement of fish by several researchers into "homelands" during apartheid years.

public awareness campaigns that should be backed by effective legislation. The lack of personnel and resources in many South African nature conservation agencies will hinder implementation of legislation.

What Vectors and Pathways of Introduction Are Likely to Exist in the Future?

Pathways provided by anglers and engineering works remain a serious problem. There may also be an increase in the hazard from the aquarium trade, which will include the possible introduction of alien pathogens, as well as additional invasive fish species. Every section of river or water body in South Africa that has not been invaded by alien/translocated fish species should be identified and given protection. There should also be river rehabilitation in areas where endemic species can still be saved from past invasions. Without proper legislation, heavy fines, and implementation by public agencies, the conservation of the native freshwater fish fauna—notably, the smaller species—looks bleak, especially in the Eastern and Western Cape provinces. Awareness campaigns need to be intensified and environmental groups co-opted to conserve what remains of the highly threatened freshwater fish fauna of South Africa.

TABLE 12.8. Pathways of alien (al) and translocated (tr) freshwater fishes in South Africa that are considered to have had major impacts in the systems in which they occur.

				Pathways				
Species	Nature Conservation Agencies	Anglers	Private Hatcheries	University Hatcheries	Aquarium Trade	Private	Farmers	Interbasin Transfers
Oncorhynchus mykiss (al)								
Rainbow trout	P,Pr(fewer),F?	P,Pr,F	P,Pr,F	P,Pr,F?			P,Pr,F	
Salmo trutta (al)								
Brown trout	P,Pr,F?	P,Pr,F	P,Pr,F	P,Pr,F?			Pr,F	
Carassius auratus (al)								
Goldfish	P		P,Pr,F	Pr,F	P,Pr,F	P,Pr,F	P,Pr,F	
Ctenopharyngodon idella (al)								
Grass carp	P,Pr,F?	P,Pr,F	Pr,F	Pr,F			P,Pr,F	
Cyprinus carpio (al)								
Common carp	P,Pr,F	P,Pr,F	P,Pr,F	??	P,Pr,F	P,Pr,F		P,Pr,F
Lepomis macrochirus (al)								
Bluegill sunfish	P	P,Pr,F					P,Pr,F	
Micropterus dolomieu (al)								
Smallmouth bass	P	P,Pr,F					P,Pr,F	
Micropterus punctulatus (al)								
Spotted bass	P	P,Pr,F					P,Pr,F	

continues

TABLE 12.8. Continued

				Pathways				
Species	Nature Conservation Agencies	Anglers	Private Hatcheries	University Hatcheries	Aquarium Trade	Private	Farmers	Interbasin Transfers
Micropterus salmoides (al)								
Largemouth bass	P,Pr(fewer),F?	P,Pr,F					P,Pr,F	
Labeo umbratus (tr)								
Moggel	Pr,F(Free State Conservation)							P,Pr,F
Clarias gariepinus (tr)								
Sharptooth catfish	P,Pr,F	P,Pr,F	P,Pr,F	P,Pr,F			P,Pr,F	P,Pr,F
Oreochromis mossambicus (tr)								
Mozambique tilapia	P	P,Pr,F	P,Pr,F				P,Pr,F	
Tilapia rendalli swierstrae (tr)								
tilapia Southern redbreast	P,Pr?	P,Pr,F	P,Pr,F				P,Pr,F	
Tilapia sparrmanii (tr)								
Vlei tilapia	P	P,Pr,F	P,Pr,F				P,Pr,F	
TOTAL PAST	13	12	8	3	2	2	14	3
TOTAL PRESENT	7 (fewer agencies)	12	9	5	2	2	14	3
TOTAL FUTURE	6?	12	9	5?	2	2	14	3

(P = past; Pr = present; F = future) Based on information reviewed in this chapter.

TABLE 12.9. Management strategies currently in place and recommendations to reduce future spread along identified invasion pathways for freshwater fishes in South Africa.

Pathway	Present Management Strategy	Recommended Management Strategy
Nature conservation agencies	Internal management decisions for each province vary; e.g., some provinces still officially produce invasive alien fish species for distribution.	National panel to investigate and coordinate provinces
Anglers (game fish)	Nature conservation agencies ordinances	National panel to investigate and coordinate legislation
Anglers (food)	Nature conservation agencies ordinances	Public education on the ecological dangers of moving fish and legal implications; e.g., vehicle confiscation, heavy fines, cleanup costs
Aquaculture (game fish)	Nature conservation agencies ordinances	As for anglers (food)
Aquaculture (food fish)	Nature conservation agencies ordinances	Should be controlled by one department, not shared by Departments of Agriculture and Conservation (which leads to uncoordinated decisions). Public education on the ecological dangers of moving fish and legal implications, e.g., vehicle confiscation, heavy fines, cleanup costs; i.e., effective legislation.
Aquarium trade	From outside: customs; from within country: Nature conservation agencies ordinances	Assess whether there is effective customs control for incoming fish. Public education on the care and control of aquarium fish.
Public aquaria	Nature conservation agencies ordinances	As for aquarium trade
Private individuals (garden ponds)	Nature conservation agencies ordinances	As for aquarium trade
Interbasin transfers	Impact Assessment Regulations of 1996	Revisit existing IBTs to investigate problems. Future IBTs: hold engineers accountable for any introductions.
Researchers	Nature conservation agencies ordinances	Alien invasive awareness at tertiary institutions
International rivers	?	Political agreements with neighboring countries

MARINE ORGANISMS

The only alien marine species known to have been deliberately introduced to South Africa are commercially cultured mollusks, including small experimental shipments of California red abalone (*Haliotis rufescens*), Manila clams (*Tapes phillipinarum*), and several species of oyster (Griffiths et al. 1992). Despite the continuing introduction of millions of oyster spat each year by the mariculture industry, there have been no published records of any of these species becoming naturalized (De Moor and Bruton 1988). However, at the time of writing naturalized populations of *Crassostrea gigas* have just been discovered in four permanently open estuaries along the Southern Cape coast (T. Robinson and C. L. Griffiths unpublished data).

Accidental introductions have taken two main pathways of transport: species attached to the hulls of ships or oil rigs, or on dry ballast, and planktonic and larval forms in the ballast water of ships (see Carlton 1989). The former mechanism has operated for centuries and may have been responsible for the early introduction of species now regarded as part of the natural biota. The use of ballast water is a more recent phenomenon and is now almost certainly the principal vector of new marine introductions.

Griffiths et al. (1992) listed seven mollusks, one polychaete, and two crab species either known or suspected to have been accidentally introduced to the South African coastline. One of these, the crab *Pilumnoides perlatus* (then thought to be an introduction from South America), has subsequently been reclassified and is now recognized as a separate endemic species, *P. rubus*. Several additional introductions have also come to light since 1992, including the anemone *Metridium senile* (Griffiths et al. 1996) and a number of ascidian species (Monniot et al. 2001). All of these are confined to present or historical harbor areas. Only two introduced species have become invasive and established widespread populations along the South African coastline: the Mediterranean mussel, *Mytilus galloprovincialis*, and the European shore crab, *Carcinus maenas*.

Mytilus galloprovincialis was first recorded in South Africa by Grant et al. (1984), by which time it had already established large populations along the west coast between Cape Point and Lüderitz in Namibia. The exact date and site of introduction are unknown, but circumstantial evidence suggests that the introduction was recent, perhaps in the late 1970s or early 1980s, and mediated by humans (Hockey and van Erkom Schurink 1992, Griffiths et al. 1992). The species was not recognized as an alien before becoming so widespread, as it is superficially similar to the indigenous *Choromytilus meridionalis.* Initial recognition relied on genetic techniques (Grant and Cherry 1985), although methods of separating the species in the field are now established (van Erkom Schurink and Griffiths 1990). By the early 1990s *Mytilus galloprovincialis* had spread as far east as Port Alfred, on the southeastern Cape, and was the dominant intertidal organism along the entire west coast, with an estimated biomass of 194 tons of wet mass per kilometer of rocky coast (van Erkom Schurink and Griffiths 1990).

The other invasive alien on South African shores, the European shore crab, *Carcinus maenas*, was first recorded from Table Bay Docks in 1983 (Joska and Branch 1986) and by 1990 had spread from Camps Bay to Saldanha Bay (le Roux et al. 1990), a distance of some 100 kilometers. No subsequent expansion has been noted, and all existing records are from wave-sheltered sites, suggesting that this species has difficulty establishing on the exposed open coast of South Africa. For this reason, and because of dietary differences between the species (le Roux et al. 1990), it seems unlikely that *C. maenas* will compete with or displace indigenous crab species colonizing open rocky shores to any significant extent (le Roux et al. 1990, Griffiths et al. 1992). This species is, however, a voracious, hardy, and adaptable predator that has become a serious pest in other regions where it has been introduced (<http://www.natureserve.org/publications/leastwanted/crab.html>). There is concern that it may pose a threat if it becomes established in the Saldanha Bay–Langebaan Lagoon region, which is both the site of an important marine national park and the center of the mariculture industry in South Africa. Although a single mating pair was collected at this site in the late 1980s (le Roux et al. 1990), there have been no subsequent records from this site.

No invasive alien seaweeds have been reported from South African waters (J. J. Bolton, pers. comm.) despite the fact that several species of large brown algae, notably *Undaria* spp. and *Sargassum muticum*, have spread around the world, apparently with the oyster industry (e.g., Rueness 1989).

The Changing Importance of Vectors and Pathways over Time

Oceangoing vessels capable of introducing marine aliens have been visiting the Cape for about five centuries. The nature and speed of these vessels, and hence their efficacy as vectors, have changed over time. Early vessels were wooden and slow; they were efficient transporters of sessile species, such as barnacles, and of wood-boring forms—notably, shipworms and gribbles. Many such species have been moved all over the globe, and it is difficult or impossible to determine the natural distributions for many taxa. The steel vessels used in later years are unsuitable for borers and, with the introduction of antifouling paints, now carry reduced loads of fouling organisms. Although there have been some changes over time in the relative importance of trading routes (e.g., with opening of the Suez Canal and increased movement of oil around the Cape with the subsequent introduction of supertankers unable to use the canal), the rate of introduction of new external fouling organisms is probably declining. At the same time, the use of ballast water by oil tankers and bulk carriers has come into common use. Because of the increased speed of these vessels, this mechanism is capable of transporting both adult planktonic and larval benthic forms. The general trend has been a decrease in the transport of sessile invertebrates and an increase in the introduction of larval and planktonic forms, which can include pathogens and toxic dinoflagellate spores. We know of two possible introductions of harmful algae via ballast water to South African waters. *Aureococcus anophagefferens* is a small coccoid, nonmotile pelagophyte. Since 1996 it has formed high-biomass blooms in Saldanha Bay that harm the mariculture industry. *Gymnodinium mikimotoi*, a dinoflagellate, forms blooms in False Bay. These taxa probably arrived in ballast water because they appeared suddenly, not having been previously recorded, and then bloomed regularly. The former appeared in a port, whereas the appearance of the latter has been linked to a tanker that underwent repairs in False Bay (G. Pitcher, pers. comm.). The construction of new harbors has clearly established new foci for introduction. A prime example has been the construction of new harbors in Richards Bay and Saldanha Bay in the 1970s. Both these sites lie in ecologically sensitive and previously pristine bays. The volume of ballast water dumped into these systems is huge; in 1999, 805 ships deballasted 19.5 million tons of water into Richards Bay, and 403 ships deposited 9.1 million tons into Saldanha Bay (Portnet, unpublished records).

Other more modern vectors include the import of spat of organisms used for mariculture. Most of the oysters reared in South Africa are raised from spat imported from Chile. Such imports always carry the risk that other alien species, especially pathogens, may be introduced along with the spat. Some translocation of pest species within South African waters has also been mediated by mariculture. For example, the movement of mussel ropes stocked with the invasive *Mytilus galloprovincialis* from Saldanha Bay to Algoa Bay established a new focus from which this species expanded up the east coast. Other potential modes of introduction are the release of organisms from aquaria and

TABLE 12.10. The relative importance of known pathways for the introduction of alien marine organisms to South African waters: past, present, and future. Based on information reviewed in this chapter.

Pathway	Past	Present	Future
ACCIDENTAL			
Attachment to hulls of ships and rigs	Moderate/high	Moderate	Moderate
Dry ballast	Moderate	Very low	Very low
Ballast water	None	Moderate/high	High
DELIBERATE			
Mariculture	Low	Moderate/high	Moderate/high
Aquaria, curio, and pet trade	None	Low	Moderate

research laboratories. Several such introductions were documented in other parts of the world (Carlton 1989), but none are known from South Africa. The changing importance of vectors/pathways is summarized in Table 12.10.

What Vectors and Pathways of Introduction Are Likely to Exist in the Future?

Despite the fact that ships have been calling at the Cape for hundreds of years, several new introductions have been documented over the past two decades, as mentioned earlier. The reasons for these new introductions are not clear. They may have arrived in the ballast water of tankers. Alternatively, they may have been introduced on the hulls of oil rigs, which made their first appearance in the Cape in the 1970s; the rigs may act as vectors for species not carried by, or unable to survive on, the faster-moving tankers. The principal vector now is thought to be ballast water, and this is likely to remain the main vector in the future. Deliberate importation of species for aquaculture and the aquarium trade also needs to be tightly controlled.

Carter (1996) lists recommendations for the mitigation of the ballast water pathway. They include the chemical or heat treatment of ballast water, the storage of such water in onshore tanks, or open sea exchanges of ballast water. The most appropriate of these methods appears to be open-sea exchange, although vessels that lack segregated ballast tanks (and thus carry contaminated ballast) could discharge into holding tanks onshore.

This water can be either treated and released, or loaded onto other vessels leaving the harbor.

LEGISLATION

Few laws in South Africa regulate invasive alien organisms. Most environmental legislation deals primarily with the protection of a natural resource, such as water, or with the conservation of agricultural resources. In some instances, the lack of legal controls molds the pathway of invasion. South African law, including the regulations that affect alien species, is generally command-and-control based (i.e., prohibited acts are declared and penalties are set for contraventions). However, in some cases, laws also provide incentives that may shape invasion pathways.

Laws Affecting the Movement of Alien Species

Several international conventions and national laws are broadly applicable to invasive alien taxa in that they operate (at least potentially) to block pathways. It is important to note that there are activities or invasive taxa that are not controlled; for these, it is the absence of legislation that shapes the pathway.

International Controls

Several international conventions affect pathways for invasion directly or indirectly. South Africa is party to some of these, including the Convention Relative to the Preservation of Fauna and Flora in Their Natural State, the International Plant Protection Convention, the Convention on Wetlands of International Importance Especially as Waterfowl Habitat, the Convention on International Trade in Endangered Species of Wild Fauna and Flora, the United Nations Convention on the Law of the Sea, and the Convention on Biological Diversity. Other international legal instruments are the World Trade Organization's Agreement on Sanitary and Phytosanitary Measures and the U.N. Food and Agriculture Organization's Codes of Practice on the Introduction of Marine Organisms and on the Import and Release of Exotic Biological Control Agents. It is important to realize that international law obligations become binding on individuals only after they have been embodied in national law. Because South Africa, in many instances, has not enacted appropriate domestic legislation, membership of these conventions and treaties is largely of academic interest. This is particularly problematic because it is the spread of invasive aliens within South Africa, as much as the initial introductions, which constitutes the threat to the region's biodiversity. The imperative to enact laws that control such spread is therefore great. The following are the most significant international legal instruments to which South Africa is party:

- The International Plant Protection Convention is primarily concerned with plant quarantine in the context of international trade (<http://www.fao.org/ag/agp/agpp/pq/en/ippce.htm>). To this end, it envisages a system of phytosanitary certificates based on a set of international standards. These requirements have, to some extent, been enacted in South Africa's domestic legislation through the Agricultural Pests Act (see page 336). However, their primary focus is on organisms likely to endanger agricultural resources rather than biodiversity.
- The World Trade Organization's Agreement on Sanitary and Phytosanitary Measures is also based on phytosanitary standards (<http://www.wto.org/english/tratop_e/sps_e/spsund_e.htm#Intro>). However, it recognizes the potential that such requirements may have for creating a disguised restriction on international trade and the difficulty of developing countries in complying with these standards. It therefore seeks to achieve a balance between the need for uniform phytosanitary requirements and the inability of all countries to implement the same measures. The indirect effect of this balance may be to create pathways for invasive aliens, but we have no evidence of this.

The Convention on Biological Diversity (the CBD; <http://www.biodiv.org/>) requires signatories to "prevent the introduction of, control or eradicate those alien species that threaten ecosystem, habitats or species." The CBD also requires signatory states to provide financial support and incentives for national activities to achieve the objectives of the CBD, including the control of invasive species. Conventions are binding on nations rather than individuals, and they give rise to legal controls that are enforceable against individuals only when the obligations contained in these conventions have been enacted in domestic legislation. South Africa has yet to do this in the case of the CBD, but the National Environmental Management: Biodiversity Bill on the conservation and sustainable use of South Africa's biodiversity has been published for comment as B30-2003 (in the *Government Gazette* 24935 30 May 2003). This Bill recognizes that past efforts of control were unsuccessful, primarily because responses were reactive. The Bill proposed strengthening existing legislation controlling the introduction and spread of potentially harmful organisms and improved enforcement. It also proposes the development of regulatory procedures for comprehensively assessing the potential risks of the introduction of alien organisms before their introduction. Initiatives to encourage landowners to control or eradicate alien organisms that threaten biodiversity were also proposed (<http://easd.org.za/sapol/diversity.htm>).

CITES (the Convention on International Trade in Endangered Species of Wild Fauna and Flora; <http://www.cites.org/>) regulates the cross-boundary movement of organisms and imposes an obligation on parties to penalize those trading or in possession of taxa protected by the convention (article viii). Obviously, the primary aim of the convention is the protection of endangered taxa, but it does regulate the importation of alien organisms.

National Controls Affecting the Movement of Alien Species

South African common law provides no direct mechanism for addressing the spread of invasive alien organisms. The law of nuisance, based on property rights, prohibits anyone from interfering unreasonably with another person's right to use or enjoy his/her property. Unless it could be demonstrated that the spread of invasive aliens is infringing on this right, the common law would not be a useful instrument of control. One circumstance that could conceivably give rise to an action at common law is where invading aliens can be shown to be detrimentally affecting the specific purpose for which the property is being used.

The Constitution of the Republic of South Africa Act (108 of 1996; <http://www.gov.za/constitution/>) guarantees everyone the right to an environment that is not harmful to their health or well-being (section 24[a]). The same section also imposes an obligation of the State to protect the environment for the benefit of future generations through reasonable legislative and other measures that prevent "ecological degradation, promote conservation, and secure ecologically sustainable development." The failure of the State to introduce legislative and other measures that might promote biodiversity and adequately regulate invasive aliens is arguably a contravention of this right. It may well be possible to enforce control of invasive aliens through the Constitution.

Few statutory controls have as their primary purpose the control of invasive aliens. Those that do exist either operate to serve some other goal, such as the protection of water, or are of a broad nature. Plant invaders are more rigorously controlled by legislation than invasive animals, except when the animals constitute an agricultural pest. The existing controls are as follow:

REGULATIONS PROMULGATED UNDER THE CONSERVATION OF AGRICULTURAL RESOURCES ACT (CARA, 43 OF 1983):
The Conservation of Agricultural Resources Act allows the Minister of Agriculture to declare any plant species an "invader plant." In terms of CARA, regulations have been passed specifying plants and identifying

areas in which each is declared an invader plant. Of particular relevance is the provision that provides for the listing of alien plant taxa in three categories (<http://www.gov.za/gazette/notices/1999/2485.htm>). The proposed amendments include two possible formulations, both of which regulate declared weeds and invader plants, and distinguish between (1) plants that are harmful to humans, animals, or the environment and serve no obvious economic purpose; (2) plants that are commercially useful but invasive; and (3) invasive plants that have value as ornamentals. According to this proposal, category 1 species are not permitted in any area under any conditions, whereas species in the other two categories may be allowed. Category 2 species may occur or be established in areas demarcated for that purpose, subject to certain conditions. Category 3 species may be established on any land as long as they are controlled where they are a threat to the natural resources or where they invade the natural vegetation. Seeds of species in category 3 may not be imported and used in trading. The consequence of declarations in CARA is that land users in affected areas are required to control listed invader plants if they occur to such an extent that they are, or could be, detrimental to the production potential of natural agricultural resources. According to the Act, "control" means the combating of invader plants, but not at the expense of the production potential of natural agricultural resources. It is therefore possible that in the interests of agricultural resources, these regulations will shape invasion pathways. To ensure that the necessary measures are being effected, the Minister (of Agriculture) is empowered to carry out activities to control invader plants on private or public land and recover the proportion of the costs that relate to any beneficial effect on the land from the landowner. If there are no benefits to the owner, then the minister may be reluctant to take these steps, since he or she will not be able to recover any of the costs.

ENVIRONMENT CONSERVATION ACT (73 OF 1989)

Although largely repealed by the National Environmental Management Act (NEMA, 107 of 1998), the Environment Conservation Act (ECA) contains some provisions that are still in force and that may be used to affect pathways of invasion:

- Section 31A allows a competent authority to stop an activity that may seriously damage or detrimentally affect the environment, or to direct an individual to take the steps necessary to eliminate or reduce the harm. Such activities may include the spread (or failure to prevent the spread) of invasive alien plants or animals. NEMA contains a very similar provision and may be put to similar use.

- The Environmental Impact Assessment Regulations: The Minister of Environmental Affairs and Tourism is empowered to identify activities that "may have a substantial detrimental effect on the environment." These activities may not be undertaken except with the written authorization of the minister, or by a competent authority. Such an authorization may be issued only after compliance with the regulations, which may include the submission of an environmental impact assessment report. For our purposes, the most important activity identified is the "intensive husbandry of, or importation of, any plant or animal that has been declared a weed or an invasive alien species." However, the regulation of concentrating livestock in a confined structure for the purpose of mass commercial production and the release of any organism for the purpose of biological pest control may also have a bearing on pathways.

AGRICULTURAL PESTS ACT (36 OF 1983)

This act prohibits the importation of plants, pathogens, exotic animals, growth medium, "infectious things" (any thing, except a plant, which may serve as a medium for the importation or spreading of any pathogen, insect, or alien animal), honey, beeswax, or used apiary equipment into the country without a permit and then only through a prescribed port. Significantly, a permit will be granted only if the importer is able to furnish a phytosanitary certificate. Such a certificate will be acceptable to the competent authorities if it can be declared that the plants in question were inspected at some point during their growth and found to be visually free of disease. It is possible, however, that after the inspection, the material to be imported will contract a disease. In this way, notwithstanding the controls, invasive alien species may be spread. The act also regulates the existence of defined agricultural pests on land. Again, the primary aim of this act is to preserve agricultural resources. In practice, this act is effective only in limiting the introduction of pests (especially parasites and pathogens) of plants and animals. Provisions in the act for screening vertebrate animals and higher plants for their potential invasiveness are, in our view, ineffective. This shortcoming stems from the lack of an objective and practical means for deciding which taxa to allow in and which to exclude.

Provincial Controls Affecting the Movement of Alien Taxa

Each of the four former provinces enacted a Nature Conservation Ordinance. All have similar provisions with regard to the control of invasive aliens. For the sake of brevity, only the Nature and Environment Conservation Ordinance, enacted by the former Cape Province, is discussed here. It controls alien invaders by regulating aquatic vegetation and establishes a system of provincial, local authority, and private nature reserves within which measures must be taken to protect indigenous flora. It also protects some animal species. However, these protections relate to the prevention of damaging plants and directly harming animals, rather than the introduction of other (invasive) species that may have the same effect. The introduction into any inland waters of any aquatic growth is prohibited without a permit. The Nature Conservation Board is required to take the necessary measures to control aquatic growths in inland waters. The Director of the Nature Conservation Board may order owners of inland waters to destroy aquatic growth considered injurious to other aquatic growth or to the water. The cultivation, possession, transportation, sale, donation, purchase, acquisition, or importation into the province of a noxious aquatic growth is prohibited.

The keeping of nonnative organisms is also governed under this ordinance and is prohibited without the permission of the Nature Conservation Board. The Board's policy is to grant permits if it is satisfied that the specimen came from a legitimate source, that the organism is kept in adequate facilities, and that it will not be released into the wild. There is apparently no national policy on this issue.

Laws Impacting on the Economic, Social, and Political Forces That Drive Invasions

There are several laws that, although not directly concerned with invasive aliens, impact on them indirectly. Underpinning each are different imperatives, mostly economic. For example, CARA is primarily concerned with preserving agricultural resources, thereby ensuring national food security. Forestry represents an important source of revenue for the country, and the National Forests Act has to some extent facilitated the spread of invasive aliens because of the (short-term) economic benefit it afforded. The following are pathways afforded (or denied) to the spread of invasive aliens by different national acts.

Food Security

CARA is the only national legislation that regulates invasive species, but its primary focus is conserving agricultural resources. CARA allows the minister to provide financial assistance to land users for combating invasive plants, and allows for the provision of departmental officials who may give advice to land users on the control of invasive plants and may enter private land to do so.

Water Security

NATIONAL WATER ACT (36 OF 1998)

The National Water Act (NWA), which aims to ensure water security, exemplifies the way in which social and economic pressures have indirectly led to control over the spread of invasive aliens. To the extent that invasive aliens may threaten water security, they were included in NWA to deal with the threat. These powers allow the Minister (of Water Affairs and Forestry) to declare "streamflow reduction activities" (SRAs), over which specified controls may be exerted (<http://www.dwaf.gov.za/ sfra/>). The only activity currently specified in NWA is commercial afforestation, but the Minister can declare other forms of land use to be SRAs if they are "likely to reduce the availability of water in a watercourse to the Reserve, to meet international obligations, or to other water users." It is possible that other activities involving invasive alien plants will be declared streamflow-reducing activities if they impact on water security.

The Minister may also grant financial assistance for clearing invasive alien plants that impact negatively on water resources. Furthermore, NWA provides that water use charges may be imposed to cover the costs of water resource protection. If such protection includes the clearing of invasive alien plants that impact negatively on the water resource, water users might be charged for that, particularly if it was their act or omission that caused the spread of the invasive alien species.

MOUNTAIN CATCHMENT AREAS ACT (63 OF 1970)

The Mountain Catchment Areas Act (MCAA) provides for the conservation, use, management, and control of land situated in mountain catchment areas. Underlying this act is a need to protect the quality of headwaters in watersheds. The MCAA allows competent authorities (designated in each province) to declare any area as a mountain catchment area. For such areas, the competent authority may direct that specified control measures be taken (such measures also apply within 5 kilometers of the boundary of the mountain catchment area). Such directives may include the protection and treatment of natural vegetation within the specified area and the destruction of "intruding vegetation."

Economic Security

NATIONAL FORESTS ACT (84 OF 1998)

The National Forests Act (NFA) promotes the sustainable management and development of forests. It specifies principles that must guide decisions affecting forests, including the requirement that "forests must be developed and managed so as to conserve biological diversity, ecosystems and habitats . . . [and to] conserve natural resources, especially soil and water."

The minister is required to monitor forests, but not specifically with reference to their impact on the spread of invasive aliens. Although there are currently no requirements contained in the NFA or regulations for the control of alien invaders, the minister may develop criteria, indicators, and standards regarding sustainable forest management on the advice of the Committee for Sustainable Forest Management. These could include management practices that reduce the spread of invasive alien plants (Lawes et al. 1999, p. 462). The minister may create or promote certification programs and other incentives to encourage sustainable forest management, which may include management of the spread of invasive alien plants.

The Marine Environment

Regulation 68 of the Marine Living Resources Act (18 of 1998) prohibits anyone from releasing into South African waters any fish without the written permission of the minister (indigenous wild fish caught in the Republic are excluded). "Fish" is defined to include "the marine living resources of the sea and the seashore, including any aquatic plant or animal, whether piscine or not and any mollusk, crustacean, coral, sponge, holothurian or echinoderm, reptile and marine animal, and includes their eggs, larva and all juvenile stages, but does not include seabirds and seals." Accordingly, aside from birds and sea mammals, this prohibition is extensive. It is, however, not well policed, and thus remains an ineffective regulator of pathways for the introduction and spread of alien organisms.

SUMMARY OF LEGISLATION

No laws specifically control invasive alien species, except to the extent that these serve other interests, such as the preservation of agricultural resources. Accordingly, there are legal instruments that may be used to prevent the spread of invasive aliens, but that is not their primary aim. Table 12.11 summarizes the various legal instruments that regulate (at least potentially) invasion pathways in South Africa. As in most other

TABLE 12.11. Summary of international and national controls that affect (or potentially affect) pathways for biological invasions in South Africa. Many other international controls exist that could potentially influence pathways for invasions (see text), but appropriate domestic legislation has yet to be enacted.

Laws	Applicable Controls	Effectiveness	Indicators
INTERNATIONAL LAWS			
CITES	Regulates the importation and movement of specified animals.	Not effective in South Africa because there is inadequate domestic legislation to enforce this international law obligation.	Increase or decrease in numbers of alien animals that are not controlled and that are regulated under CITES.
Convention on Biological Diversity	Requires member states to control spread of alien invasives.	Not effective in South Africa because there is inadequate domestic legislation to enforce this international law obligation.	Increase or decrease in numbers of invasive aliens.
DOMESTIC LAWS			
Conservation of Agricultural Resources Act	Regulations passed under the act control particular plants in identified areas.	Combating of invaders is not required at the expense of agricultural resources, and this measure may therefore, indirectly, become a facilitative vector.	Increase or decrease in numbers of alien invasives in agricultural areas.
	Draft regulations have been proposed requiring the control of plants that are not useful to humans, are commercially useful, or are only ornamentally useful but are invasives.	If these are implemented, they will regulate previously unregulated invasive species (e.g., many ornamental plants), but the regulations are reactive in nature.	Increase or decrease in numbers of invasives imported for ornamental reasons.
Environment Conservation Act	Allows a competent authority to direct a person to stop an activity that is causing serious damage to the environment.	Reactive in nature.	Number of directives sent under this legislation.
	Regulations requiring environmental impact assessments on defined activities that may cause the spread of alien invasives.	These regulations are proactive, but apply only in specified circumstances, not all of which cover potential pathways.	Number of environmental impact assessments carried out in respect of potentially threatening alien invasive species.

Laws	Applicable Controls	Effectiveness	Indicators
Agricultural Pests Act	Prohibits the importation of plants, animals, and pathogens without a permit, which will not be granted without a phytosanitary certificate.	The phytosanitary certificate is often granted by means of a visual inspection or during a period during the plant's lifetime, prior to it becoming an organism of the invasive species.	Increase or decrease in numbers of new species of aliens.
National Water Act	Regulates commercial afforestation that has been declared to be a streamflow reduction activity and that therefore requires a license in terms of this act.	The requirement of the spread of alien invasives is not included as a criterion for granting a license under this act.	Increase or decrease in uncontrolled migration of alien invasive plants from forestry areas.
Mountain Catchment Areas Act	Allows competent authorities to define an area as a mountain catchment area and to control activities within that area and within 5 km of the boundary of that area.	It is potentially a proactive measure but is not well enforced.	Increase or decrease of the spread of alien invasive species in mountain catchment areas.
National Forests Act	Requires that forests must be developed and managed so as to conserve biologically diverse ecosystems and habitats.	Does not specifically require that alien invasives be contained within forestry area boundaries.	Increase or decrease in uncontrolled migration of alien invasive plants from forestry areas.
Marine Living Resources Act	Regulations promulgated under this act prohibit the release of nonindigenous organisms into the sea without a permit.	Difficult to enforce.	Increase or decrease in number of new and existing marine alien invasive species.
PROVINCIAL LAWS			
Nature Conservation Ordinances	Primarily protects specified indigenous flora and fauna rather than introducing measures that control exotic species. The introduction into inland waters of any aquatic growth is prohibited without a permit.	Focus is on the protection of indigenous species rather than the control of exotic ones.	Increase or decrease in spread of alien invasive species, particularly in protected areas.

countries, a major problem facing those tasked with managing invasions is the fragmented nature of existing regulations. This problem may be addressed when the White Paper on Biodiversity becomes an act.

The very absence of laws constitutes a pathway by which invasive aliens may be spread.

Some existing controls, such as those regulating the import of species that may be or may carry invasive aliens, are implemented in such a way that they may facilitate the movement of invasive aliens. This point is particularly significant for South Africa since it rejoined the global market. Increased importation of goods and species that are likely to carry (or be) invasive aliens, coupled with the manner of implementing controls, constitutes a significant vector.

Invasive alien plants are more strictly controlled than invasive alien animals, except when the latter fall within the definition of agricultural pests. The only other method by which invasive alien animals may be regulated is through broad controls, such as those contained in the ECA or in NEMA.

Economic and other imperatives give rise to pathways, which in some instances discourage the spread of invasive alien species, but in other instances encourage them. The need for water security is an example of the former. Examples of the latter include the incentives for commercial forestry and agriculture that have led to legislation primarily concerned with fostering those interests, sometimes at the expense of controlling invasive aliens.

CONCLUSIONS

This chapter has considered invasion pathways for selected taxonomic groups and spheres of human activity that rely to varying degrees on non-native organisms. Each case highlights different problems facing those responsible for managing biological invasions at local, regional, national, and international scales. Pathways of invasion for different taxa in South Africa have developed over decades and centuries, their dimensions being molded by different suites of factors.

The extent to which invasion pathways can be managed to reduce the deleterious effects of invasions depends largely on the value (financial or other) attached to the goods and services supplied by the various alien taxa. In all the cases examined here, pathways have been dynamic, with major trends clearly associated with sociopolitical events and, increasingly, with innovations in human economies at the global scale. For all our examples, invasion pathways that have operated in the past have left us with a troublesome legacy. We need management interventions to deal with invasions resulting from the pool of alien organisms already introduced. Innovations are also needed to modify pathways to reduce the likelihood of future problems.

The forestry pathway is clearly delineated. Innovative approaches have been implemented to make the existing configuration of species in plantations less of a threat to biodiversity outside areas demarcated for forestry. These approaches were driven by the forest industry, at least partly in direct response to international pressure from current and potential customers. Various measures are needed to further restructure forestry practice to reduce future impacts. For the other human endeavors discussed here, different approaches are urgently needed to reduce the impacts of already introduced aliens and to prevent/reduce impacts from further introductions. Education, the entrenchment of environmental ethics, the implementation of financial incentives, and legal controls are key considerations. A hierarchy of assessment tools, including objectively defined criteria and indicators (as discussed by Lawes et al. 1999 for sustainable plantation management), needs to be developed.

Legislation has played a relatively minor role in shaping the pathways for the introduction of alien taxa that have led to South Africa's current problem with invasive alien species. On paper, South Africa's current environmental legislation is among the best in the world, particularly with regard to water use and pollution. Biodiversity concerns are, however, not adequately addressed. Few regulations relate directly to invasions, but several are in preparation. The proposed National Environmental Management Act: Biodiversity will assist in ensuring the improved control of invasions, especially for plants. Legal instruments governing the future importation of alien taxa are similar to those in many first-world countries. However, serious problems with implementation and enforcement due to a shortage of resources greatly reduce the effectiveness of legislation. An important shortcoming in existing legislation is that different provinces have ordinances stipulating different requirements for import permits for alien vertebrates. Taxa that may be validly flagged as "safe" introductions for one province may well constitute a high risk in another province. Once these taxa are in the country, no means are in place for controlling their further movement across provincial boundaries. National and regional (involving all countries in the Southern African Development Community [SADC]) guidelines are therefore urgently required. Urgent attention should be given to formulating objective criteria for assessing the potential risks of members of all taxonomic groups becoming invasive in the region.

ACKNOWLEDGMENTS

We thank Karen Kirkman (South African Forestry Company Limited), Richard Mack (Washington State University), Guy Preston (Working for Water Programme, Department of Water Affairs and Forestry), and Shirley Pierce for valuable comments on parts of the chapter. Work on this chapter was funded by the Institute for Plant Conservation, University of Cape Town.

REFERENCES

Anonymous. 1902. *Reports of the Conservators of Forests for the Year 1901.* Cape Town: Department of Agriculture, Cape of Good Hope.

Anonymous. 1999. *The Forestry Industry. South Africa's Green Giant.* Newspaper Supplement, 30 March 1999, to newspapers published by the Independent Newspaper Group.

Berruti, A. 1997. House crow. In *The Atlas of Southern African Birds*, Vol. 2, J. A. Harrison, D. G. Allan, L. G. Underhill, M. Herremans, A. J. Tree, V. Parker, and C. J. Brown, eds., p. 108. Johannesburg: Bird Life South Africa.

Bigalke, R. C. 1990. *Guidelines for the Application of Conservation Practices in Production Forestry.* Pretoria: Forestry Council.

Boshoff, A., G. Kerley, and G. Castley. 1997. *Who's Who in the Open Air Zoo?* Poster presented at the Arid Zone Ecology Forum, Prince Albert, September 1997.

Botha, C. and J. Botha. 2000. *Bring Back Nature to Your Garden.* Durban: Wildlife and Environmental Society of South Africa, KwaZulu-Natal Region.

Brooke, R. K., P. H. Lloyd, and A. L. de Villiers. 1986. Alien and translocated terrestrial vertebrates in South Africa. In *The Ecology and Management of Biological Invasions in Southern Africa*, I. A. W. Macdonald, F. J. Kruger, and A. A. Ferrar, eds., pp. 63–74. Cape Town: Oxford University Press.

Brown-Rossouw, R. and D. M. Richardson. 2000. The Role of Nurseries in Distributing Invasive Alien Plants in the Western Cape, South Africa. CSIR Report ENV-S-S 2000-021. Stellenbosch: CSIR Division of Water, Environment, and Forestry Technology.

Cambray, J. A. 2003. Opinion paper: The need for research on the impacts of translocated sharptooth catfish, *Clarias gariepinus*, in South Africa. *African Journal of Aquatic Sciences* 28: 2: in press.

Cambray, J. A. and P. G. Bianco. 1998. Freshwater fish crisis: A Blue Planet perspective. *Italian Journal of Zoology (Supplemental)* 65: 345–356.

Cambray, J. A. and R. Jubb. 1977. Dispersal of fishes via the Orange-Fish Tunnel, South Africa. *Journal of the Limnological Society of Southern Africa* 3: 33–35.

Cameron, T., ed. 1991. *A New Illustrated History of South Africa.* Johannesburg: Southern Book Publishers.

Carlton, J. T. 1989. Man's role in changing the face of the ocean: biological invasions and implications for conservation of the near-shore environment. *Conservation Biology* 3: 452–465.

Carter, R. A. 1996. The potential ecological impacts of ballast water discharge by oil tankers in the Saldanha Bay/Langebaan Lagoon system. Specialist Study SII. In *Environmental Impact Assessment—Strategic Fuel Fund Saldanha.* CSIR Report EMAS-VC 96005D. Stellenbosch: CSIR Division of Water, Environment, and Forestry Technology.

Cellier, G. A. 1991. The potential for economic development in rural Kwazulu through the use of commercial *Eucalyptus* woodlots. In *Intensive Forestry: The Role of Eucalypts,* A. P. G. Schonau, ed., pp. 834–835. Durban: International Union of Forest Research Organizations.

Cellier, G. A. 1993. The changing landscape—is there room for forestry in the new South Africa? *South African Forestry Journal* 167: 57–61.

Cellier, G. A. 1994. Are all trees green? The forestry industry replies. *Africa Environment and Wildlife* 2(1): 79–85.

Coetzee, H. 1985. Provenance research on Mexican pines. *South African Forestry Journal* 135: 68–73.

Davies, B. and J. Day. 1998. *Vanishing Waters.* Cape Town: University of Cape Town Press.

Deacon, J. 1986. Human settlement in South Africa and archaeological evidence for alien plants and animals. In *The Ecology and Management of Biological Invasions in Southern Africa,* I. A. W. Macdonald, F. J. Kruger, and A. A. Ferrar, eds., pp. 3–19. Cape Town: Oxford University Press.

Deacon, H. J. and J. Deacon. 1999. *Human Beginnings in South Africa. Uncovering the Secrets of the Stone Age.* Cape Town: David Philip.

Dean, W. R. J. 2000. Alien avifauna in southern Africa: what factors determine success? *South African Journal of Science* 96: 9–14.

De Moor, I. J. and M. N. Bruton. 1988. *Atlas of Alien and Translocated Indigenous Aquatic Animals in Southern Africa.* South African National Scientific Programmes Report 144. Pretoria: Council for Scientific and Industrial Research.

Denny, R. P. and A. W. Schumann. 1993. Weed control. In *Forestry Handbook,* H. J. van der Sijde, ed., pp. 219–230. Pretoria: Southern African Institute of Forestry.

De Wit, M., D. Crookes, and B. W. van Wilgen. 2001. Conflicts of interest in environmental management: estimating the costs and benefits of a tree invasion. *Biological Invasions* 3: 167–178.

Donald, D. G. M. 1994. *Paulownia*—the tree of the future? *South African Forestry Journal* 154: 94–98.

DWAF. 1996. *Sustainable Forest Development in South Africa.* White Paper. Pretoria: Department of Water Affairs and Forestry

DWAF. 1997. *South Africa's National Forestry Action Plan.* Pretoria: Department of Water Affairs and Forestry.

Esterhuyse, C. J. 1989. *Agroforestry.* Pretoria: Department of Environment Affairs.

FIEC. 1995. *Guidelines for Environmental Conservation Management in Com-*

mercial Forests in South Africa. Johannesburg: Forestry Industry Environmental Council.

Fox, R. and K. Rowntree. 2000. *The Geography of South Africa in a Changing World*. Cape Town: Oxford University Press.

FSC. 1999. *Principles and Criteria*. Document 1.2. London: Forestry Stewardship Council. Available at <http://www.fscoax.org/principal.htm>.

Gandar, M. V. 1984. Wood as a source of fuel in South Africa. *South African Forestry Journal* 129: 1–9.

Grant, W. S. and M. I. Cherry. 1985. *Mytilus galloprovincialis* Lmk. in southern Africa. *Journal of Experimental Marine Biology and Ecology* 90: 179–191.

Grant, W. S., M. I. Cherry, and A. T. Lombard. 1984. A cryptic species of *Mytilus* (Mollusca: Bivalvia) on the west coast of South Africa. *South African Journal of Marine Science* 2: 149–162.

Green, R. E. 1997. The influence of numbers released on the outcome of attempts to introduce exotic bird species to New Zealand. *Journal of Animal Ecology* 66: 25–35.

Griffiths, C. L., P. A. R. Hockey, C. van Erkom Schurink, and P. J. le Roux. 1992. Marine invasive aliens on South African shores: implications for community structure and trophic functioning. *South African Journal of Marine Science* 12: 713–722.

Griffiths, C. L., L. M. Kruger, and C. Ewart Smith. 1996. First record of the sea anemone *Metridium senile* from South Africa. *South African Journal of Zoology* 31: 157–158.

Ham, C. and J. M. Theron. 1999. Community forestry and woodlot development in South Africa: the past, present and future. *South African Forestry Journal* 184: 71–79.

Harrison, J. A., D. G. Allan, L. G. Underhill, M. Herremans, A. J. Tree, V. Parker, and C. J. Brown, eds. 1997. *The Atlas of Southern African Birds. Vol. 1: Non-passerines, Vol. 2: Passerines*. Johannesburg: Bird Life South Africa.

Hockey, P. A. R. and C. van Erkom Schurink. 1992. The invasive biology of the mussel *Mytilus galloprovincialis* on the southern African coast. *Transactions of the Royal Society of South Africa* 48: 124–139.

Impson, N. D., I. R. Bills, and J. A. Cambray. 1999. *The Primary Freshwater Fishes of the Cape Floristic Region: Conservation Needs for a Unique and Highly Threatened Fauna*. Internal publication. Cape Town: Cape Nature Conservation.

International Standards Organization (ISO). 1996. *Environmental Management Systems—Specification with Guidance for Use*. Document ISO 14001: 1996 (E).

Ivey, P. and B. J. Heydenrych, eds. 1995. *Invasive Plants in Nurseries*. Flora Conservation Committee Report. 95/1. Cape Town: Flora Conservation Committee, Botanical Society of South Africa.

Joska, M. A. P. and G. M. Branch. 1986. The European shore crab—another alien invader? *African Wildlife* 40: 63–65.

Keet, J. D. M. 1929. *Tree Planting in Orange Free State, Griqualand West, Bechuanaland, and North-eastern Districts of the Cape Province*. Bulletin No. 24. Pretoria: Department of Forestry.

Keet, J. D. M. 1936. *Report on Drift Sands in South Africa. Forestry Series No. 9.* Pretoria: Department of Agriculture and Forestry.

Keet, J. D. M. 1974. Historical review of the development of forestry in South Africa. Unpublished report. Pretoria: Department of Water Affairs and Forestry.

Kietzka, J. 1988. *Pinus maximinoi*: a promising species in South Africa. *South African Forestry Journal* 145: 33–38.

King, N. L. 1943. Historical sketch of the development of forestry in South Africa. *Journal of the South African Forestry Association* 1: 4–16.

Lawes, M. J., D. Everard, and H. A. C. Eeley. 1999. Developing environmental criteria and indicators for sustainable plantation management: the South African perspective. *South African Journal of Science* 95: 461–469.

Legat, C. E. 1917. *Annual Report of the Forest Department for the Year Ended 31 March 1915.* Pretoria: Department of Forestry.

le Roux, J. P., G. M. Branch, and M. A. P. Joska. 1990. On the distribution, diet and possible impact of the invasive European shore crab *Carcinus maenas* (L.) along the South African coast. *South African Journal of Marine Science* 9: 85–93.

Lever, C. 1987. *Naturalized Birds of the World.* London: Longman Scientific and Technical.

Liversidge, R. 1985. Alien bird species introduced into southern Africa. In *Proceedings of the Birds and Man Symposium,* L. J. Bunning, ed., pp. 31–44. Johannesburg: Witwatersrand Bird Club.

Lückhoff, H. A. 1973. The story of forestry and its people. In *Our Green Heritage,* W. F. E. Immelman, C. L. Wicht, and D. P. Ackerman, eds., pp. 20–32. Cape Town: Tafelberg.

Macdonald, I. A. W., F. J. Kruger, and A. A. Ferrar. 1986. *The Ecology and Management of Biological Invasions in Southern Africa.* Cape Town: Oxford University Press.

Mack, R. N. 2001. Motivations and consequences of the human dispersal of plants. In *The Great Reshuffling: Human Dimensions in Invasive Alien Species,* J. A. McNeely, ed,. pp. 23–34. Gland, Switzerland, and Cambridge: International Union for the Conservation of Nature.

Maclean, G. L. 1993. *Roberts' Birds of Southern Africa.* Cape Town: John Voelcker Bird Book Fund.

Monniot, F., C. Monniot, C. L. Griffiths, and M. Schleyer. 2001. South African ascidians. *Annals of the South African Museum* 108: 1–141.

Mulder, P. F. S. and G. D. Engelbrecht. 1998. Watershed exchange across the Orange-Vaal and Tugela River systems: implications for fish fauna. In *FISA/PARADI,* L. Coetzee, J. Gon, and C. Kulongowski, eds., p. 250. Grahamstown: African Fishes and Fisheries Diversity and Utilisation.

Olckers, T. and M. P. Hill, eds. 1999. Biological control of weeds in South Africa. *African Entomology Memoir* 1: 1–182.

Pieterse, P. J. and C. Boucher. 1997. A case against controlling introduced Acacias—19 years later. *South African Forestry Journal* 180: 37–44.

Poynton, R. J. 1979a. *Tree Planting in Southern Africa. Volume 1. The Eucalypts.* Pretoria: Department of Forestry.

Poynton, R. J. 1979b. *Tree Planting in Southern Africa. Volume 2. The Pines.* Pretoria: Department of Forestry.

Poynton, R. J. 1984. *Characteristics and Uses of Selected Trees and Shrubs Cultivated in South Africa.* Pretoria: Directorate of Forestry.

Poynton, R. J. 1990. The genus *Prosopis* in southern Africa. *South African Forestry Journal* 152: 62–66.

Poynton, R. J. Ms. *Acacia* Miller. Part of unpublished manuscript for the genus *Acacia.* Copy supplied by CSIR Division of Water, Environment, and Forestry Technology, Pretoria, South Africa.

Rejmánek, M., D. M. Richardson, S. I. Higgins, M. Pitcairn, and E. Grotkopp. 2004. Ecology of invasive plants: state of the art. In *Invasive Alien Species,* H. A. Mooney, ed. Washington, DC: Island Press. In press.

Richardson, D. M. and S. I. Higgins. 1998. Pines as invaders in the Southern Hemisphere. In *Ecology and Biogeography of* Pinus, D. M. Richardson, ed., pp. 450–473. Cambridge: Cambridge University Press.

Richardson, D. M., I. A. W. Macdonald, J. H. Hoffmann, and L. Henderson. 1997. Alien plant invasions. In *Vegetation of Southern Africa,* R. M. Cowling, D. M. Richardson, and S. M. Pierce, eds., pp. 535–570. Cambridge: Cambridge University Press.

Richardson, D. M., P. Pyšek, M. Rejmánek, M. G. Barbour, F. D. Panetta, and C. J. West. 2000. Naturalization and invasion of alien plants: concepts and definitions. *Diversity and Distributions* 6: 93–107.

Rouget, M., D. M. Richardson, J. A. Nel, and B. W. van Wilgen. 2002. Invasion of *Acacia mearnsii, Eucalyptus* spp. and *Pinus* spp. in South Africa—the role of commercial forestry and environmental factors. *Biological Invasions* 4: 397–412.

Rueness, J. 1989. *Sargassum muticum* and other introduced Japanese macroalgae: biological pollution of European coasts. *Marine Pollution Bulletin* 20: 173–176.

Shaughnessy, G. L. 1986. A case study of some woody plant introductions to the Cape Town area. In *The Ecology and Management of Biological Invasions in Southern Africa,* I. A. W. Macdonald, F. J. Kruger, and A. A. Ferrar, eds., pp. 37–43. Cape Town: Oxford University Press.

Sheat, W. G. 1982. *The A–Z of Gardening in South Africa.* Cape Town: Struik.

Siegfried, W. R. 1962. *Introduced Vertebrates in the Cape Province.* Cape Department of Nature Conservation. Investigation. Report 4: 1–16. Cape Town: Cape Department of Nature Conservation.

Simon, D. and M. Ramutsindela. 2000. Political geographies of change in southern Africa. In *The Geography of South Africa in a Changing World,* R. Fox and K. Rowntree, eds., pp. 89–113. Cape Town: Oxford University Press.

Skelton, P. H. 1987. South African Red Data Book—Fishes. *South African National Scientific Programmes Report* 137: 1–199.

Skinner, J. and R. H. N. Smithers. 1990. *The Mammals of the Southern African Subregion.* 2nd ed. Pretoria: University of Pretoria.

Smithers, R. H. N. 1983. *The Mammals of the Southern African Subregion.* Pretoria: University of Pretoria.

Stirton, C. H. 1978. *Plant Invaders. Beautiful, but Dangerous: A Guide to the Identification of Twenty-Six Plant Invaders of the Province of the Cape of Good Hope.* Cape Town: Department of Nature and Environmental Conservation of the Cape Provincial Administration.

van der Zel, D. W. 1990. *The Afforestation Permit System.* Information Leaflet No. 1. Pretoria: Department of Environment Affairs.

van der Zel, D. W. 1995. Accomplishments and dynamics of the South African afforestation permit system. *South African Forestry Journal* 172: 49–57.

van der Zel, D. W. and A. J. Brink. 1980. Die geskiedenis van bosbou in Suider-Afrika. Deel II: Plantasiebosbou. *South African Journal of Forestry* 115: 17–27.

van Erkom Schurink, C. and C. L. Griffiths. 1990. Marine mussels in South Africa—their distribution patterns, standing stocks, exploitation and culture. *Journal of Shellfisheries Research* 9: 75–85.

van Wilgen, B. W., D. M. Richardson, C. Marais, D. Magadlela, and D. C. Le Maitre. 2001. The economic consequences of alien plant invasions: examples of impacts and approaches for sustainable management in South Africa. *Environment, Development and Sustainability* 3: 145–168.

van Wilgen, B. W., F. van der Heyden, H. G. Zimmermann, D. Magadlela, and T. Willems. 2000. Big returns from small organisms: developing a strategy for the biological control of invasive alien plants in South Africa. *South African Journal of Science* 96: 148–152.

Versfeld, D. B., D. C. Le Maitre, and R. A. Chapman. 1998. *Alien Invading Plants and Water Resources in South Africa: A Preliminary Assessment.* WRC report: TT99/98. Pretoria: Water Research Commission.

Vitousek, P. M., H. A. Mooney, J. Lubchenco, and J. M. Melillo. 1997. Human domination of earth's ecosystems. *Science* 277: 494–499.

Wells, M. J., A. A. Balsinhas, H. Joffe, V. M. Engelbrecht, G. Harding, and C. H. Stirton. 1986. A catalogue of problem plants in southern Africa. *Memoirs of the Botanical Survey of South Africa* 53: 1–658.

Williamson, M. 1996. *Biological Invasions.* London: Chapman and Hall.

Working for Water Programme. 1998. *Annual Report 1997/98. Working for Water Programme.* Pretoria: Department of Water Affairs and Forestry.

Chapter 13

Mitigating Introduction of Invasive Plant Pests in the United States

Joseph F. Cavey

The United States Department of Agriculture, Animal and Plant Health Inspection Service (USDA, APHIS) is responsible for the difficult and critical task of preventing the entry of invasive and injurious pest organisms into the United States. The agency's Veterinary Services division prevents entry of pests affecting domesticated animals, and the Plant Protection and Quarantine (PPQ) division prevents introduction of organisms that adversely affect plants (i.e., plant pests). The Plant Protection Act gives APHIS the authority to regulate international arrivals of cargo, vessels, and travelers (U.S. Congress 2000). APHIS uses this authority to support an integrated safeguarding system involving numerous and varied pest exclusion activities (USDA 1996, National Plant Board 1999).

APHIS uses diverse analysis, inspection, and commodity permitting methods to mitigate the routes, or modes, of pest entry known as pathways. These pathways carry numerous plant pests, including insects, plant pathogens, nematodes, snails, mites, and noxious weeds. APHIS defines pathways as narrowly as a single plant product from a single country (e.g., tomatoes from the Netherlands) or more broadly.

For illustrative purposes, the following is a list of seven major plant pest pathways monitored by APHIS:

- plants for propagation as cargo (imported nursery stock)
- plant products not for propagation as cargo (fruit and vegetables)
- miscellaneous cargo (machinery, ceramic tiles)

- plant material carried by international travelers
- conveyances (ships, aircraft, or railroad cars)
- mail
- organism imports (live insects for butterfly houses)

For these pathways, APHIS uses different but related methods to prevent the entry of invasive plant pests depending on requirements of the law and international phytosanitary standards (Code of Federal Regulations 2002, Food and Agriculture Organization 1996, National Plant Board 1999). APHIS's methods differ for each pathway, but the following description of the process for regulating the pathway of imported fruits and vegetables intended for consumption (i.e., not propagation) provides the most inclusive and illustrative example of the varied methods and tools employed by the agency. Interdiction methods APHIS uses to detect and prevent smuggled agricultural products are not addressed in the description of this process.

PROCESS FOR IMPORTING FRUITS AND VEGETABLES FOR CONSUMPTION

The process APHIS uses to determine conditions for importing fruits and vegetables for consumption is illustrated in Figure 13.1. Commodities not currently imported to the United States must be specifically requested for entry. Risk analysis by the agency for each requested commodity determines if entry will be allowed and under what conditions. Finally, through the federal regulatory process, a regulation allowing importation is usually produced, as outlined below. Parties wishing to import fruits and vegetables not presently allowed into the United States must submit a written request for a permit to import a specific product from a specific country of origin (USDA 2002). APHIS first determines if a previous decision was made to allow entry for a product from an origin. If a previous decision to allow the commodity was made, new requests from additional importers are allowed entry under the same conditions. If APHIS previously decided against entry for that product, the commodity is again refused entry or the new request may trigger reconsideration (e.g., if new mitigation methods effective against critical pests in the pathway have been developed).

APHIS receives approximately fifty importation requests per year for fruits and vegetables that are currently not permitted entry (Imai 2001). Because APHIS cannot process so many requests immediately, the agency must prioritize them. Date of submission is the primary criterion used to

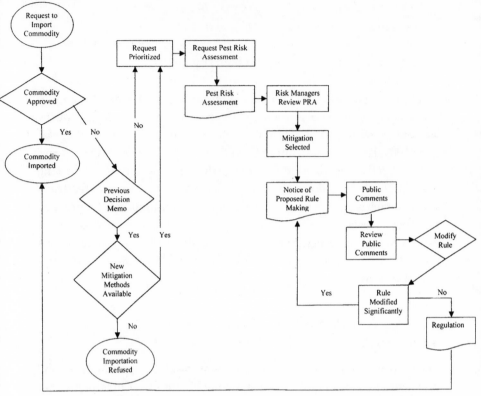

FIGURE 13.1. The process APHIS used to determine conditions for importing fruits and vegetables for consumption.

rank import requests for processing. Bilateral agreements with foreign governments may adjust rank order. APHIS provides guidelines for conducting commodity pest risk analyses to the public and foreign customers (USDA 2000). If the requestor uses these guidelines to provide a pest risk assessment with the importation request, APHIS raises priority for that request.

A risk analysis process begins once an import request reaches the top of the priority list. The risk analysis process consists of three elements: risk assessment, risk communication, and risk management (Orr and Cohen 1991). Risk assessment occurs from the onset and throughout the process of risk analysis among risk assessors, risk managers, stakeholders, and other involved parties. This communication attempts to assure a thorough evaluation of pest risk and accurate interpretation of the assessment necessary for appropriate management of risk.

Risk Assessment

A commodity risk assessment (RA) typically answers three questions:

1) Which pests could be imported if the commodity is imported?

2) What are the consequences if the pests are imported?

3) What is the likelihood that the adverse event will happen?

The first question identifies the pests that are associated with the plant producing the product that will be imported to the United States and, especially, the pests most likely to follow the pathway, or arrive with the product. The next question asks what consequences (e.g., economic impact) to agriculture, silviculture, and the environment would result if the plant pests associated with the commodity became established in the United States. Finally, the RA investigates the likelihood that these plant pests will be introduced by the imported commodity. The answers to these questions characterize the risk associated with importing the requested commodity (Orr and Cohen 1991).

APHIS conducts both qualitative and quantitative commodity risk assessments (National Plant Board 1999). Qualitative assessments, which are more common, estimate risk in qualitative terms, such as "high," "medium," or "low." Quantitative assessments often express risk as numerical probability estimates. APHIS reserves more rigorous quantitative risk assessments for particularly controversial and high-profile import requests.

Preparing commodity risk assessments takes time. Assessors attempt to determine all organisms known to be associated with plant species in question, through searches of the scientific literature and pest interception data. Each identified organism is considered a potential plant pest. From this usually long list of species, assessors determine which organisms meet the definition of a quarantine pest (Food and Agriculture Organization 1996). Finally, assessors edit the reduced list of quarantine pests to include only those pests likely to follow the commodity pathway. For example, if tomato fruit is the requested commodity, and one of the listed pests is an exotic nematode that attacks only tomato roots, that pest would not likely accompany the shipment of fruit and would not be considered further in assessing importation risk. At this point, assessors evaluate and rate each remaining pest on the list for the likelihood that it would establish in the United States and for potential environmental and economic impact. Each pest receives an overall risk rating (USDA 2000).

Although these risk assessments can be resource intensive, thorough analysis is often necessary to mitigate the arrival of unwanted pests effectively. For example, little or no biological data exists for many of the organisms associated with most commodities. When exhaustive literature searches fail, risk assessors extrapolate expected biological characteristics from related, more extensively documented organisms. Depending on the characteristics attributed to the assessed organism, recommended mitigation measures could vary from commodity inspection at arrival in the United States (e.g., for an easily detected, leaf-feeding caterpillar) to required treatment of the commodity (e.g., for more cryptic, fruit-boring caterpillars).

Risk Management: Mitigation Options

After completion of the commodity RA, risk mitigation teams determine options for safely importing the commodity in light of RA findings. Mitigation options include prohibited entry (most restrictive mitigation), pest-free concepts, systems approaches, treatment, or entry with inspection of the commodity (least restrictive mitigation). In selecting the appropriate option, APHIS evaluates the various options for efficacy and impact, considering factors such as biological effectiveness, cost/benefit of implementation, impact of existing regulations, commercial impact, social impact, environmental impact, phytosanitary policy considerations, time required to implement a new regulation, and efficacy of the option against other quarantine pests. APHIS adheres to the international guideline principle of "minimal impact" set by the Food and Agricultural Organization (FAO) when choosing appropriate mitigation: "Phytosanitary measures shall be consistent with the pest risk involved, and shall represent the least restrictive measures available which result in the minimum impediment to the international movement of people, commodities and conveyances" (FAO 1995).

Inspection

Inspection remains a primary mitigation tool for APHIS. More than 2,000 APHIS inspectors and 1,100 technicians work at international airports, seaports, land border crossings, and mail facilities in the United States (Smith 2002). Inspectors use some 75 X-ray machines and 50 trained dogs to screen passenger baggage and mail for prohibited plants and plant products, and a variety of sampling and inspection techniques for cargo. Each year, APHIS-PPQ intercepts more than 1.5 million prohibited or infested plants and plant products and 40,000 to 50,000 plant pests arriving at U.S. borders (USDA 1999). Each of the pest records constitutes an interception of a single pest taxon on a single imported commodity shipment, commodity carried by a traveler, or item associated with a conveyance such as an airplane or railcar.

For some commodities shipped as cargo to the United States, inspection at the country of origin prior to departure is most practical and efficient. APHIS conducts these "preclearance" inspection programs in cooperation with twenty-eight countries for numerous commodities, including mangoes, apples, and flower bulbs (USDA 1999). But, for the large majority of imported products, APHIS conducts inspections at ports of entry.

Entry inspections generally conform to a 1–2 percent sampling standard. Inspectors often focus their efforts on pests identified by the risk assessment as likely to arrive with the commodity and on pests previously intercepted from that commodity. However, they also search for any live organisms that arrive with shipments. Traditionally, port inspectors learned which shipments presented highest pest risk through shared experience within and among ports and by analyzing APHIS pest interception data. Today, APHIS augments these methods with a random sample survey designed to monitor the amount and kind of quarantine materials and pests approaching ports through known pest entry pathways. The results of this Agricultural Quarantine Inspection Monitoring (AQIM) program are used to estimate objectively the pest risk present in commodity pathways, by comparing AQIM results to risk illustrated by actual pest interceptions at a port. Resources can be shifted to better manage high-risk pathways, and new pathways may occasionally be discovered by this analysis.

Inspection: Pest Identification Support

Inspections of cargo, conveyances, and travelers produce interceptions of potentially invasive pest organisms. In some cases, finding pests will not significantly delay international commerce or tourism. For example, if a traveler carries a piece of fruit that is normally allowed entry but on which a pest is found, the fruit is confiscated and destroyed and the traveler may proceed. Likewise, when pests are found in cold storage food lockers on board ships, an APHIS requirement that the ship's captain safely retain all food stores on board until the ship leaves U.S. waters prevents delay that may result from imposing other quarantine measures. APHIS processes pest interceptions such as these in a timely manner, by identifying the intercepted organism and adding records of intercepted quarantine pests to a pest interception database.

If organisms are found with imported cargo, APHIS-PPQ detains the shipment and must quickly ascertain whether the shipment represents a pest risk. Usually, the decision is based on whether the intercepted organism is a quarantine pest. The organism must be identified before APHIS decides whether quarantine action is necessary prior to releasing the shipment.

Inspection programs for invasive species require credible, reliable, and responsive taxonomic support. APHIS provides such support through a nationwide infrastructure that includes personnel trained in taxonomic identification at ports of arrival and taxonomic specialists in some ports and at centralized laboratories, such as the USDA Agricultural Research Service's Systematic Entomology Laboratory in the National Museum of Natural History in Washington, D.C.

Port inspectors receive pest identification training designed primarily to aid in screening organisms clearly not posing a plant health threat. In addition, APHIS places nearly seventy Area Identifiers at large and moderate-sized ports of entry. These identifiers are trained in the taxonomy of either plants, arthropods, plant pathogens, or a combination of these disciplines. They identify organisms intercepted in their port and other ports in their area of coverage. APHIS National Identification Services staff at headquarters in Riverdale, Maryland, monitors and issues authority to Area Identifiers to identify intercepted organisms. If an Area Identifier does not have authority to identify a particular organism, he/she forwards the interception to an appropriate taxonomic specialist. APHIS employs some taxonomic specialists and obtains the services of others through cooperative agreements and contracts, including those with the USDA Agricultural Research Service for insects, nematodes, and pathogens, and with the Academy of Natural Sciences in Philadelphia for mollusks.

When cargo is held pending identification of an intercepted potential pest, the interception receives an "urgent" designation, which means that APHIS attempts to complete the needed identification and decide upon disposition of the cargo within one working day. APHIS uses overnight mail service to forward urgent interceptions to taxonomic specialists when Area Identifiers do not have delegated authority to recognize the organism. Specialists communicate their identifications to APHIS headquarters by email, phone, or facsimile. APHIS headquarters then communicates the identification and recommended cargo disposition (mitigate for infestation or release shipment) to the port.

Recently, APHIS adopted digital imaging technology for pest identification use. Ports prepare digital images of intercepted organisms through microscopes and email the images to appropriate specialists. APHIS currently uses this technology in more than forty U.S. ports of entry. Specialists identify more than 70 percent of these images with greater than 99 percent accuracy but request overnight forwarding of actual specimens when unsure. Remote identification technology often reduces the time required to complete identification of intercepted organisms from a working day to several hours.

Pest-free Area Concepts

Pest-free area concepts recognize biological, physical, or other natural limiting factors that help to assure freedom from specified pests of concern. Exporting countries must demonstrate that production areas are free of the specified pest (usually by comprehensive survey), continue monitoring and surveillance, and provide contingency plans for excluding the pest from the area should future surveys detect the pest there. This option has been used more frequently in recent years.

Systems Approaches

Systems approaches use a series of risk mitigation measures that individually and cumulatively reduce pest risk. These are defined as "[A] defined set of phytosanitary procedures, at least two of which have an independent effect in mitigating pest risk associated with the movement of commodities" (National Plant Board 2002). Complex systems of growing, packing, and shipping commodities can increase the probability that targeted pests will fail to travel with the commodity and survive the journey to the United States. Some examples of mitigation measures that may be used in combination to form a systems approach for a commodity include the following:

- surveying, trapping, and controlling pests in growing areas
- employing regular field sanitation procedures
- using pest-resistant hosts
- safeguarding the commodity from infestation after harvest
- allowing shipment only during low-risk seasons, e.g., winter in the United States
- requiring inspection at the packing house and/or port of arrival
- limiting distribution of the product to areas of the United States not suitable to pest establishment

Systems approaches encourage industry/grower participation and innovation in the safeguarding process and offer alternatives to chemical treatments (National Plant Board 2002). However, because many of the quarantine measures applied to the commodity occur in the country of origin, monitoring systems approaches can be difficult and resource intensive for APHIS. Monitoring activities are often conducted by the exporting country's plant protection organization in compliance with APHIS. APHIS may also conduct periodic audit inspections in the United States to monitor program effectiveness. As with pest-free area concepts, the use of systems approaches has increased in recent years.

Treatment

APHIS requires treatment of commodities if associated pests are difficult to detect, are likely to be exported with the commodity, and can easily become established through the pathway. Depending on the pest, APHIS may prescribe heat or cold treatment, hot water or chemical dips, fumigation, or other effective methods.

APHIS requires chemical treatments only when deemed necessary to prevent pest introduction. Inspection is considered an adequate mitigation measure in lieu of mandatory treatment for the following:

- pests not usually associated with the exported plant part, e.g., fruit weevils that attack immature but not mature fruit
- pests with limited mobility
- large external feeders, e.g., grasshoppers or scarab beetles
- immature stages of especially large moths and some internal feeding caterpillars if exterior damage to the commodity is evident
- some pathogens if they produce obvious and characteristic symptoms and/or likelihood of transmission from the imported plant part is low
- some mites if they are minor pests, usually do not feed or occur on the imported plant part, or do not prefer the commodity as a host

Prohibition

APHIS prohibits importation of plant products when effective treatments would be required but do not exist or are not feasible to mitigate risk presented by pests likely to arrive with the product. For example, numerous diseases of potatoes still exotic to the United States do not consistently manifest symptoms on tubers (thereby disqualifying inspection as an option) and cannot be effectively eliminated by treatments without destroying the commodity.

Risk Management: Rule Making

APHIS must promulgate a federal regulation to allow importation of an unprecedented commodity/origin combination. After analyzing available mitigation options, APHIS chooses a preferred option and begins the rule making process. A notice of proposed rule making is published in the Federal Register to solicit public comment, usually for a sixty-day period. The notice lists the mitigation options considered and explains why APHIS chose the preferred option. APHIS then reviews submitted comments and may reconsider import requirements in light of comment.

If APHIS decides to proceed with the proposed rule, or to modify it slightly, the rule becomes regulation and commodity importation is allowed under the specified protocol. But if the agency modifies the proposed rule significantly, additional public comment on the modified rule may be warranted. In this case, a second notice is published, comments are reviewed, and needed modifications are made before publishing the final regulation. The regulation specifies entry requirements for the fruit or vegetable commodity and is published in the Code of Federal Regulations (CFR 2002).

MITIGATING OTHER PATHWAYS

APHIS uses risk analysis to help prevent pests from entering the United States with fruits and vegetables (USDA 2000). The agency uses different approaches for different pathways as stipulated by existing regulations and traditional procedures. The following briefly describes approaches APHIS uses for other major pathways.

Commodities allowed entry may be imported by mail or carried by international travelers if declared upon entry and found free of pests during inspection. Other regulated plant products, such as cut flowers and nonregulated miscellaneous cargo (e.g., machinery or clothing), are also inspected by APHIS at ports of entry. Commodities not allowed entry or found infested are seized and destroyed, refused entry and exported, or treated to eliminate pests.

APHIS prohibits entry of noncommercial plant products carried by conveyances, such as ships' stores (food items for the crew), decorative flower arrangements in airplanes, or regulated items belonging to foreign crewmembers. On commercial vessels, these items must be retained and safeguarded on board to prevent pest introduction while in the United States.

Propagative material such as nursery stock has been imported into the United States since the first settlers (Sailer 1978, U.S. Congress 1993). The APHIS regulation on nursery stock prohibits certain species (e.g., gladiolus from Africa) and requires post-entry quarantine or treatment for others (e.g., chestnuts other than from North America) (CFR 2002). All others are allowed to enter with inspection. Because risk of pest establishment is especially high for organisms arriving in this pathway, inspection requirements are more restrictive than for nonpropagative plant products. Propagative material can enter only through plant inspection stations located in 15 of the approximately 100 U.S. ports of entry. Importers must bring their shipments to these stations for clearance, where optimized conditions allow for efficient commodity inspection by highly experienced personnel.

Requests for importing organisms, primarily for biological control or research, are subjected to the APHIS permit process. Analysis of each request determines if the organism may be imported and under what conditions.

USDA, APHIS works to prevent the introduction of exotic plant pests from a variety of major pest pathways. The agency works within its authority designated by the Plant Pest Act (U.S. Congress 2000) and international trade and phytosanitary agreements (Food and Agriculture Organization 1995), using risk assessment to tailor appropriate management measures to mitigate risk in specific pathways. The APHIS procedure for mitigating the pest risk associated with the broad pathway for plant pests offered by imported fruits and vegetables illustrates application of the agency's authority in accord with international principles. This process safeguards agricultural systems and the environment in the United States from invasions of exotic plant pests by appropriately regulating commodities in international commerce that present pest risk.

Acknowledgments

I am most grateful to Charles E. Miller and Narcy Klag, APHIS-PPQ, Riverdale, Maryland, for contributing information on mitigation measures and import processes, respectively; and to Jane Levy, Russell Stewart (retired), Wayne Burnett, Narcy Klag, Joel Floyd, and Charles E. Miller, APHIS-PPQ, Riverdale, Maryland, for reviewing drafts of the manuscript.

References

Code of Federal Regulations (CFR). 2002. *Code of Federal Regulations, Title: 7, Agriculture, Parts 300 to 399, a Special Edition of the Federal Register, Revised as of January 1, 2002*. Office of the Federal Register, National Archives and Records Administration. Washington, DC: U.S. Government Printing Office.

Food and Agriculture Organization (FAO). 1995. *International Standards for Phytosanitary Measures, Principles of Plant Quarantine as Related to International Trade*. Rome: Secretariat of the International Plant Protection Convention, Food and Agriculture Organization of the United Nations.

Food and Agriculture Organization (FAO). 1996. *International Standards for Phytosanitary Measures, Part 1—Import Regulations: Guidelines for Pest*

Risk Analysis. Rome: Secretariat of the International Plant Protection Convention, Food and Agriculture Organization of the United Nations.

Imai, E. 2001. USDA, APHIS, PPQ. Personal communication with J. F. Cavey.

National Plant Board. 1999. *Safeguarding American plant resources: a stakeholder review of the APHIS-PPQ safeguarding system.* <http://www.aphis.usda.gov/npb/safegard.html>.

National Plant Board. 2002. *Preventing the introduction of plant pathogens into the United States: the role and application of the systems approach.* <http://www.aphis.usda.gov/ppq/systemsapproach/sysapp.pdf>.

Orr, R. L. and S. D. Cohen. 1991. *Generic pest risk assessment process.* Washington, DC: Department of Agriculture, Animal and Plant Health Inspection Service.

Sailer, R. 1978. Our immigrant insect fauna. *Bulletin of the Entomology Society of America* 24(1): 3–11.

Smith, J. 2002. USDA, APHIS-PPQ. Personal communication with J. F. Cavey, October 3.

U.S. Congress, House. 2000. *106th Congress, 2nd Session, Congressional Record: conference report on HR2559, Agricultural Risk Protection Act of 2000* (24 May 2000), pp. H3763–3804.

U.S. Congress, Office of Technology Assessment. 1993. *Harmful Non-Indigenous Species in the United States,* OTA-F-565. Washington, D.C.: U.S. Government Printing Office.

U.S. Department of Agriculture, APHIS. 1996. Delivery of plant protection programs in the United States—the role of the Animal and Plant Health Inspection Service PPQ concept paper. Unpublished document.

U.S. Department of Agriculture, APHIS. 1999. *Facts about APHIS: excluding foreign pests and diseases.* <http://www.aphis.usda.gov/oa/exclude.html>.

U.S. Department of Agriculture, APHIS. 2000. *Guidelines for pathway-initiated pest risk assessments, Version 5.02.* <http://www.aphis.usda.gov/ppq/pra/commodity/cpraguide.pdf>.

U.S. Department of Agriculture, APHIS. 2002. *Agricultural Permits, Fruits and Vegetables.* <http://www.aphis.usda.gov/ppq/permits/fruits_veg/index.html>.

Chapter 14

Environmental Diplomacy and the Global Movement of Invasive Alien Species: A U.S. Perspective

Jamie K. Reaser, Brooks B. Yeager, Paul R. Phifer,
Alicia K. Hancock, and Alexis T. Gutierrez

Due, at least in part, to policies that promote economic growth through globalization, the world is now crisscrossed with an increasingly expanding network of commercial "expressways." As a result, we are both witnessing and participating in a global reshuffling of biological organisms: some that move as commodities, others that "hitchhike" with people, products, and services. Facilitated by the rapid expansion of international trade, travel, and transport, as well as ongoing changes in land use and climate, the rate of biological invasion, as well as diversity and volume of invaders, has never been so high (Vitousek et al. 1997, McNeely et al. 2001, Westbrooks et al. 2001).

In most cases, the translocation of biological organisms does not pose a problem; either the organisms do not survive in their new conditions without deliberate cultivation and husbandry or their populations are small and easily managed (Bright 1998, 1999, Mack 2000, Mack et al. 2000). However, an estimated 1 out of every 1,000 organisms is introduced into a new environment where it thrives (Williamson and Brown 1986, Williamson 1996). These organisms, collectively known as "invasive alien species," spread, proliferate, and cause serious harm to the environment, the economy, or human health.

Invasive alien species are one of the most significant drivers of environmental change worldwide (Mooney and Hobbs 2000, Sala et al. 2000). They contribute to social instability and economic hardship, placing constraints on sustainable development, economic growth, and environmental conservation (McNeely et al. 2001). Society pays a great price for invasive alien species—costs measured not just in currency but also in unemployment, damaged goods and equipment, power failures, food and water shortages, environmental degradation, loss of biodiversity, increased rates and severity of natural disasters, disease epidemics, and lost lives (Bright 1998, Carlton 2001, McNeely et al. 2001).

This chapter provides an overview of the U.S. government's response to the international aspects of the movement of invasive alien species. In particular, the authors focus on the challenges and opportunities for the U.S. government to engage in diplomacy with other governments. In order to effectively minimize the spread and impact of invasive alien species, the United States needs to use international agreements and voluntary programs of cooperation to raise awareness of the invasive alien species issue, including the ecological, economic, and social costs; implement effective international policies and partnerships; share information and technologies; and provide technical and financial assistance worldwide.

BACKGROUND

In 1993, the U.S. Office of Technology Assessment (OTA) concluded that the number of invasive alien species and their cumulative impacts were a growing burden for the United States, and that federal policies and programs to protect the country from the most harmful invasive alien species were inadequate (U.S. Congress 1993). Four years later, more than 500 scientists and natural resource managers from across the country wrote the federal administration to express their deep concern about the damage being done globally by invasive alien species, estimated at more than $100 billion annually in environmental losses and control costs to the United States alone (Pimentel et al. 2000) and at least this much in six other countries combined (Pimentel et al. 2001). Authors of these letters called for a coordinated federal strategy to prevent the introduction and spread of invasive alien species.

The president of the United States responded with the Invasive Species Executive Order (13112) on February 3, 1999. The Executive Order established the interdepartmental National Invasive Species Council (NISC) to direct and coordinate the work of the U.S. government agencies on invasive alien species. It also empowered a nonfederal Invasive Species Advisory Committee (ISAC) of relevant experts from academia, industries, and nongovernmental organizations, states, and tribes with the duty to inform and review the activities of NISC. Furthermore, the order called for NISC, with input from ISAC, to prepare an Invasive Species Management Plan that would identify and direct opportunities for the U.S. government to implement federal policies and programs that could adequately prevent and control the spread of invasive alien species.

The first National Invasive Species Management Plan was adopted by NISC on January 18, 2001 (www.invasivespecies.gov). The plan includes fifty-seven action items (many with multiple components) to be implemented within approximately two years by the U.S. government in conjunction with relevant partners. Eighteen of the action items are explicitly international in scope and require the support and cooperation of other governments for full implementation.

An International Problem

In a large continental country like the United States, organisms sometimes cause harm when they are relocated within national borders. For example, crayfish and other freshwater organisms native to the southeastern United States have been relocated to the western United States to serve as game species or food for game species. In many instances, these newcomers have proven to be voracious predators, competitors, and/or vectors of disease and parasites. The long-term result has been a significant reduction in the freshwater biodiversity of western watersheds (Fuller et al. 1999, Claudi and Leach 2000, Carlton 2001). Nevertheless, the United States' most challenging invasive alien species problems occur when nonnative species are brought into the country intentionally or unintentionally (e.g., Fuller et al. 1999, Carlton 2001). The country is learning the hard way that international trade is not "free" and that with increases in commerce come greater risks and costs—to human health, economies, and the environment. The United States is not alone in having learned this lesson. Every country has been invaded by nonnative species, and many literally cannot afford the costs (Box 14.1).

The United States' ability to prevent invasive alien species from crossing its borders depends not just on its policies and capabilities, but also on the policies and capabilities of other countries to effectively manage and control the domestic movement of potentially invasive alien species and invasion pathways. If a potentially invasive alien species never leaves its native range, it will never become a problem in another country. Of course, invasive alien species do not respect jurisdictional boundaries; they can walk, fly, swim, crawl, and hitchhike from one country to another, threat-

Box 14.1. Costs of Invasion (U.S. dollars)*

- U.S. losses due to invasive plant pathogens total approximately $23 billion per year.
- Invasion of the Black Sea by the Leidy's comb jellyfish, introduced through ballast water, cost already stressed regional fisheries more than $350 million and affected the livelihoods of approximately 2 million people.
- Water hyacinth in Lake Victoria costs Uganda, Tanzania, and Kenya approximately $150 million annually for control and removal. In addition, water hyacinth mats are threatening the Lake Victoria fishing industry, which provides half of the protein supply for this area's 30 million residents.
- Introduced donkeys and goats on the Galápagos Islands are trampling native vegetation and outcompeting endemic tortoises and iguanas for food. The presence of these invasive alien species threatens the highly profitable ecotourist industry. The costs of eradication to the Ecuadorian National Park system are estimated at more than $8 million.
- A 1991 epidemic of cholera, possibly arriving in Peruvian ports through ballast discharge from South Asian ships, resulted in losses of more than $1 billion in exports of potentially contaminated seafood and in tourist revenues. Latin America spent more than $200 billion over the next four years repairing sewage and drinking water systems in an attempt to stem the epidemic.
- Epidemics of introduced viruses, bacteria, and protozoa cost the Ecuadorian shrimp industry $200 million in 1993. A viral outbreak destroyed 80 percent of Taiwan's shrimp harvest in 1987, and India's industry crashed in 1994 when disease eliminated $63.8 million worth of shrimp in Adhra Pradesh and Tamil Nadu.
- Managing invasive sea lampreys in the Great Lakes costs the U.S. Department of State at least $10 million per year.

* Adapted from Bensted-Smith 1997, Bright 1998, Pimentel et al. 2000, NISC unpublished.

ening an entire region once they become established (Mack et al. 2000, National Invasive Species Council 2001).

As countries become more aware of the implications of invasive alien species, they tend to look inward and focus on protecting themselves. To the contrary, the National Invasive Species Management Plan clearly reflects an understanding by the U.S. government that the invasive alien species problems faced by other countries can be the direct result of U.S. actions. As a result of the U.S. government's intentions to be a world leader in the provision of goods and services, it has introduced invasive alien species to other countries through development assistance programs, military operations, famine relief projects, and other government-sponsored activities (Box 14.2). As demands for U.S. exports grow, the potential increases for organisms that are invasive in the United States to be shipped

Box 14.2. International Assistance

AN INVASIVE PATHWAY

Invasive alien species are sometimes transported around the world through development assistance programs, famine relief projects, and military operations. Well-intentioned efforts can have unexpected and persistent, negative consequences. Unfortunately, U.S. overseas activities offer multiple examples of both intentional and unintentional introductions.

Intentional Introductions

The United States and other developed nations have, for aquaculture purposes, introduced Madagascar and Nile tilapia into lakes and ponds all over the world, as well as carp, bass, trout, and other invasive fish species (FAO Database on Introductions of Invasive Species). The golden apple snail, a South American species, was introduced into East and Southeast Asia as a food development project, and has proved devastating to Asian rice fields. The snail has already cost Philippine rice farmers approximately $1 billion in crop losses (http://www.fao.org/NEWS/1998/RIFILI-E.HTM).

The U.S. Agency for International Development has historically introduced eucalypts and many other nonnative forestry species in many developing countries. Several international forestry projects promote the use of *Leucaena leucocephala*, a fast-growing Central American leguminous tree, which can be used as "green fertilizer" and a forage crop but can also be invasive (Bright 1998).

Unintentional Introductions

It appears that U.S. military involvement in the Balkans brought a corn borer into the area along with food and assistance shipments. The borer is now affecting corn production throughout the area and may spread through central Europe.

U.S. military transports after World War II introduced the brown tree snake, a native of Papua New Guinea, to Guam. The snake has since eliminated nine of Guam's eleven native bird species and causes more than $1 million in damages annually (J. Waage, pers. comm.).

A MEANS TO HALT AND MITIGATE INVASIONS

The United States also offers a wide variety of international assistance programs that are specifically aimed at preventing the introduction and controlling the spread of invasive alien species. The following are some examples from the U.S. Agency for International Development:

- A program in Ecuador to study control techniques for invasive alien species in the Galápagos (Ospina 1997).
- A project in Bangladesh to reestablish native fish in the local canals (http://www.doi.gov/intl/bangladesha.html).
- An effort in Uganda to conduct environmental impact studies on water hyacinth controls (Bright 1998).
- Both the U.S. Department of Interior and the U.S. Agency for Development have provided resources to support the Working for Water Programme in South Africa. This project is widely regarded as one of the leading models, worldwide, of "best practice" in the control and eradication of invasive alien species (http://www.doi.gov/intl/southafricaa.html).

overseas purposely (e.g., American bullfrog, *Rana catesbieana*, for food markets) or unintentionally (e.g., pine wood nematode, *Bursaphelenchus lignicolus*, as a hitchhiker in solid wood packaging materials). Lastly, as international travel increases, so does the likelihood that U.S. citizens may unwittingly transmit invasive alien species (especially seeds, insects, and pathogens) on their clothing or in their possessions (National Invasive Species Council 2001, National Research Council 2002).

THE DIPLOMATIC LANDSCAPE

Ultimately, a nation's ability to effectively address the problems posed by invasive alien species depends on the government's understanding of the problem, its willingness to protect national security (e.g., human and environmental health, agricultural production), and its various institutional and scientific capabilities.

In December 1999, the U.S. Department of State undertook a comprehensive survey to better understand the priorities and policies held by other governments with regard to invasive alien species. The survey's preliminary results revealed that the United States faces many challenges to addressing the invasive alien species issue internationally.

Findings from the survey indicated that few countries (Australia, New Zealand, South Africa, and Norway, for example) considered invasive alien species a high priority, had coordinated policies and plans in place specifically aimed at minimizing the problem, and were dedicating substantial resources to prevent and control the spread of invasive alien species. In many other countries, however, high-level government officials were largely unaware of the threats posed by invasive alien species. Further, government officials that were aware of the problem frequently considered it a low priority. Among those developing countries that recognized the gravity of the situation, there were many hampered by a lack of scientific, technological, and financial resources. For example, island nations, such as those in the Caribbean region, considered invasive alien species a high priority but were restricted by a lack of resources to address the issue adequately.

The survey also revealed that, within and among governments, efforts to address invasive alien species are typically not well coordinated. In most governments, aspects of the invasive alien species problem are under the jurisdiction of multiple ministries and, in some cases, subnational governing units. Mexico, for example, reported that eleven agencies have responsibilities for different aspects of invasive alien species prevention and control. Governing units often have different priorities, goals, and objectives, and most governments lacked a mechanism for coordination and cooperation.

The assessment also indicated that it is not uncommon for neighboring countries to be unaware of each other's policies and practices to prevent and control the spread of invasive alien species. In some cases, the priorities of one country might actually be facilitating invasions across country borders; one country might be importing what it considers to be valuable forage grasses, while its neighbor is aggressively trying to eradicate the same plant as an invasive alien species. Clearly, such a "migration of problems" has the potential to create regional friction.

The U.S. Department of State repeated the survey in 2001. Although it did indicate a moderate increase in the number of countries aware of the problems posed by invasive alien species and a slight increase in the level of priority among previously aware governments (likely as a result of increased attention to the issue by the Convention on Biological Diversity and work of the Global Invasive Species Programme), other patterns had not significantly changed.

INTERNATIONAL OPPORTUNITIES FOR ENGAGEMENT

The U.S. Department of State used the information obtained in its 1999 and 2001 surveys to set its priorities for diplomatic engagement on the issue of invasive alien species. Clearly, before effective international policies could be established, significant strides needed to be made to increase awareness of the issue; communicate its implications for economic growth and sustainable development; and motivate governments to implement national strategies to minimize the spread and impact of invasive alien species.

The following is an overview of the primary mechanisms through which the United States can raise awareness of the threats posed by invasive alien species and engage with other governments to implement effective solutions to the problem on regional and global scales.

Influencing International Law and Policy

Governments and international organizations are beginning to use conventions, treaties, and other agreements more effectively to limit the spread of invasive alien species. The Global Invasive Species Programme (GISP) and NISC have identified more than fifty international agreements that specifically address invasive alien species (Shine et al. 2000; www.invasivespecies.gov). Some of these international agreements are between two countries (bilateral), while others are regional or global (multilateral). These international agreements may have broad scope, or may be specific to certain types of invasive organisms, ecosystems, or invasion pathways (e.g., see Box 14.3). For example, the United States and Canada cooperate to control sea lampreys and ballast water through the Convention on Great Lakes Fisheries and Great Lakes Water Quality Agreement, respectively. The South Pacific Regional Environment Programme (SPREP) has developed a Regional Invasive Species Strategy for the Pacific Islands. Article 8(h) under the Convention on Biological Diversity (CBD) calls for governments to "prevent the introduction of, control or eradicate those alien (nonnative) species which threaten ecosystems, habitats or species," and it has adopted fifteen nonbinding Guiding Principles and a work program on invasive alien species. The International Plant Protection Convention (IPPC) is currently developing standards on a variety of issues relevant to invasive alien species, including the environmental impact of quarantine pests. The Convention on Wetlands of International Importance especially as Waterfowl Habitat (Ramsar Convention) addresses issues relating to the impacts of invasive alien species of freshwater, estu-

Box 14.3. International Legal Instruments Pertaining to Invasive Alien Species

Convention on Biological Diversity: Article 8(h) states that contracting parties, as far as possible and appropriate, are to "prevent the introduction of, control or eradicate those alien species which threaten ecosystems, habitats or species." Signed by the United States on June 4, 1993; not yet ratified. <http://www.biodiv.org>.

International Plant Protection Convention: Applies primarily to quarantine pests in international trade. The convention creates an international regime to prevent the spread and introduction of plant and plant product pests premised on the exchange of phytosanitary certificates between importing and exporting countries' national plant protection offices. Ratified by the United States on August 18, 1972. <http://www.fao.org/legal/treaties/004t-e.htm>.

Convention on Wetlands of International Importance especially as Waterfowl Habitat (Ramsar Convention): The Ramsar Resolution on Invasive Species and Wetlands: 7.14 (May 1999) urges countries to prepare an inventory and risk assessment of alien species in wetlands and to establish programs to target priority invasive species for control or eradication. It seeks to address environmental, economic, and social impacts of the transport of alien species on the global spread of invasive wetland species. Improvements in technology and information sharing are also featured. It now has a joint work program with the CBD. The United States ratified the Ramsar Convention on December 18, 1986. <http://www.ramsar.org/index.html>.

International Maritime Organization (IMO): The IMO has in place voluntary guidelines for the control and management of ship's ballast water and sediment discharges and is currently working to develop a legally binding instrument that would effectively minimize the spread of invasive alien species through ballast water exchange. The IMO might also be well positioned to address the invasion of nonnative species through hull fouling and the relocation of marine structures (e.g., oil rig platforms). <http://www.imo.org>.

Convention on Great Lakes Fisheries between the United States and Canada: This convention established the Great Lakes Fisheries Commission, whose primary purpose is the control and eradication of nonnative, highly invasive species, such as the Atlantic sea lamprey, from the Great Lakes. Ratified by the United States in 1955. <http://www.glfc.org/pubs/conv.htm>.

South Pacific Regional Environment Programme (SPREP) Convention: SPREP's Invasive Species Strategy for the Pacific Islands Region, developed in 1999, is meant to promote the efforts of Pacific island countries in protecting their natural heritage from the impacts of invasive species. It encourages and facilitates cooperative measures to develop and maintain a coordinated network of information, prevent the introduction of new invasive species, reduce the impact of existing invasive species, raise awareness, and build the capacity required to manage invasive threats (Sherley 2000).

(For additional information, see <http://www.invasivespecies.gov>.)

arine, and coastal ecosystems, including coral reefs. And the United States and other governments are participating in negotiations on ballast water management under the International Maritime Organization (IMO).

Recognizing the intrinsic difficulty of regulating the many forms of trade, tourism, and transport that facilitate the movement of invasive alien species, industries, international organizations, and governments have begun to use codes of conduct and other "soft law" tools voluntarily to minimize the spread of invasive alien species (Box 14.4). Ideally, these codes of conduct and guidelines will set forth the practices and policies that, if adopted, will minimize the spread and impact of invasive alien species. There are many areas in which the U.S. government could promote and employ codes of conduct to help prevent the movement of invasive alien species. Arguably, the most significant opportunities exist in pathways management. By voluntarily agreeing to develop and adopt technological or policy changes to various commercial practices and establishing these as "best practices," the United States could "close" or at least "clean up" several major pathways of invasion (Table 14.1).

Because invasive alien species issues influence such a wide variety of sectors (e.g., agriculture, human health, environment), each with diverse interests, no single international agreement will provide comprehensive policy coverage. The first step to strengthening policy frameworks internationally is for governments and other bodies to assess the various existing international regimes and institutions that are relevant to the invasive alien species problem and design a strategy to utilize each of these regimes fully. This action might involve increasing a government's ability to enforce existing policies, amend agreements, and harmonize agreements. Governments should then consider new international agreements (binding or nonbinding) to fill the gaps in coverage that cannot be accomplished through changes to existing legal regimes.

TABLE 14.1. Examples of unintentional pathways for the invasion of alien species and measures that could be taken to minimize their spread and impact (see Chapters 7 and 18, this volume, for discussions relevant to ballast water).

Pathway/Vector	Major Types of Invasives	Recommended Change*
Solid wood packaging materials	Pathogens and insects that impact forests and wood product industries Note: Use of wood products for this purpose can also result in losses of economically and ecologically valuable timber.	Use materials made of recycled plastics, steam clean after use, and store in sealed environment.
Seaweed for bait and seafood packaging (provides moisture)	Wide variety of marine biota that cause losses of biodiversity and impact infrastructure	Use damp recycled paper.
Tires	Mosquitoes that carry pathogens that transmit human and/or animal diseases	Store dry and steam clean before shipping.
Seeds	Invasive plants can out-compete native plants and noninvasive forage species. They may have indirect impacts on wildlife, especially pollinators and herbivore populations. Some species are known to alter water and fire cycles.	Label all packages with scientific and common names so they can be checked against locally relevant lists of invasive alien species. Establish penalties for false labeling and exceeding thresholds of contamination with extraneous seeds.
Military vehicles and equipment	Seeds, insects, insect eggs, and other organisms can be lodged in or on vehicles, equipment, and supplies.	Steam clean vehicles and equipment and inspect all property before return to country of origin.
Used cars	Seeds, insects, insect eggs, and other organisms can be lodged in or on these vehicles.	Steam clean all undercarriages and inspect interiors.
Clay tiles (for roofing, flooring, decoration, etc.)	Snails that can carry human and/or animal diseases, seeds, insects, insect eggs, and other organisms can inhabit tiles, pallets, and shipping containers.	Steam clean tiles, pallets, and shipping containers immediately before shipment and store in sealed containers.

* = The "recommended changes" are meant to be indicative of possible solutions and are not intended to represent the full suite of alternative practices that are technically feasible.

Box 14.4. Codes of Conduct Pertaining to Invasive Alien Species

Selected codes of conduct or guidelines now being utilized internationally:

- *FAO Code of Conduct for Responsible Fisheries:* This code requests that governments consult with their neighboring states before introducing nonindigenous species into transboundary aquatic ecosystems. Article 9.3.1 asks that efforts be undertaken by national governments, fishing entities, and regional and subregional organizations to minimize the harmful effects of introducing nonnative species. The code went into effect on October 31, 1995. <http://www.fao.org/fi/agreem/codecond/ficonde.asp>.

- *IUCN Guidelines for the Prevention of Biodiversity Loss Caused by Alien Invasive Species:* These guidelines for national governments and management agencies are meant to increase awareness and understanding of the impact of invasive species. The document includes guidelines on prevention, eradication, control, and species reintroduction. The guidelines went into effect in February 2000. <http://www.iucn.org/themes/ssc/pubs/policy/invasivesEng.htm>.

- *IMO Guidelines for the Control and Management of Ships' Ballast Water to Minimize the Transfer of Harmful Aquatic Organisms and Pathogens:* This document provides guidelines for the uptake and release of ballast water so as to reduce the transfer of invasive aquatic species. It is recommended that every ship that carries ballast water be provided with a ballast water management plan to assist in the fulfillment of these guidelines. The guidelines went into effect on November 27, 1997. <http://www.imo.org>.

- *FAO Code of Conduct for the Import and Release of Exotic Biological Control Agents:* This code is meant to help national governments, international organizations, industry, and research institutes facilitate the safe import, export, and release of nonnative biological control agents. The code went into effect on November 1, 1995.

(For additional information, see http://www.invasivespecies.gov.)

Sharing Information and Technology and Providing Technical Assistance

Policymakers need up-to-date scientific and technical information if they are going to develop and adopt decisions and take actions to effectively limit the spread of invasive alien species. Information critical to one country's success in controlling an invasive alien species may already be available in the country where the organism originated or in another country where it has already been a problem. Countries that openly exchange information and technologies relevant to invasive alien species are best poised to prevent the introduction of invasive alien species and respond rapidly when invasions do occur. In order to be readily useable by all countries, scientific and technical information needs to be collected and shared according to the same standards and protocols. Unfortunately, most countries do not inventory or monitor the invasive alien species within their borders (see National Biodiversity Strategy and Action Plans, www.biodiv.org). If relevant information does exist, it is typically scattered among collections and databases. Most of these collections have different management standards and few are widely accessible to the public (Ricciardi et al. 2000, Wittenberg and Cock 2001).

The United States and some other governments recognize that international programs that openly share information and technologies can greatly reduce the risks of invasions by nonnative species and lower the costs of controlling these pests. For many years, various agencies of the U.S. government (e.g., U.S. Department of Agriculture's Agriculture Research Services, the U.S. Geological Survey, the Centers for Disease Control) have assisted countries with scientific information on the invasive alien species that threaten their economies and human health. The United States has also provided technologies, such as biocontrol agents, that have helped countries manage invasive alien species (Mack et al. 2000). While most of this information and technology sharing has been from the United States to a specific developing country, the United States is now working with other governments in the Western Hemisphere to develop regional networks that will enable exchange of information in a timely and effective manner (Box 14.5). Furthermore, the United States has provided technical guidance and seed grants to facilitate diplomatic activities to address invasive alien species issues and in doing so has taken a lead in GISP's efforts to, among other things, establish a Global Invasive Species Information Network that will effectively link databases relevant to invasive alien species worldwide (GISP 2002).

Building the GISP Partnership Network

Intergovernmental cooperation has proven to be an effective tool for addressing environmental problems, particularly when the threats are global in scale. For example, in 1994 the governments of the United States, Japan, Australia, France, Jamaica, the Philippines, Sweden, and the United Kingdom founded the International Coral Reef Initiative (ICRI), an infor-

Box 14.5. Regional Networks for Information Exchange

THE INTER-AMERICAN BIODIVERSITY INFORMATION
NETWORK (IABIN)

IABIN, the result of an initiative of the November 1996 Santa Cruz Summit for Sustainable Development, is a forum through which countries and institutions collaborate to identify and meet information needs for conserving and sustainably using biological resources. IABIN's goal is to increase the amount of information on biodiversity shared among the nations of the Western Hemisphere. Guided by an intergovernmental working group, in concert with other ongoing initiatives, the network shares biological information relevant to decision making, science, and education. IABIN's Web site provides U.S.-sourced data, but also has links to networks in Brazil, Canada, Costa Rica, and Mexico. The U.S. Geological Survey works closely with IABIN, whose objectives are shared by the United States. This network seeks to build on, not substitute for, other information-sharing initiatives. With technical and financial support from the United States, IABIN has established the IABIN Invasives Information Network (I3N) in an effort to establish databases on invasive alien species throughout Latin America. <http://www.nbii.gov/>.

THE NORTH AMERICAN BIODIVERSITY INFORMATION
NETWORK (NABIN)

NABIN, an initiative of the North American Commission for Environmental Cooperation (CEC), seeks to promote open access to biodiversity data and collaboration among biodiversity scientists in North America (Canada, Mexico, and the United States). The NABIN data network is developing an infrastructure for search and retrieval of information available electronically on biological collections, and has developed a tool, Species Analyst, for predicting species distributions based on information drawn from distributed databases. These are already being applied to invasive alien species issues. <http://www.cec.org>.

mal, voluntary network dedicated to the conservation of coral reefs (ICRI's founding partners also consisted of such agencies as the World Bank and UNEP). Since its conception, more than forty countries have adopted ICRI's "Call to Action" and developed a Global Coral Reef Monitoring Network (GCRMN), International Coral Reef Information Network (ICRIN), and an International Coral Reef Action Network (ICRAN).

The 5th Conference of the Parties of the CBD passed a decision (V/8) that "strongly encourages Parties [governments] to develop mechanisms for transboundary cooperation and regional and multilateral cooperation in order to deal with the invasive alien species issue, including the exchange of best practices." The United States, along with other governments such as South Africa, turned this directive into an opportunity to encourage GISP to evolve from its origins as a consortium of three international organizations into a partnership network that also engages governments, international governmental organizations, nongovernmental organizations, academia, the private sector, and other relevant stakeholders.

In March 2002 the CBD's Subsidiary Body on Scientific, Technical and Technological Advice (SBSTTA), GISP released a Call to Action, inviting all stakeholders to become members of a "GISP Partnership Network." More than fifty governments, as well as numerous industries, scientific institutes, nongovernmental organizations, and intergovernmental organizations have signed the Call to Action, making GISP a truly cooperative program of global scale. In its Phase II Implementation Plan (released at CBD COP6, April 2002), GISP outlines a leadership structure and program of work that clearly provides opportunities for governments to work together, and with other bodies, to ensure that decisions made by relevant policy frameworks can be translated into productive, on-the-ground action (see http://www.jasper.stanford.edu/GISP).

The United States' most significant contribution to the GISP Partnership Network so far has been the financial support for five regional workshops on invasive alien species (Baltic-Nordic, South America, Southern Africa, South-Southeast Asia, and the Austral-Pacific). The purpose of these workshops was to (1) raise awareness of the invasive alien species problem among policymakers, (2) forge cooperation across environmental and agricultural sectors, and (3) lay the foundation for regional action plans.

CHALLENGES AND OPPORTUNITIES AHEAD

Even with state-of-the-art economic and demographic projections, the direction and degree of future social, economic, and environmental changes will be somewhat unpredictable, and they may contain yet unrecognized threats (Mack et al. 2000). Governments need to develop strategies to prevent and control invasive alien species that (1) clearly recognize the need to manage dynamic landscapes rich in local variability, (2) employ a variety of approaches, (3) promote cautious decision making, (4) conduct experiments and monitor the results, and (5) continually adapt policy and management decisions to reflect the best available information.

In order to be effective over the long term, the United States' policies on invasive alien species and its diplomatic efforts overseas need to consider emerging social, economic, and environmental trends carefully and adapt accordingly. Humans are changing the global environment, the biosphere, in unprecedented ways (Mooney and Hobbs 2000, Sala et al. 2000). As a consequence, preventing and controlling invasions of nonnative species will become an even more challenging task in the future.

The U.S. government and its citizens have an opportunity to provide leadership by making choices that will ultimately prevent the spread of invasive alien species. Unless the country changes some of its business practices and the character of its consumer patterns, it will face greater uncertainties and even more significant risks in managing invasive alien species. Table 14.2 provides a summary of some of the most evident and globally important social, economic, and environmental trends relevant to

TABLE 14.2. Social, economic, and environmental trends that will influence U.S. efforts to minimize the spread and impact of invasive alien species (IAS).

Category	*Issue*	*Trends*
Social and economic changes	Commerce—trade, tourism, and transport	Continued expansion. Moving larger volumes of people and goods to greater numbers of destinations. Increasing number and diversity of IAS moved as a result (Bright 1998, McNeely et al. 2001).
	Technology	Unforeseen technological breakthroughs will have profound and unpredictable impacts. Some technologies will prevent the introduction and spread of IAS; others will enhance the transport of organisms. The Internet and genetic modification are examples of technologies that could be both "good" and "bad" where IAS are concerned (Bright 1998, McNeely et al. 2001).
	Wealth	Trade- and technology-based contributions to the economy will enable many people to demand more new types of goods from the marketplace. As more people are able to travel the world, there is greater opportunity for transport of a cornucopia of organisms. On the other hand, many developing countries already cannot afford the costs of IAS. Developed countries will need to support a larger share of the burden to keep international commerce "clean" of IAS and will need to target their trading to partners with policies that effectively limit the spread of IAS (McNeely 2001).

continues

TABLE 14.2. *Continued*

Category	Issue	Trends
	Biosecurity	The anthrax attacks in the United States have raised concern over prospects for use of biological agents in acts of terrorism. Increases in regional and global political conflicts will surely translate into greater risks for the deliberate introduction of invasive pathogens that affect humans and livestock, as well as agricultural and ecological systems (Meyerson and Reaser 2002a, 2002b, 2003).
Environmental changes	Land use	Land uses (agricultural production, etc.) are likely to intensify. Typically, human-induced disturbances increase the opportunities for invasion. Land use changes can themselves be brought about by the purposeful introduction of nonnative organisms, e.g., new forage or plantation species. Land use change thus works in two directions to promote IAS: it can promote changes in the landscape that increase the opportunities for invasion, and it can purposefully bring nonnative species into altered ecosystems (Bright 1998, Mooney and Hobbs 2000).
	Climate change	The changing global climate is likely to offer IAS even more opportunities to establish and spread. Changes in temperature, moisture, and seasonality can effectively create "new" habitats and dramatically modify the relationships among species, opening up new opportunities for IAS (Mooney 1996, Sutherst 2000).
	Nitrogen deposition	Atmospheric nitrogen levels have significantly risen as a result of industrial pollution, automobile emissions, and increases in the use of agricultural fertilizers. Especially in naturally nutrient-poor soils, nitrogen deposition can accelerate the spread of fast-growing grasses and other species. Elevated levels of nitrogen deposition have already been implicated in the invasion of grassland ecosystems in California and central Europe (Weiss 1999, Scherer-Lorenzen et al. 2000).
	Invasive alien species	IAS are themselves agents of global change. They can alter the dynamics of ecosystems, force land use changes, and even influence local and regional climatic conditions. For example, invading trees can transform grassland into forest and thus greatly reduce the supply of surface water available for drinking and irrigation. Invading grasses can increase rates of fire disturbances, adding more carbon

Invasive alien species *continued*	into the atmosphere, changing a forest into a grassland, and placing private property at greater risks (D'Antonio 2000, Mack et al. 2000).
Carbon dioxide	The burning of fossil fuels and destruction (especially burning) of forests release large amounts of carbon dioxide (CO_2) into the atmosphere. Increasing carbon dioxide concentrations may enable some plants, particularly annual grasses, to use water more efficiently and extend their ranges into more arid landscapes. Leguminous shrubs might be able to process nitrogen more efficiently and thus grow faster. On the other hand, perennial grasses might perform less well. All other factors aside, increased levels of carbon dioxide are expected to cause annual grasses and legumes to become more invasive, while making grasslands more susceptible to invasion by other plant species (Dukes 2000, 2002 and Mooney 1999).

invasive alien species that the United States and other governments must consider if its future policy decisions are to be well informed. Ideally, the invasive alien species issue will eventually become integrated into other relevant thematic areas (e.g., poverty elimination, deforestation, climate change, anti-desertification, water conservation), rather than be solely addressed as an independent issue.

Acknowledgments

We are grateful to the many people who made this review possible through support of the Global Invasive Species Programme. In particular, we thank William Gregg and Dick Mack for providing valuable contributions to the development and technical review of the manuscript.

References

Bensted-Smith, R. 1997. The war against aliens in Galapagos. *IUCN World Conservation* 28: 40.

Bright, C. 1998. *Life Out of Bounds.* New York: W. W. Norton and Company.

Bright, C. 1999. Invasive species: pathogens of globalization. *Foreign Policy* 1999: 50–64.

Carlton, J. T. 2001. *Introduced Species in U.S. Coastal Waters: Environmental Impacts and Management Priorities.* Arlington: Pew Oceans Commission.

Claudi, R. and J. H. Leach. 2000. *Nonindigenous Freshwater Organisms: Vectors, Biology, and Impacts.* Salem: CRC Press.

D'Antonio, C. M. 2000. Fire, plant invasions, and global changes. In *Invasive Species in a Changing World,* H. A. Mooney and R. J. Hobbs, eds., pp. 65–94. Washington, DC: Island Press.

Dukes, J. S. 2000. Will the increasing atmospheric CO_2 concentration affect the success of invasive species? In *Invasive Species in a Changing World,* H. A. Mooney and R. J. Hobbs, eds., pp. 95–114. Washington, DC: Island Press.

Dukes, J. S. 2002. Comparison of the effect of elevated CO_2 on invasive species (*Centaurea solstitialis*) in monoculture and community settings. *Plant Ecology* 160: 225–234.

Dukes, J. S. and H. A. Mooney. 1999. Does global change increase the success of biological invaders? *Trends in Ecology and Evolution* 14: 135–139.

Fuller, P. L., L. G. Nico, and J. D. Williams. 1999. *Nonindigenous Fishes Introduced into Inland Waters of the United States.* Special Publication 27: 1–2. Bethesda: American Fisheries Society.

Global Invasive Species Programme. 2002. *Phase II Implementation Plan.* Palo Alto, California.

Mack, R. N. 2000. Cultivation fosters plant naturalization by reducing environmental stochasticity. *Biological Invasions* 2(2): 111–122.

Mack, R. N., D. Simberloff, W. M. Lonsdale, H. Evans, M. Clout, and F. A. Bazzaz. 2000. Biotic invasions: causes, epidemiology, global consequences and control. *Ecological Applications* 10: 689–710.

McNeely, J., ed. 2001. The great reshuffling: human dimensions of invasive alien species. Gland, Switzerland: IUCN.

McNeely, J. A., H. A. Mooney, L. E. Neville, P. J. Schei, and J. K. Waage, eds. 2001. *Global Strategy on Invasive Alien Species.* Cambridge: IUCN in collaboration with the Global Invasive Species Programme.

Meyerson, L. A. and J. K. Reaser. 2002a. A unified definition of biosecurity. *Science* 295: 44.

Meyerson, L. A. and J. K. Reaser. 2002b. A comprehensive approach to biosecurity. *BioScience* 52: 593–600.

Meyerson, L. A. and J. K. Reaser. 2003. Bioinvasions, bioterrorism, and biosecurity. *Front. Ecol. Environ.* 1: 307–314.

Mooney, H. A. 1996. Biological invasions and global change. In *Proceedings of the Norway/UN Conference on Alien Species,* O. T. Sandlund, P. J. Schei, and A. S. Viken, eds., pp. 123–126. Trondheim: Directorate for Nature Management.

Mooney, H. A. and R. J. Hobbs, eds. 2000. *Invasive Species in a Changing World.* Washington, DC: Island Press.

National Invasive Species Council. 2001. *Meeting the Invasive Species Challenge: The U.S.'s First National Invasive Species Management Plan.* Washington, DC: National Invasive Species Council. www.invasivespecies.gov.

National Research Council. 2002. *Predicting Invasions by Nonindigenous Plants and Plant Pests.* Washington, DC: National Academy of Sciences.

Ospina, P. 1997. Eradication and quarantine: two ways to save the islands. *IUCN-World Conservation* 28: 43.

Pimentel, D., L. Lach, R. Zuniga, and D. Morrison. 2000. Environmental and economic costs of nonindigenous species in the United States. *BioScience* 50: 53–65.

Pimentel, D., S. McNair, J. Janecka, J. Wightman, C. Simmonds, C. O'Connell, E. Wong, L. Russel, J. Zern, T. Aquino, and T. Tsomondo. 2001. Economic and environmental threats of alien plant, animal, and microbe invasions. *Agriculture, Ecosystems & Environment* 84: 1–20.

Ricciardi, A., W. W. M. Steiner, R. N. Mack, and D. Simberloff. 2000. Towards a global information system for invasive species. *BioScience* 50: 239–244.

Sala, O. E., F. S. Chapin III, J. J. Armesto, E. Berlow, J. Bloomfield, R. Dirzo, E. Huber-Sanwald, L. F. Huenneke, R. B. Jackson, A. Kinzig, R. Leemans, D. M. Lodge, H. A. Mooney, M. Oesterheld, N. L. Poff, M. T. Sykes, B. H. Walker, M. Walker, and D. H. Hall. 2000. Global biodiversity scenarios for the year 2100. *Science* 287: 1770–1774.

Scherer-Lorenzen, M., A. Elend, S. Nollert, and E. Schulze. 2000. Plant invasions in Germany: general aspects and impact of nitrogen deposition. In *Invasive Species in a Changing World*, H. A. Mooney and R. J. Hobbs, eds., pp. 351–368. Washington, DC: Island Press.

Sherley, G. 2000. *Invasive Species in the Pacific: A Technical Review and Draft Regional Strategy.* Apia: South Pacific Regional Environment Programme.

Shine, C., N. Williams, and L. Gundling. 2000. *A guide to designing legal and institutional frameworks on alien invasive species.* Cambridge, UK: IUCN.

Sutherst, R. W. 2000. Climate change and invasive species—a conceptual framework. In *Invasive Species in a Changing World*, H. A. Mooney and R. J. Hobbs, eds., pp. 211–240. Washington, DC: Island Press.

U.S. Congress, Office of Technology Assessment. 1993. *Harmful Non-Indigenous Species in the United States*, OTA-F-565. Washington, D.C.: U.S. Government Printing Office.

Vitousek, P. M., H. A. Mooney, J. Lubchenco, and J. M. Melillo. 1997. Human domination of Earth's ecosystems. *Science* 277: 494–499.

Weiss, S. 1999. Cars, cows, and checkerspot butterflies: nitrogen deposition and management of nutrient-poor grasslands for a threatened species. *Conservation Biology* 13: 1476–1486.

Westbrooks, R. G., W. P. Gregg, and R. E. Eplee. 2001. My view. *Weed Science* 49: 303–304.

Williamson, M. 1996. *Biological Invasions.* London: Chapman and Hall.

Williamson, M. and K. C. Brown. 1986. The analysis and modeling in British invasions. *Philosophical Transactions of the Royal Society* B 314: 505–522.

Wittenberg, R. and M. J. W. Cock. 2001. *Invasive Alien Species: A Toolkit of Best Prevention and Management Practices.* Oxon: CAB International.

Chapter 15

Biosecurity and the Role of Risk Assessment

Keith R. Hayes

In December 1999 the world's economic ministers met in Seattle, Washington, at the third ministerial meeting of the World Trade Organization (WTO), to develop the latest trade rules for the global world economy. The meeting failed to achieve this, in part because of the unprecedented protests from a variety of interest groups worried about the social, economic, and environmental impacts of free trade.

One of the most immediate environmental impacts of trade is biological pollution—the introduction of nonnative or "exotic" plants and animals. Since 1950, world trade has increased fourteenfold (Nordstrom and Vaughan 1999). In the same period the number of biological invasions in terrestrial, freshwater, and marine habitats has increased exponentially (Ruesink et al. 1995, Ruiz et al. 1997). Biological invasions are one of the most serious ecological problems of the early twenty-first century—and the trade policies of the new global economy are an unwitting contributor to this problem.

The environmental implications of global trade were recognized, albeit indirectly, at the inception of the General Agreement on Tariffs and Trade (GATT) in 1947. The WTO recently revisited the question of trade and the environment in its fourth special studies report (Nordstrom and Vaughan 1999). The report notes that trade threatens biodiversity because of habitat destruction (principally logging) and biological pollution. Curiously, the report ignores biological pollution in its subsequent deliberations. Furthermore, it does not examine the conflict that arises between global trade agreements and a nation's environmental autonomy—its ability to protect its own native fauna and flora—or how this conflict is best managed.

Article XX of GATT 1947 allows countries to ignore normal trading rules in order to protect human, animal, or plant life or health, provided that such measures do not discriminate between sources of imports or constitute a disguised restriction on international trade. These provisions have been replaced, at least for biosecurity purposes, by the agreement on Sanitary and Phytosanitary Measures (the "SPS agreement"), negotiated in 1994 at the end of the Uruguay round of multilateral negotiations. The SPS agreement allows countries to enforce measures to prevent the spread of plant, animal, or other disease agents, and to prevent or control the spread of pests. These measures, however, must be based on scientific justification or on an "objective" assessment of the risks to human, animal, or plant health (Doyle et al. 1996). In 1997 the International Plant Protection Convention (IPPC) was revised in line with the SPS agreement.

The SPS agreement and the IPPC, however, are just two—albeit prominent—examples of how risk assessment is being used to manage biosecurity issues. Various other policies and agreements also rely on risk assessment to manage the bioinvasion risk associated with the movement of goods and people, both internationally and domestically. The purpose of this chapter is to examine the process of bioinvasion risk assessment and some of the problems associated with this process. The chapter looks at ten examples of bioinvasion risk assessment that are, or have been, used to manage the translocation of various commodities, and it ends by asking whether bioinvasion risk assessment, as currently practiced, is up to the job.

RISK ASSESSMENT AND PATHWAY ANALYSIS

Risk assessment should contribute to pathway analysis and biosecurity in three ways: it should identify hazards, quantify risks, and help identify management options and strategies.

First Steps

Bioinvasion risk assessment is just one step in a wider process designed to maintain a nation's biosecurity. The assessment usually takes place before the commodity in question arrives at the border and is generally applied to the "pathway" (Fig. 15.1). Here pathway refers to the way in which biological pollutants—exotic plants and animals—are translocated within a nation, or delivered to a nation's doorstep.

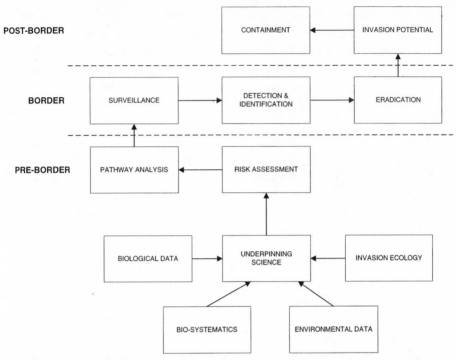

FIGURE 15.1. An example of a biosecurity strategy. Risk assessment, supported by biosystematic research and biological and environmental data, is an important pre-border component of this strategy.

The risk assessment is usually predictive (performed before the commodity is expected to arrive) but could also be applied retrospectively (subsequent to an invasion in order to identify the most likely pathway). Before starting a bioinvasion risk assessment, the analyst must decide:

- which framework to adopt—such as the "chain of event" model recommended by the Office Internationale des Epizooties (OIE) (OIE 1996, Morley 1993), the U.S. Environmental Protection Agency (U.S. EPA) framework for ecological risk assessment (U.S. EPA 1992, 1998), and an assortment of others (reviewed in Hayes 1997).
- which approach to adopt—Are there enough relevant data to take a deductive "accident statistic" approach to risk calculation? Will the analysis be inductive and model-based, or a mixture of both?
- which metric to adopt—Will the risk assessment be qualitative, semi-quantitative, fully quantitative, or a mixture of some or all of these?

- which process to adopt—Will panels of experts conduct the risk assessment, or will the risk calculation be automated?

Each of these decisions has a strong bearing on the risk assessment process. A panel of experts cannot be expected to make daily multiple assessments—these must be automated. Qualitative risk assessments, however, are not well suited to automation; there is usually too much uncertainty and too many permutations. For example, the U.S. Department of Agriculture pest risk assessment process (Orr 1995) scores seven components of the bioinvasion process as high, medium, or low, and allocates one of five uncertainty codes to each score. The total number of permutations in the risk calculation is $(3 \times 5)^7 = 1.7 \times 10^8$—clearly a difficult procedure to automate.

Identify Hazards

Hazard identification is the critical step in risk assessment; hazards that are not identified are not assessed, so risk is underestimated. Biosecurity hazards can be broadly classified into three groups: vectors, pathways, and species. In most situations one or more of these is known beforehand—usually the vector (e.g., salmon imports, shipping, live bait). The pathway(s) associated with the vector and the species hazard are usually more difficult to identify.

In the first instance, multiple pathways are often associated with single vectors both pre- and post-border (Fig. 15.1). Shipping is the principle vector responsible for transmitting marine pests around the world. The main pathways of this vector are ballast water and hull fouling (William et al. 1988, Pierce et al. 1997, Hallegraeff 1998, Gollasch et al. 1998), but there are at least nineteen possible (pre-border) pathways on a typical ship (Carlton et al. 1995). Similarly, the Australian import risk analysis for nonviable salmonids (Kahn et al. 1999) identifies at least seven possible (post-border) pathways for contaminated salmon (Fig. 15.2).

Secondly, most risk assessments define only those pathways associated with the planned event chain for the commodity concerned. Accidents during the event chain may introduce new pathways or change the significance of the usual ones. The situation is made still more complicated if multiple species are associated with a vector and/or pathway. Biosecurity hazard assessment must therefore identify the complete vector/pathway/species set. In effect the hazard assessment must define the "mechanics" of the invasion process: which species might invade and exactly how they could do this.

Leaving aside the problem of the species for the moment, there are several techniques for predicting how they could invade. These techniques

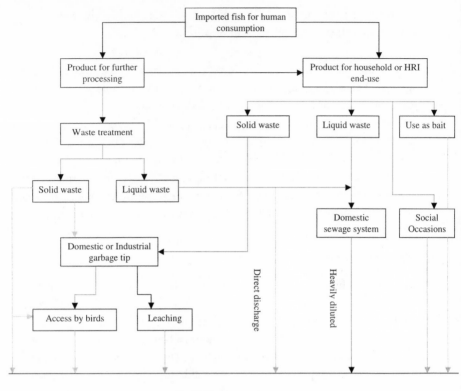

Key

Probable pathway ⟶

Less significant ┄┄➤

Exceptional ┄┄▸

FIGURE 15.2. Contaminated salmon exposure pathways (post-border) identified in the Canadian salmon import risk assessment conducted by Australia.

are, broadly, deductive or inductive. Deductive techniques simply record what has happened in the past and are usually implemented through a checklist or an unstructured brainstorming.

Deductive hazard identification appears to be the norm for biosecurity risk assessment. The techniques are simple and easy to implement, but may mislead the analyst into believing, without confirmation, that all hazards have been identified. Deductive techniques, by their very nature, will capture only those events that lie within the professional experience of the assessor(s). Alternatively, inductive techniques could include the following:

- logic tree analysis—fault tree and event trees
- hazard and operability analysis (HAZOP) (Kletz 1986)
- failure modes and effects analysis (Palady 1995)

These techniques were originally developed in an industrial context, usually as the first step in a quantitative risk assessment, and have been successfully used for many decades. They are successful because they are logical, rigorous, and systematic, and make no prior assumptions about the likelihood of hazardous scenarios. They are therefore an excellent way to identify plausible, but improbable, events that may be overlooked by deductive approaches because they are not within the analyst's professional experience. It is rare, however, to see these techniques used in a bioinvasion or ecological risk context. Notable exceptions are GENHAZ— a HAZOP analysis for genetically modified organisms (Royal Commission on Environmental Pollution 1991), the fault-tree analysis developed for the Australian Quarantine and Inspection Service's (AQIS) ballast water risk assessment (Hayes 2002a), and the infection modes and effects analysis applied to small craft (Hayes 2002b).

Quantify Risks

Risk is a function of the likelihood and consequences of undesired events. In a biosecurity context the undesired event is biological pollution. The second role of bioinvasion risk assessment is, therefore, to quantify the likelihood and consequences of biological pollution. Quantitative risk assessments for biological stressors, however, are notoriously difficult (Fiksel and Covello 1985, Simberloff and Alexander 1994), so most assessments address these issues qualitatively (see Table 15.1). Analysts usually reject quantitative techniques because the problem is too complex or because reliable data are too few.

It would be unduly pessimistic, however, to believe that qualitative risk estimates are the best that bioinvasion analysts can hope to achieve. Bioinvasion pathways usually involve several steps: infection, entry, survival, establishment and/or dispersal, and impacts. The uncertainty associated with each step increases from left to right—that is, from infection to impacts. Probabilistic techniques become increasingly difficult as one moves from low uncertainty to near ignorance (Faucheux and Froger 1995). It may not be necessary, however, to quantify all of the steps in the invasion sequence. For species that are already pests, with a well-documented impact history, quantified estimates of inoculation (i.e., all those steps up to and including survival in the recipient area) may be sufficient from a risk manager's perspective. Alternatively, the analyst could adopt quantitative methods as far as possible, and only use qualitative approaches for higher order events such as establishment and dispersal.

Quantitative risk assessment techniques can (again) be broadly classified as deductive or inductive. Deductive techniques use an "accident statistic" or frequentist approach to calculate probability. For example, Cohen and Carlton (1997) report interceptions of the Chinese mitten crab, *Eriochier sinensis*, at San Francisco's airport (hand-carried by disembarking airplane passengers). It is possible to establish the probability of entry through this pathway by comparing these figures with total passenger numbers over an equivalent period. These figures will help predict future arrival rates so long as the conditions under which they were collected (e.g., flight patterns, market price of the crabs) remain broadly the same.

The alternative inductive approach models some or all of the steps in the invasion chain. These models can range from simple point estimates, such as the efficacy of evisceration on reducing the risk of importing contaminated salmon (MacDiarmid 1994), to complex ecosystem models, such as that used to investigate the impacts of *Mnemiopsis leidyi* in the Black Sea (Berdnikov et al. 1999). Variability is propagated through these models by describing statistical distributions for important model parameters and using Monte Carlo simulation techniques to sample from these distributions (O'Neill et al. 1982, Vose 1996), or through interval analysis, fuzzy arithmetic, or probability bounding (Ferson et al. 2001).

Quantitative risk estimates have the following advantages:

- They allow interval, fuzzy, and probabilistic expressions of variability.
- They quickly identify what is unknown. By just undertaking it, the assessor(s) is forced to think very hard about what is and isn't known about the bioinvasion process.
- They are well suited to an iterative assessment cycle: calculate risk, collect data, ground-truth predictions, refine models, and recalculate risk.
- They are amenable to the scientific method.
- They can be used to compare alternative management strategies through a risk–benefit analysis.

Quantitative assessments are amenable to scientific scrutiny because their predictions are testable; for example, the predicted number of infected animal units per import ton can be tested through routine sampling. Indeed, continual monitoring, surveillance, and ground-truthing should be an essential component of any ecological risk assessment, quantitative or otherwise. Quantitative risk estimates, however, are not necessarily "objective." Subjective probability-elicitation techniques are (and should remain) an important tool in the risk analyst's toolbox (Hayes 1998). Furthermore, important subjective judgments are involved in all quantitative risk assessments—all probability-based inferences rely on a statistical model, but the choice of model is largely subjective. Even the simplest hypothesis test involves fundamentally subjective choices about the design and duration of the experiment (Berger and Berry 1988). The strength of quantitative risk assessment, as in science, lies not in its objectivity but rather in the way it exposes subjective input.

Manage Risk

Biosecurity ultimately aims to eliminate and contain (or slow the spread of) established pests and to slow the rate of new invasions. In its third role, bioinvasion risk assessment is a management tool to assist this process. In this role the risk assessment should

- help identify weak links in the invasion chain.
- identify where further data collection would improve the risk assessment.
- identify the management options with the greatest benefit–cost ratio.
- quantify the probability of the management strategy failing, perhaps as part of a formal management strategy evaluation (Smith 1993).

The weak link in an invasion chain is the step with the lowest probability of success. Management strategies aimed at barring or slowing this step are likely to return good cost benefit. In this context it is important to emphasize that, even if risk assessment fails to eliminate biological invasions (which it inevitably will do), slowing the rate of invasions or spread of an established pest has considerable value (Sharov and Liebhold 1998).

TEN EXAMPLES OF BIOINVASION RISK ASSESSMENT

Table 15.1 summarizes ten examples of bioinvasion risk assessments that are used by various regulatory agencies to manage imports or translocations of one kind or another. Most bioinvasion risk assessments are based on, or mirror, the OIE's "chain of events" framework. This framework views bioinvasion as the culmination of a series of steps, each of which must be successfully negotiated by the invading species, and to which a probability of success can be assigned. The overall probability of success—that is, invasion—is the product of the probabilities assigned to each step in the event chain. The risks associated with genetically modified organisms have been analyzed similarly (Alexander 1985) (Table 15.2).

The OIE framework is simple and effective and approaches bioinvasions from the perspective of the invading species. The framework's efficacy can be improved, however, by incorporating basic tenets of the quantitative risk assessment (QRA) paradigm. The QRA paradigm con-

TABLE 15.1. Ten examples of bioinvasion risk assessment used by various authorities to manage imports and the translocation of species.

Name	Reference	Metric	Framework
Generic nonindigenous aquatic organisms risk analysis review process	Orr (1995)	Qualitative	Broadly based on the QRA paradigm
National policy for the translocation of live aquatic organisms	Ministerial Council on Forestry, Fisheries and Aquaculture (1999)	Qualitative	Mirrors the OIE "chain of events" model and the QRA paradigm
Assessment procedure for Schedule 6 of the Wildlife Protection (Regulation of Exports and Imports) Act 1985		Qualitative	Mirrors the OIE "chain of events" model
Review and decision model for evaluating proposed introductions of aquatic organisms	Kohler and Stanley (1984)	Semiquantitative	Mirrors the OIE "chain of events" model
Expert system for screening potentially invasive alien plants in South African fynbos	Tucker and Richardson (1995)	Mixed qualitative and quantitative	No obvious precedence—specifically designed around the fynbos environment
Decision support protocol used to assess the advisability of allowing the importation of alien aquatic animals into southern Africa	De Moor and Bruton (1993)	Mixed qualitative and semiquantitative	Mirrors the OIE "chain of events" model
AQIS Weed Risk Assessment	Pheloung (1995)	Semiquantitative	No obvious precedence
Ballast Water Risk Assessment—12 Queensland Ports	Hilliard et al. (1997)	Semiquantitative	No obvious precedence
Import Health Risk Analysis: Salmonids for Human Consumption	Stone et al. (1997)	Quantitative	Follows the OIE "chain of events" model
Risk Assessment for Ballast Water Introductions	Hayes and Hewitt (2000a, 2000b)	Quantitative	Mixes the OIE "chain of events" model with techniques of the QRA paradigm

proach and Process	Calculation	Regulator
cies-specific: seven elements of the invasion process scored aigh, medium, or low, together with five uncertainty es. Decision rules combine these into overall risk rating. k = f{Pr(establishment) × consequences of establishment}}	Expert assessor or panel review	U.S. Department of Agriculture—agricultural imports into the United States
cies-specific: 44 questions address the likelihood and sequences of escape/release, survival, and establishment species. Risk = f{likelihood and consequences of escape, vival, and establishment}	Expert assessor or panel review	State fishery agencies—translocation of live aquatic organisms within Australia
cies-specific: designed to establish a "white list." 46 ques-ns that address the species life cycle, ability to survive in stralia, and potential impacts in the wild	Expert assessor or panel review	Environment Australia—animal imports into Australia
cies-specific: 10 questions related to the species' ability escape into the wild, its ability to establish in the wild, l its potential ecological impacts	Panel review	Unknown—aquaculture introductions
cies-specific: 24 questions regarding the environmental ditions in the species' native range, population charac-istics and habitat specialization, seed dispersal and pro-tion, likely seed predation patterns in the fynbos, and -history adaptations to fynbos fire conditions	Expert assessor	Unknown—ornamental plant imports into South Africa
cies-specific: 3 pruning questions followed by 9 questions arding the species' ability to survive in the recipient envi-ment, the species' ability to invade the environment, resistance to invasion, and any possible ecological impacts. es Chutter's Biotic Index to measure inter alia invasion istance.	Expert assessor	Unknown—freshwater fish imports into South Africa
cies-specific: 49 questions on the biogeography, biology/ logy, and undesirable attributes of the species. Answers s, no, or don't know) are scored—overall species score itively correlated to weediness.	Expert assessor	Australian Quarantine and Inspection Service—plant imports into Australia
rironmental similarity: Risk = (inoculation factor × rironmental similarity index × risk biota factor), where inoculation factor = (the number of vessel arrivals × the ume of ballast discharged), the environmental similarity ex = 1/[(GM similarity coefficient) × (source port group nber)], and risk biota factor = arbitrary weighting factor ween 1 and 2.	Automated	Queensland Ports Corporation—ballast water "imports"
cies-specific: Models four steps in the invasion process— prevalence of infected fish in wild stocks, number of in-ted fish per ton imported, contamination of waterways via ee exposure scenarios, and infection of native salmonids	Automated	New Zealand Ministry of Agriculture—Canadian salmon imports
rironmental similarity and species-specific: Conducts a ple environmental match hazard assessment, followed a species-specific risk assessment. The latter models four ps in the invasion process—infection status of donor port, ction of vessel, journey survival, and survival of species he recipient port.	Automated	Australian Quarantine and Inspection Service—ballast water decision support system

Table 15.2. The probability of deleterious effect from a genetically engineered organism is the product of six factors (adapted from Alexander 1985).

Probability	Event
P_1	RELEASE—will the genetically engineered organism escape?
P_2	SURVIVAL—will it survive in the natural environment?
P_3	MULTIPLICATION—will it proliferate?
P_4	DISSEMINATION—will it be dispersed to distant sites?
P_5	TRANSFER—will its genetic information be transferred to other species?
P_6	HARM—will the engineered organism be harmful?

$$\text{Risk} = P_1 \times P_2 \times P_3 \times P_4 \times P_5 \times P_6$$

sists of five steps: hazard identification, frequency assessment, consequence assessment, risk calculation, and uncertainty analysis. It was originally developed for complex industrial systems, but its techniques can be applied equally well to complex ecological systems.

For example, Australia's policy for translocating live aquatic organisms (Ministerial Council on Forestry, Fisheries and Aquaculture 1999) is largely based on the OIE framework but also borrows from the QRA paradigm, distinguishing between the frequency of undesired events and their consequences. The policy addresses three events in the bioinvasion chain: escape/release, survival, and establishment. The assessment is qualitative, at least in the first instance, and asks questions to determine the likelihood of escape or release. The assessment then considers the consequences of release. The next set of questions determines the likelihood of survival followed by the consequences of survival, and finally the likelihood and consequences of establishment. The question sequence emphasizes that the consequences at one step in the bioinvasion event chain (which species escape, at which life stages, how many individuals, when and where) are intimately linked to the likelihood of success at the next step (will these individuals survive), which in turn determines the consequences at the next step, and so forth.

With one exception, all of the bioinvasion models summarized in Table 15.1 are species-specific, meaning that each is applied on a species-by-species basis. The one exception is the Queensland Ports Corporation ballast water risk assessment (Hilliard et al. 1997). This is based on the environmental similarity between the donor and recipient ports, as measured by forty environmental variables. Carlton et al. (1995) developed a similar, but much simpler, approach based purely on comparing the salinity of ballast water to that of the recipient port.

Bioinvasion risk assessments based on environmental comparison avoid the question, which species will be the next invader? They are therefore well suited to a translocated habitat or environment containing a multitude of species (e.g., ballast water, the soil of a potted plant, consignment of Siberian timber, etc.). This approach assumes that the likelihood of invasion is directly proportional to the biophysical similarity between donor and recipient environments. As a first step in the risk assessment process this is a useful approach, but as a stand-alone assessment it is flawed. Notwithstanding the difficulties of measuring biophysical similarity, this approach is virtually impossible to improve empirically (Hayes and Hewitt 2000a), and is not conservative for species with broader environmental tolerances than their current range. The water hyacinth, *Eichhornia crassipes*, is a spectacular example of the inability of environmental-matching to predict the future range of an invading exotic (Mack 1996).

Seven out of the ten assessments summarized in Table 15.1 calculate risk with the aid of an expert assessor or panel review. Three of these assessments are qualitative: in each case the assessor(s) must answer a series of questions about the species concerned and its ability to survive and reproduce in the recipient environment. The other four are mixed qualitative/semiquantitative: the assessor assigns scores to a similar series of questions. Qualitative and semiquantitative assessments are attractive because they are flexible—they can be applied when data are scarce, and are easily modified to the particular circumstances of the assessment. Risk assessment, however, is all about uncertainty; indeed, this is what distinguishes it from environmental impact assessment. Uncertainty in a risk assessment takes many forms (Baybutt 1989, Morgan and Henrion 1990, Pate-Cornell 1996, Regan et al. 2002), the most important being variability (a characteristic's true heterogeneity) and epistemic (our incomplete knowledge of the system in question). Qualitative and semiquantitative risk assessments do not address either of these very well. Typically, one of the following three approaches is adopted:

- Uncertainty is not formally addressed in the assessment procedure. A high-risk status is (presumably) assigned to the species or commodity if there is insufficient information to allow a reasonable assessment, but this is entirely at the discretion of the assessor(s). Examples include Australia's translocation policy and the Schedule 6 assessment procedure (Table 15.1).
- The assessment is inherently conservative. A high-risk status is auto-

matically assigned to the species or commodity unless there is sufficient information (as judged by the assessor) to indicate a low invasion risk. Examples include the expert system for screening alien plants in the South African fynbos (Tucker and Richardson 1995) and the AQIS weed risk assessment (Pheloung 1995).

• The assessment includes a qualitative or semiquantitative description of uncertainty. The assessor(s) describe, score, or rank their level of uncertainty with each question, which is then accounted for in the final risk calculation. Examples include the review and decision model for introductions of aquatic organisms (Kohler and Stanley 1984) and the generic aquatic organisms risk analysis review process (Orr 1995).

Clearly none of these approaches is "risk assessment in the sense of explicitly characterizing the probability of populations or communities [of organisms] becoming impaired" (Calow 1993, p. 1519). Furthermore, the assessments are scarcely "objective" and are difficult to justify scientifically because they do not (usually) make testable predictions. The AQIS weed risk assessment is possibly an exception; the procedure was empirically tested against 370 known invasive weeds in Australia. The assessment's decision boundaries (accept/evaluate/reject) were then manipulated to ensure all serious weeds were rejected, while the number of further evaluations and useful plants rejected were minimized. Although not a scientific hypothesis test, it does introduce some empirical rigor, which is usually lacking in qualitative/semiquantitative assessments.

Three out of the five quantitative or semiquantitative assessments summarized in Table 15.1 automate the risk calculation. Of these the New Zealand import health risk analysis (Stone et al. 1997) and the AQIS ballast water risk assessment (Hayes and Hewitt 1998, 2000a, 2000b) are the most sophisticated. The former uses a mixture of deductive and inductive techniques to quantify the risk of introducing the bacterium *Aeromonas salmonicida* into New Zealand with salmon imports. Figure 15.3 summarizes the risk assessment procedure, illustrating the OIE "chain of events" model, and the analysis conducted at each step of the event chain. Figure 15.4 summarizes the AQIS ballast water risk assessment, which again emulates the OIE framework. The assessment is entirely inductive, however, because the lack of data on species assemblages in specific ballasting conditions, or on the frequency of successful inoculation or establishment, does not allow a deductive approach.

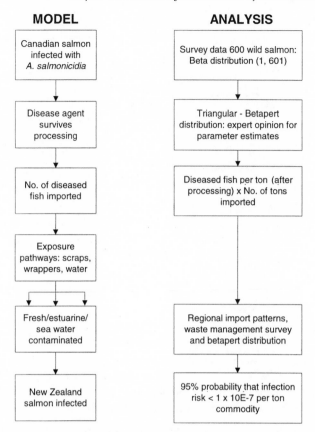

FIGURE 15.3. A summary of the New Zealand salmon import risk assessment framework showing the analyses at each step of the bioinvasion event chain.

Ballast Water Example

The AQIS ballast water risk assessment is based on the OIE framework; ballast-meditated invasions are viewed as a sequential chain of events (Fig. 15.4). The risk assessment, a central component of the decision-support system implemented by AQIS in July 2001, was designed to deliver species- and vessel-specific risk estimates, on a per vessel-visit basis. The assessment uses a relatively low-order endpoint—survival in the recipient port—to maintain a reasonable bound on uncertainty, and defines ballast water risk as

$$\text{Risk}_{species} = p(A) \cdot p(B) \cdot p(C) \cdot p(D),$$

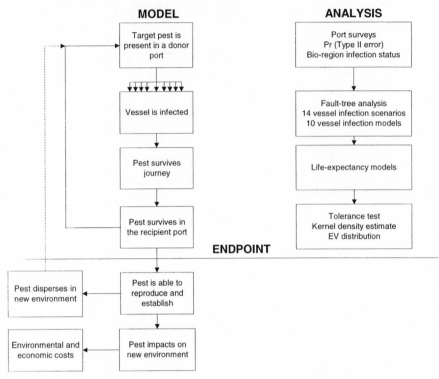

FIGURE 15.4. A summary of the AQIS ballast water risk assessment framework showing the analyses at each step of the bioinvasion event chain.

where $p(A)$ is the probability that the donor port is contaminated with the species in question, $p(B)$ the probability that the vessel becomes infected with this species, $p(C)$ the probability that the species survives the vessel's journey, and $p(D)$ the probability that the species will subsequently survive in the recipient port.

The assessment is inductive, modeling each step in the invasion chain by discrete modules. The assessment is also hierarchical, allowing increasingly accurate risk estimates to be made when more data are made available to the analysis (Hayes and Hewitt 1998, 2000a, 2000b).

Identify Hazards

The ballast water risk assessment borrows two hazard identification techniques from the QRA paradigm: fault-tree analysis and HAZOP analysis. Fault trees identify the chain of events that lead to hazardous occurrences—in this instance, vessel infection. HAZOP uses guide words to test the effects of deviations in the process of intent—in this case, port-based processes (environmental and anthropogenic) that might invalidate the predictive algorithms in the assessment.

Figure 15.5 illustrates the start of the fault-tree analysis completed for the risk assessment (but see Hayes 2002a for the complete tree). The analysis helped identify ten vessel-infection scenarios, which are mutually exclusive for the life stages of most species (i.e., each life stage falls into one or another of the infection scenarios). The infection scenarios are defined by the habitat of the life stage (water column, soft/hard substrate or epibiotic) and its infection characteristics (planktonic, tychoplanktonic, neustonic, vertical migrator, and floating detached) (Fig. 15.6).

The HAZOP analysis is designed to highlight environmental and anthropogenic activity in a port (donor and recipient) that may not be captured by the risk assessment models. Bioinvasion processes are very time-dependent. Carlton (1996) makes this point well, listing six scenarios that can change an unsuccessful process to a successful invasion, including environmental changes in existing donor regions and recipient regions (e.g., donor and recipient ports) and dispersal vector changes. A fault-tree analysis, however, does not capture time-dependent variables very well. A more creative and open-ended approach, such as HAZOP, could help here. Table 15.3, for example, shows the start of a possible HAZOP analysis for port-based processes. By applying guide words such as "more," "less," "none," and so forth to the main environmental variables, such as temperature and salinity, the analysis forces the assessor to consider whether deviations from the "typical" conditions could occur, and/or the extent to which these deviations are adequately described by existing datasets. This approach, however, is yet to be tested on an actual port.

In the AQIS ballast water risk assessment, the species hazard is defined in advance; the risk assessment is applied to a target list of species that are known to be marine pests in their native or introduced range. The assessment is currently limited to these species. It can, however, be applied to any species provided certain data requirements are met. The assessment does not, however, identify potential pests among the many hundreds of species that are daily transported around the world in ballast water, and does not therefore calculate the risk of introduction for these species. A module that addresses the residual risks associated with these is currently being developed (Barry and Bugg 2002).

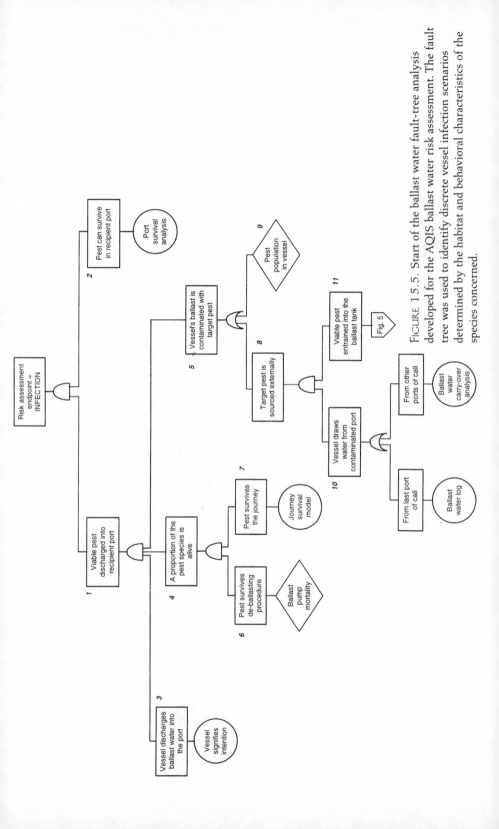

FIGURE 15.5. Start of the ballast water fault-tree analysis developed for the AQIS ballast water risk assessment. The fault tree was used to identify discrete vessel infection scenarios determined by the habitat and behavioral characteristics of the species concerned.

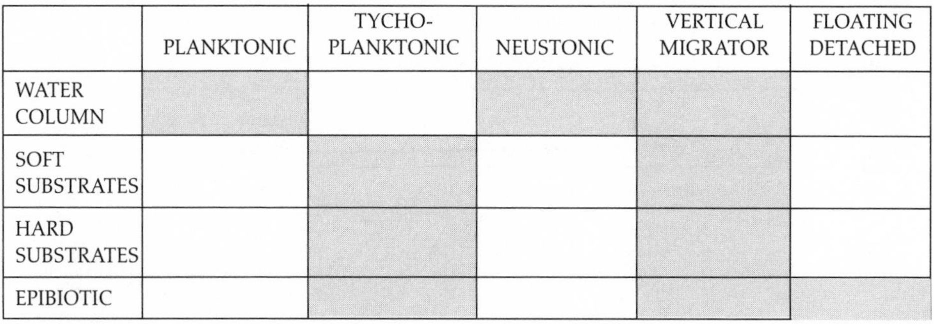

	PLANKTONIC	TYCHO-PLANKTONIC	NEUSTONIC	VERTICAL MIGRATOR	FLOATING DETACHED
WATER COLUMN					
SOFT SUBSTRATES					
HARD SUBSTRATES					
EPIBIOTIC					

FIGURE 15.6. Ballast water infection scenarios used to model vessel infection in the AQIS ballast water risk assessment.

TABLE 15.3. Example of a HAZOP analysis for port-based processes.

Guide Word	Deviation	Possible Cause	Consequences and Significance Assessment
None	Salinity—no freshwater	Drought	Consider likelihood and effects of drought conditions on port environment in relation to species tolerance for salinity, temperature, etc., and port circulation patterns.
More of	Temperature—increased	Simply a warm year	Test derivation of extreme value distribution for sea-surface temperature—time-series length relative to other meteorological records, presence of trend.
		Localized warming, e.g., industrial outfall	Check for presence of industrial outfalls at temperatures significantly higher than ambient.
		Localized warming due to lagoon effect	Check for presence of partially enclosed habitats with restricted water circulation.
	Oxygen—increased	Reduction in detritus input	Check for evidence of reduction in detritus loads and variation in oxygen minima at the freshwater–brackish water interface, relative to species tolerance.
Less of	Salinity—reduced	Increased freshwater input into areas of restricted circulation	Check for presence of partially enclosed habitats with restricted circulation and freshwater input, together with extremes in freshwater discharge history.
		Stratified flow regime within estuary	Check for evidence of salt wedge and freshwater lens, variation in relation to freshwater input, and the extent to which this is captured in port data.

continues

Table 15.3. *Continued*

Guide Word	Deviation	Possible Cause	Consequences and Significance Assessment
As well as	Bed shear stress—increase	Sympathetic effect of extreme tidal current and wind-induced shear stress	Hypothesize potential maximum shear bed stress conditions (and likely return period) on basis of sympathetic extremes of tide and wind, highlighting likelihood distribution in time.
	Target species presence—altered behavior	Predation avoidance responses between target species	Consider any evidence for behavioral interactions between target species and implications for vessel infection models—e.g., altered vertical migration patterns.
Where else	Salinity—altered circulation	Flood events change pattern of freshwater sources	Consider likelihood of new freshwater (or storm drain) inputs into port environment and likely significance with respect to circulation and salinity/temperature regime.
	Target species—altered distribution	Settlement or colonization of new areas in port	Consider availability of existing and new habitats within port and potential implications for pest distribution.
When else	Altered reproductive season	Species' hypothetical niche is broader than realized niche	Consider any evidence for species reproductive season extending either side of documented season in native or introduced range.

Quantify Risks

The ballast water risk assessment attempts to quantify four steps in the bioinvasion process, up to and including survival in the recipient port (Fig. 15.4). The probability of donor port infection is based on the results of port surveys, the attendant probability of Type II error—that is, the species is not detected by a survey but is in fact present (Hayes and Hewitt 2000a)—and the date since it was last surveyed. The probability of a port gaining species since its last survey depends on, among other things, the time since the last survey and the vector volume (e.g., ship visits, volume of ballast discharged). The infection status of ports that have not been surveyed is inferred from the infection status of the bioregion in which the port is located, and the native and/or introduced range of the species concerned.

The ballast water infection scenarios identified in Figure 15.6 are used to model the probability of vessel infection with any life stages of the species concerned. The framework is hierarchical. In the first instance it uses simple models for each step of the invasion process and maintains conservative assumptions if any one of the steps cannot be quantified. For example, it assumes that the probability of vessel infection is "1" if the requisite data are unavailable. More sophisticated models are used when more data becomes available. In this way several levels or tiers of assessment can be made, each progressively more accurate with increasing data.

A Bayesian journey-survival model is used to quantify the analyst's uncertainty about the species' life expectancy in the ballast tank, and hence the probability that the species will survive the journey (Hayes 1998). Figure 15.7, for example, shows the posterior distribution function for the life expectancy of the larval life stages of the seastar, *Asterias amurensis*. The probability of journey survival $p(C)$ is given by the probability that the species life expectancy equals or exceeds the vessel's journey duration (days).

Finally, the probability of survival in the recipient port is calculated by comparing kernel density estimates, extreme-value distributions, or time-series models of temperature and/or salinity in the recipient port with the temperature and/or salinity tolerance of the species concerned. Figure 15.8, for example, shows the kernel density estimates for maximum sea surface temperature in Hobart and Sydney in January 1994. A kernel density is a nonparametric estimate of a probability distribution function—in

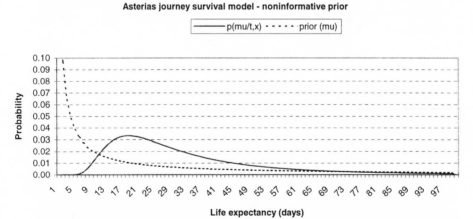

FIGURE 15.7. The posterior distribution function for the life expectancy of the larval life stages of the seastar *Asterias amurensis*. The distribution function is used by a demonstration project of the AQIS ballast water risk assessment to model journey survival. It is based on a noninformative prior distribution (that assumes a constant daily rate of mortality) and the results of four surveys aboard the MV *Iron Sturt*. The surveys recorded larval life stages of *Asterias* as dead after two journeys of 12 days and one journey of 33 days. The last survey recorded larvae alive after a journey of 16 days.

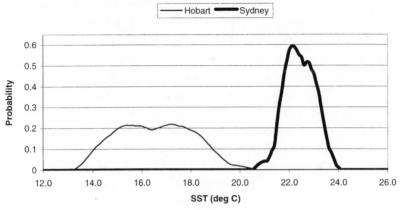

FIGURE 15.8. Kernel density estimates of maximum sea-surface temperature in Hobart and Sydney in January 1994. Probability density estimates such as these are used by a demonstration project of the AQIS ballast water risk assessment to calculate the probability that the species concerned will survive if discharged into.

this instance, the probability of sea surface temperature—and can be used to provide a direct estimate of survival probability $p(D)$. The kernel densities shown in Figure 15.8 were constructed using the Epanechnikov kernel and an "automatic" bandwidth (see Hayes and Hewitt 2000b for further details).

Manage Risk

Of thirty-two ballast water and sediment management options, only one—ballast water exchange—is widely practiced by the shipping industry (Carlton et al. 1995). If alternative management strategies are practiced in the future, then it is unlikely that any one option would be the most appropriate for all vessels on all routes. Future management options might be adopted in one of two ways:

- on a bioregion-to-bioregion basis. For example, all ballast translocated between bioregions north and south of the equator, or on the western/eastern margins of continents, that are environmentally similar but biologically distinct, might be more stringently controlled.
- on a vessel-, tank-, and species-specific basis. For example, some of the ballast translocated on a domestic voyage might be controlled to avoid the further spread of an established marine pest.

The AQIS ballast water risk assessment allows management options to be directed at specific vessel- or route-hazards, and also identifies where best to break the event chain. For example, a demonstration project (Hayes and Hewitt 2000b) illustrates that with species that have resistant diapause life stages (such as the cysts of the dinoflagellate *Gymnodinium catenatum*), appreciable risk reductions can be achieved only by avoiding vessel infection. In this case the management strategies applied at the donor port are likely to provide the best cost–benefit ratio. Alternatively, with species that have relatively delicate larval life stages (such as *Asterias amurensis*), appreciable risk reductions can be achieved by enhancing the natural mortality that occurs in the ballast tank, particularly on long journeys. In this instance management strategies applied en route might prove the most cost-efficient.

Discussion

Biosecurity risk assessment faces at least three important problems: the "which species" problem, the "firefight" problem, and the problems associated with a lack of data.

Which Species

Environmental-match assessments are a useful nonspecific measure of bioinvasion hazard. Bioinvasion risk assessment for species that are known to be harmful must be species-specific (Hayes and Hewitt 2000a). For pathways that can deliver multiple harmful species, this raises two questions:

• Which species might be on the path at any given time?
• On which species should the analyst conduct the risk assessment?

The easiest way to identify the species on a path is to wait, observe, and record what happens. This is simple, rigorous, and effective, but only proactive and precautionary if the species are not released from the pathway (e.g., ballast water that is not discharged into a port). Invasion biologists have traditionally listed the characteristics of successful invaders (those species that make it to the end of the bioinvasion event sequence; see for example, Holdgate 1986 or Lodge 1993). Recording the species that are present at each step of the bioinvasion event sequence (not just the end), and listing their biological characteristics relative to the characteristics of the pathway, would undoubtedly shed some more light on this problem. The analyst could gain further predictive insights by also exploring the invasion "mechanics" of a particular pathway, using the hazard analysis tools discussed earlier. Taken together, these approaches would help identify what types of species are likely to be present in a given pathway and why.

The next question is, on which species should the risk assessment be conducted? The risk analysis guidelines published by the OIE (Office International des Epizooties 1996) give no guidance here. Some authors suggest that the risk assessment need be conducted only on the species that is most likely to be introduced, reasoning that if the risk is negligible for this species, then the overall risk is negligible (MacDiarmid 1994). In practice, however, this approach is likely to be quickly rejected by stakeholders opposed to the pathway.

The FAO guidelines for pest risk analysis (Food and Agricultural Organization 1996) state that all pests that are likely to follow a pathway should be listed, but only "quarantine pests" should be subject to a detailed risk assessment. Quarantine pests are defined as a pest of potential economic importance to an area, and either not yet present, or present but not widely distributed (and subject to official control).

If there are a large number of species that satisfy these criteria, then they are usually prioritized in some fashion. The AQIS salmon-import risk assessment, for example, considers only the disease agents of salmon with a Humphrey score greater than 21 (with a few exceptions). The Humphrey score is the sum of the individual scores allocated by the analyst to the pathogenic significance, potential for international spread, potential for entry and establishment in Australia, socioeconomic and ecological consequences, and difficulty of control or eradication of each disease agent (AQIS 1999). It is important to note here that different, and potentially more damaging, genetic strains of species already in Australia, but not subject to official control (because, for example, they are already widespread and/or have little impact), must still be targeted for risk assessment.

The FAO guidelines clearly envisage risk assessments for multiple species. In practice, analysts may have difficulty in deciding whether a species is a quarantine pest; for example, they may not know whether the species is exotic or endemic (see "Lack of Data" on page 406). But the fundamental problem with this approach is that qualitative (and semiquantitative) risk assessments cannot provide overall measures of pathway risk for multiple species. If the probability of success at each step in the bioinvasion event chain is independent (in a statistical sense) from species to species, then the overall pathway risk is given by

$$\text{Risk}_{\text{pathway}} = 1 - \prod_{i=1}^{n}[1 - \text{Risk}_{\text{species } i}] \, ,$$

which is approximately equal to

$$\text{Risk}_{\text{pathway}} = \sum_{i=1}^{n} \text{Risk}_{\text{species } i}$$

for species $i = 1$ to n when the species risk is small. But how does one sum qualitative or semiquantitative expressions of species risk in a meaningful way? How many "negligible risks" sum to a "moderate risk"? This question cannot be answered in a qualitative fashion and therefore is usually ignored in most bioinvasion risk assessments.

The U.S. Department of Agriculture adopted an interesting approach to the problem. The import risk assessment for Monterey pine from Chile (Orr 1993) recognizes that basic biological information is unavailable for many of the insects and fungi associated with the Chilean timber. However, by developing a detailed risk assessment for pests that inhabit each of the timber's niches (the bark surface, the inner bark, the sapwood, and the heartwood), the analysts hoped to develop effective mitigation measures that would eliminate the known pests and the unknown organisms that inhabit the same niche. In effect, the analysts advocate a guild or functional group approach to risk assessment, which may help with the "which species" problem.

A guild is a group of species that exploit the same resource in a similar way (Root 1973). Guild theory suggests that species that utilize the same resource have similar diets, are found in similar habitats, and behave in a similar manner. The theory has been advocated as a useful tool for environmental impact assessment—actions that affect one member of a guild should affect all members similarly (Severinghaus 1981), and as a means of defining exposure in (chemical) ecological risk assessment (Suter 1993).

A functional group is similar to, but broader than, a guild. A functional group is made up of species categorized by body plan (anatomical and morphological), behavior, or life history, whereas a guild is defined purely on the basis of resource use. Functional group ecology has been used to understand and predict how terrestrial plants disperse, establish, and persist (Wiether et al. 1999), to categorize the response of marine algae to disturbance (Steneck and Dethier 1994), and to define the dynamics of freshwater streams (Cummins 1974). As a risk assessment tool, however, this approach does not appear to have been explored much.

A risk analyst might use functional groups or guilds in two roles: to identify the types of organisms that might occur in a pathway, and to manage these organisms. Figure 15.6, for example, identifies ballast water infection scenarios based largely on the habitat and behavior of the life stage concerned. Species that share these habitat and behavioral characteristics will infect vessels in a similar way. One might therefore expect similar infection probabilities for all species within the functional group. Thus the analyst might define functional groups on the basis of the infection characteristics of a species—how does a species join a pathway and where in that pathway does it reside? From here the analyst might define management strategies aimed at preventing a functional group from either joining the pathway or remaining viable within it, as suggested in the import risk assessment for Monterey pine.

The functional group or guild approach is probably less useful, however, when it comes to quantifying the likelihood or consequences of invasion. The probability of invasion is intimately linked to very specific species/life stage characteristics. Subsuming these characteristics within a higher ecological unit (functional group or guild) may not allow quantitative expressions of invasion probability, but should at least identify effective risk mitigation strategies.

Lack of Data

Good data are fundamental to bioinvasion risk assessment. Invasion success is a function of species, pathway, and site attributes. Any assessment of bioinvasion risk, particularly quantitative, is unlikely to be successful in the absence of detailed information on each of these attributes.

Species information is particularly problematic in this regard. Cummins (1974) noted that so long as the species is the fundamental ecological unit, the perpetually incomplete state of our taxonomic knowledge would constitute a major constraint for the development of ecological theory. Twenty-nine years later, incomplete biosystematic data still remain a major constraint for bioinvasion risk assessment. The problem is particularly acute for microbial species such as bacteria, fungi, and viruses. For example, fewer than 10 percent of the native Australian mycoflora have been identified and described (AQIS 2000). A functional group or guild approach might help here, but the probability of invasion success, and its consequences, remains fundamentally a species-specific problem. Biosystematic research is still, therefore, an important component of a good biosecurity strategy (Fig. 15.1).

To some extent the analyst can handle data availability, or the lack of it, through a hierarchical or tiered risk assessment framework. The lower tiers of a risk assessment should be protective (risk averse) and use conservative assumptions; the higher tiers should not be accessed unless certain data requirements are met, as in the ballast water risk assessment described above. This approach allows a progressively more "accurate" assessment of risk as more data become available to the analyst, thereby embodying the precautionary principle (Fairbrother and Bennett 1999). A hierarchical risk assessment penalizes data gaps, but provides risk reduction rewards for additional data collection costs. This should provide an incentive for the proponent of the activity or import in question to shoulder this cost and supply the requisite data—a sort of "polluter pays" principle.

It is appropriate to note here that Article 5.7 of the SPS agreement allows members to adopt provisional sanitary measures on the basis of available information if the scientific evidence is insufficient. Members must, however, obtain the necessary information for a more "objective assessment of risk" within a reasonable period of time. In practice, trade sanctions imposed on the basis of a low-tier conservative risk assessment are likely to be quickly challenged within the WTO as being unscientific. Furthermore, the wording of the SPS agreement is not entirely clear about who should pay to obtain the necessary information.

De facto *Fire Fighting*

Risk assessment, as with any management strategy, will not completely eliminate biological invasions. Species will continue to slip through the net. Environmental managers will inevitably find themselves "fighting fires"—having to deal with exotic species within their national boundaries. The solution here is to fight spot-fires, that is, to manage the invasion as early as possible. Border surveillance, rapid response, and predefined eradication/containment strategies are therefore critical to successful biosecurity—a fact long recognized by veterinary science. For example, in Australia, the procedures, management structures, and job descriptions in the event of a terrestrial or aquatic animal disease emergency are already well established within AUSVETPLAN (Agriculture and Resource Management Council of Australia and New Zealand 1996) and AQUAPLAN (National Office of Animal and Plant Health 1999), respectively. An emergency marine pest plan is currently being drafted along similar lines in Australia.

It is also important to record carefully which species slip through the net, how, and why. This information will eventually form the empirical databases that allow deductive approaches to risk assessment (and should provide important clues to the functional group analysis discussed above). Again, veterinary science provides a precedent; for example, the *World Animal Health* database, maintained by the OIE, contains information on the number of herds and average herd size of domesticated cattle, etc., in nations around the world, together with reported outbreaks of certain notifiable diseases. This type of information is an essential component of the quantitative import risk assessments for animals and animal products (Morley 1993).

SUMMARY

A good bioinvasion risk assessment will have the following characteristics:

- clear endpoints and well-defined boundaries that are sufficiently relevant from a policy perspective, but simple enough to minimize uncertainty in the overall risk estimate
- rigorous inductive and deductive risk/hazard assessment techniques, particularly as applied to hazard identification
- a hierarchical or tiered structure to allow increasingly accurate risk estimates as more information becomes available
- the proficiency to make predictions throughout the bioinvasion process that can be scientifically tested
- the ability to include a good analysis of variability and epistemic uncertainty

It appears, however, from a limited review of the literature, that few bioinvasion risk assessments exhibit all (or indeed any) of these characteristics. A large number of bioinvasion risk assessments are qualitative, often because there are insufficient data to conduct a quantitative assessment. These assessments usually have the following characteristics:

- They are subjective.
- They do not tackle uncertainty, or they tackle it poorly.
- They cannot address multiple risk sources.
- They do not make predictions that can be scientifically tested.

As a result, these assessments are easy to challenge and are very vulnerable to other political or economic imperatives. In the Canadian salmon case, Australia's ban on salmon imports was deemed unjustified on the basis of the available scientific evidence (World Trade Organization 1998). The qualitative risk assessment that subsequently allowed salmon imports (Kahn et al. 1999), however, is no more "objective" or scientific than the qualitative assessment (Department of Primary Industry and Energy 1996) that originally supported the ban, and does not appear to contain any substantial new evidence. It appears the decision to allow Canadian salmon imports into Australia was based on economic or political imperatives rather than on an assessment of the biological risk. The quantitative assessment conducted by New Zealand (Stone et al. 1997) may have influenced this decision.

Qualitative assessments are not bad, but often they are not sufficiently rigorous or scientific to stand up to the economic and political realities of the global world economy. Quantitative risk assessments are generally more robust, but concomitantly much more difficult to perform, particularly in data-scarce situations. If relevant data are truly scarce, however, how can any risk assessment, qualitative or otherwise, be seen as objective and scientific? In other words, for some pathways and commodities, bioinvasion risk assessment, as a discipline, is not sufficiently mature for the role that, for example, the WTO demands of it. There are two solutions to this problem: manage biological pollution in another way (for example, adopt a zero-risk approach), or invest in the science of bioinvasion risk assessment.

With respect to the latter solution, there are a number of useful avenues that might be explored:

• Develop the deductive science. Surveillance, monitoring, and reporting of species at all stages of the bioinvasion process should be made part of the iterative cycle of bioinvasion risk assessment.

• Develop the inductive science. Quantitative models that make scientifically testable predictions should be encouraged, together with rigorous hazard identification techniques such as those developed for complex industrial systems.

• Individual nations should share species and environmental data in databases that can be accessed internationally, specifically for the purpose of assisting quantitative bioinvasion risk assessment.

• Biosystematic research should continue to be funded as an important component of any biosecurity strategy.

• Rapid-response strategies should be developed to deal with the inevitable incursions of pests.

There are precedents in each of these areas, particularly in the databases and emergency response strategies of veterinary science. The challenge is to develop these areas in a systematic and coordinated fashion for all commodities and processes where risk assessment is used to manage biological pollution.

REFERENCES

Agriculture and Resource Management Council of Australia and New Zealand. 1996. *AUSVETPLAN: The Australian Veterinary Emergency Plan—Sum-*

mary. Canberra, Australia: Livestock and Pastoral Division, Department of Primary Industries and Energy.

Alexander, M. 1985. Ecological consequences: Reducing the uncertainty. *Issues in Science and Technology* Spring: 57–68.

Australian Quarantine and Inspection Service. 1999. *John Humphrey Review.* <http://www.aqis.gov.au/docs/qdu/jhreview.htm>.

Australian Quarantine and Inspection Service. 2000. *Issues Paper—Global Import Risk Analysis of the Importation of Fresh and Dried, Cultivated and Wild, Edible Mushrooms.* <http://www.aqis.gov.au/docs/plpolicy/mushTIP.doc>.

Barry, S. C. and A. L. Bugg. 2002. *Assessment of Options for Development of the Ballast Water Decision Support System.* Canberra: Bureau of Rural Sciences.

Baybutt, P. 1989. Uncertainty in risk analysis. In R. A. Cox, ed., *Mathematics in Major Accident Risk Assessment,* pp. 247–261. Oxford: Clarendon Press.

Berdnikov, S. V., V. V. Selyutin, V. V. Vasilchenko, and J. F. Caddy. 1999. Trophodynamic model of the Black and Azov Sea pelagic ecosystem: Consequences of the comb jelly *Mnemiopsis leidyi* invasion. *Fisheries Research* 42: 261–289.

Berger, J. O. and D. A. Berry. 1988. Statistical analysis and the illusion of objectivity. *American Scientist* 76: 159–165.

Calow, P. 1993. Hazards and risks in Europe: Challenges for ecotoxicology. *Environmental Toxicology and Chemistry* 12: 1519–1520.

Carlton, J. T. 1996. Pattern, process and prediction in marine invasion ecology. *Biological Conservation* 78: 97–106.

Carlton, J. T., D. M. Reid, and H. van Leeuwen. 1995. *The Role of Shipping in the Introduction of Nonindigenous Aquatic Organisms to the Coastal Waters of the United States (Other Than the Great Lakes) and an Analysis of Control Options.* Report No. CG-D-11-95, Government Accession Number AD-A294809. Springfield: National Technical Information Service.

Cohen, A. N. and J. T. Carlton. 1997. Transoceanic transport mechanisms: Introduction of the Chinese mitten crab, *Eriochier sinensis,* to California. *Pacific Science* 1: 1–11.

Cummins, K. W. 1974. Structure and function of stream ecosystems. *BioScience* 24(11): 631–641.

De Moor, I. J. and Bruton, M. N. 1993. *Decision Support Protocol Used to Assess the Advisability of Allowing the Importation of Alien Aquatic Animals into Southern Africa.* Investigational Report No. 42. Grahamstown, South Africa: J. L. B. Smith Institute of Ichthyology.

Department of Primary Industry and Energy. 1996. *Salmon Import Risk Analysis: An Assessment by the Australian Government of Quarantine Controls of Uncooked, Wild, Adult Ocean Caught Pacific Salmonid Product Sourced from the United States of America and Canada.* Final Report. Commonwealth of Canberra, Australia.

Doyle, K. A., P. T. Beers, and D. W. Wilson. 1996. Quarantine of aquatic animals in Australia. *Revue scientifique et technique de l'Office international des epizooties* 15(2): 659–673.

Fairbrother, A. and R. S. Bennett. 1999. Ecological risk assessment and the precautionary principle. *Human and Ecological Risk Assessment* 5(5): 943–949.

Faucheux, S. and G. Froger. 1995. Decision making under environmental uncertainty. *Ecological Economics* 15: 29–42.

Ferson, S., J. A. Cooper, and D. Myers. 2001. *Beyond Point Estimates: Risk Assessment Using Interval, Fuzzy and Probabilistic Arithmetic.* Society for Risk Analysis Workshop, 2 December 2001. McLean, VA: Society for Risk Analysis.

Fiksel, J. R. and V. T. Covello. 1985. The suitability and applicability of risk assessment methods for environmental application of biotechnology. In J. Fiksel and V. Covello, eds., *Biotechnology Risk Assessment: Issues and Methods for Environmental Introductions*, pp. 1–34. Final Report to the Office of Science and Technology Policy, Executive Office of the President, report No. NSF/PRA 8502286. New York: Pergamon Press.

Food and Agricultural Organization. 1996. *International Standards for Phytosanitary Measures: Part 1 Import Regulations—Guidelines for Pest Risk Analysis.* Publication No. 2. Rome: Secretariat of the International Plant Protection Convention, Food and Agriculture Organization.

Gollasch, S., M. Dammer, J. Lenz, and H. G. Andres. 1998. Non-indigenous organisms introduced via ships into German waters. In J. T. Carlton, ed., *ICES Cooperative Research Report No. 224—Ballast Water: Ecological and Fisheries Implications*, pp. 50–64. Copenhagen: International Council for the Exploration of the Sea.

Hallegraeff, G. M. 1998. Transport of toxic dinoflagellates via ships' ballast water: Bioeconomic risk assessment and the efficacy of possible ballast water management strategies. *Marine Ecology Progress Series* 168: 297–309.

Hayes, K. R. 1997. *A Review of Ecological Risk Assessment Methodologies.* CRIMP Technical Report No. 13. Hobart: CSIRO Division of Marine Research.

Hayes, K. R. 1998. *Bayesian Statistical Inference in Ecological Risk Assessment.* CRIMP Technical Report No. 17. Hobart: CSIRO Division of Marine Research.

Hayes, K. R. 2002a. Identifying hazards in complex ecological systems—Part I: Fault tree analysis for biological invasions. *Biological Invasions* 4(3): 235–249.

Hayes, K. R. 2002b. Identifying hazards in complex ecological systems—Part II: Infection modes and effects analysis for biological invasions. *Biological Invasions* 4(3): 251–261.

Hayes, K. R. and C. L. Hewitt. 1998. *Risk Assessment Framework for Ballast Water Introductions.* CRIMP Technical Report No. 14. Hobart: CSIRO Division of Marine Research.

Hayes, K. R. and C. L. Hewitt. 2000a. Quantitative biological risk assessment of the ballast water vector: An Australian approach. In J. Pederson, ed., *Marine Bioinvasions: Proceedings of the First National Conference*, pp. 370–386. Cambridge, MA: MIT Sea Grant College Program, MITSG 00-2.

Hayes, K. R. and C. L. Hewitt. 2000b. *Risk Assessment for Ballast Water Introductions*, Vol. II. CRIMP Technical Report No. 21. Hobart: CSIRO Division of Marine Research.

Hilliard, R. W., S. Walker, and S. Raaymakers. 1997. *Ballast Water Risk Assessment—12 Queensland Ports: Stage 5 Report—Executive Summary & Synthesis of Stages 1–4.* EcoPorts Monograph Series No. 14. Brisbane: Ports Corporation of Queensland.

Holdgate, M. W. 1986. Summary and conclusions: Characteristics and consequences of biological invasions. *Philosophical Transactions of the Royal Society of London. Series B: Biological Sciences* 314: 733–742.

Kahn, S. A., P. T. Beers, V. L. Findlay, I. R. Peebles, P. J. Durham, D. W. Wilson, and S. E. Gerrity. 1999. *Import Risk Analysis on Non-Viable Salmon and Non-Salmonid Marine Finfish.* Canberra: Australian Quarantine and Inspection Service.

Kletz, T. A. 1986. *HAZOP & HAZAN: Notes on the Identification and Assessment of Hazards.* Warwickshire: The Institution of Chemical Engineers.

Kohler, C. C. and J. G. Stanley. 1984. *Implementation of a Review and Decision Model for Evaluating Proposed Exotic Fish Introductions in Europe and North America.* Technical Paper 42(1): 541–549. European Inland Fisheries Advisory Commission.

Lodge, D. M. 1993. Biological invasions: Lessons for ecology. *Tree* 8(4): 133–137.

MacDiarmid, S. C. 1994. *The Risk of Introducing Exotic Diseases of Fish into New Zealand Through the Importation of Ocean-Caught Pacific Salmon from Canada.* I-CAN-135. Wellington, New Zealand: Ministry of Agriculture and Fisheries.

Mack, R. N. 1996. Predicting the identity and fate of plant invaders: Emergent and emerging approaches. *Biological Conservation* 78: 107–121.

Ministerial Council on Forestry, Fisheries and Aquaculture. 1999. *National Policy for the Translocation of Live Aquatic Organisms—Issues, Principles and Guidelines for Implementation.* Canberra: Bureau of Rural Sciences.

Morgan, M. G. and M. Henrion. 1990. *Uncertainty: A Guide to Dealing with Uncertainty in Quantitative Risk and Policy Analysis.* Cambridge: Cambridge University Press.

Morley, R. S. 1993. A model for the assessment of the animal disease risks associated with the importation of animals and animal products. *Revue scientifique et technique de l'Office international des epizooties* 12: 1055–1092.

National Office of Animal and Plant Health. 1999. *AQUAPLAN: Australia's National Strategic Plan for Aquatic Animal Health, 1998–2003.* Canberra: Agriculture Fisheries and Forestry.

Nordstrom, H. and S. Vaughan. 1999. *Trade and Environment.* Special Studies 4. Geneva: World Trade Organization.

O'Neill, R. V., R. H. Gardner, L. W. Barnthouse, G. W. Suter, S. G. Hildebrand, and C. W. Gehrs. 1982. Ecosystem risk analysis: A new methodology. *Environmental Toxicology and Chemistry* 1: 167–177.

Office International des Epizooties. 1996. *International Animal Health Code (Mammals, Birds and Bees).* Paris: OIE.

Orr, R. L. 1993. *Pest Risk Assessment of the Importation of* Pinus radiata, Nothofagus dombeyi *and* Laurelia philippiana *Logs from Chile.* Miscellaneous Publication No. 1517[CS2]. U.S. Department of Agriculture.

Orr, R. L. 1995. *Generic Nonindigenous Aquatic Organisms Risk Analysis Review Process.* Aquatic Nuisance Species Task Force USA: U.S. Department of Agriculture.

Palady, P. 1995. *Failure Modes and Effects Analysis: Predicting Problems Before They Occur.* West Palm Beach, FL: PT Publications Inc.

Pate-Cornell, M. E. 1996. Uncertainties in risk analysis: Six levels of treatment. *Reliability Engineering and System Safety* 54: 95–111.

Pheloung, P. C. 1995. *Determining the Weed Potential of New Plant Introductions to Australia*. A report on the development of a weed risk assessment system commissioned by the Australian weeds committee and the plant industries committee. Western Australia: Agriculture Protection Board.

Pierce, R. W., J. T. Carlton, D. A. Carlton, and J. B. Geller. 1997. Ballast water as a vector for tintinnid transport. *Marine Ecology Progress Series* 149: 295–297.

Regan, H. M., M. Colyvan, and M. A. Burgman. 2002. A taxonomy and treatment of uncertainty for ecology and conservation biology. *Ecological Applications* 12(2): 618–628.

Root, R. B. 1973. Organisation of a plant-arthropod association in simple and diverse habitats: The fauna of collards (*Brassica oleracea*). *Ecological Monographs* 43(1): 95–124.

Royal Commission on Environmental Pollution. 1991. *GENHAZ: A System for the Critical Appraisal of Proposals to Release Genetically Modified Organisms into the Environment*. London: HMSO.

Ruesink, J. L., I. M. Parker, M. J. Groom, and P. M. Kareiva. 1995. Reducing the risks of nonindigenous species introductions: Guilty until proven innocent. *BioScience* 45(7): 465–477.

Ruiz, G. M., J. T. Carlton, E. D. Grosholz, and A. H. Hines. 1997. Global invasions of marine and estuarine habitats by non-indigenous species: Mechanisms, extent and consequences. *American Zoology* 37: 621–632.

Severinghaus, W. D. 1981. Guild theory development as a mechanism for assessing environmental impact. *Environmental Management* 5(3): 187–190.

Sharov, A. A. and A. M. Liebhold. 1998. Bio-economics of managing the spread of exotic pest species with barrier zones. *Ecological Applications* 8(3): 833–845.

Simberloff, D. and M. Alexander. 1994. Issue paper on biological stressors. *Ecological Risk Assessment Issue Papers*. Report reference EPA/630/R-94/009. Washington, DC: U.S. Environmental Protection Agency, Office of Research and Development.

Smith, A. D. M. 1993. Management strategy evaluation—The light on the hill. In D. A. Hancock, ed., *Population Dynamics for Fishery Management*, pp. 249–253. Australian Society for Fish Biology Workshop Proceedings, Perth, 24–25 August 1993. Perth: Australian Society for Fish Biology.

Steneck, R. S. and M. N. Dethier. 1994. A functional group approach to the structure of algal-dominated communities. *Oikos* 69: 476–498.

Stone, M. A. B., S. C. MacDiarmid, and H. J. Pharo. 1997. *Import Health Risk Analysis: Salmonids for Human Consumption*. Wellington, New Zealand: Ministry of Agriculture Regulatory Authority.

Suter, G. W. 1993. *Ecological Risk Assessment*. Chelsea, MI: Lewis Publishers.

Tucker, K. C. and D. M. Richardson. 1995. An expert system for screening potentially invasive alien plants in South African fynbos. *Journal of Environmental Management* 44: 309–338.

U.S. Environmental Protection Agency. 1992. *Framework for Ecological Risk*

Assessment. Report reference EPA/630/R-92/001. Washington, DC: U.S. EPA, Office of Research and Development.

U.S. Environmental Protection Agency. 1998. *Guidelines for Ecological Risk Assessment.* Report reference EPA/630/R-95/002F, Risk Assessment Forum. Washington, DC: U.S. EPA.

Vose, D. 1996. *Quantitative Risk Assessment: A Guide to Monte Carlo Simulation Modelling.* New York: John Wiley & Sons Inc.

Wiether, E., A. van der Warf, K. Thompson, M. Roderick, E. Garnier, and E. Ove. 1999. Challenging Theophrastus: A common core list of plant traits for functional ecology. *Journal of Vegetation Science* 10: 609–620.

William, R. J., F. B. Griffiths, E. J. Van der Wal, and J. Kelly. 1988. Cargo vessel ballast water as a vector for the transport of non-indigenous marine species. *Estuarine, Coastal and Shelf Science* 26: 409–420.

World Trade Organization. 1998. *Australia—Measures Affecting Importation of Salmon.* Report of the Panel, WT/DS18/R 1998, 12 June 1998.

Chapter 16

Generic Nonindigenous Aquatic Organisms Risk Analysis Review Process

Richard Orr

In this chapter, I present a risk analysis process that was developed by committee and submitted as a report to the U.S. Aquatic Nuisance Species Task Force in 1996. The resulting report, *Generic Nonindigenous Aquatic Organisms Risk Analysis Review Process* (hereafter, Review Process), was intended to provide a coherent process for evaluation of risks associated with invasions and for consideration of management actions.

To date, the Review Process (in various permutations) has been used as a basis for more than two dozen completed risk assessments, covering a wide range of taxa from plant diseases and insects to fish and shrimp viruses. Further applications have included a number of pathway (vector) analyses on wood importations. The Review Process has proved an effective and useful tool for both aquatic and terrestrial risk evaluations.

The Review Process presented here is a slightly modified version of the original report, which was edited to reduce space. Although the original report can be obtained, it is not widely circulated or recognized for its application to vectors. For these reasons, and its general relevance to vector management, I present the edited version. Definitions of terms used throughout the document are provided in Box 16.1; copies of the original report are available from the author upon request.

Box 16.1. Risk Terms Used in the Review Process

RISK—The likelihood and magnitude of an adverse event.
RISK ANALYSIS—The process that includes both risk assessment and risk management.
RISK ASSESSMENT—The estimation of risk.
RISK COMMUNICATION—The act or process of exchanging information concerning risk.
RISK MANAGEMENT—The pragmatic decision-making process concerned with what to do about the risk.

REVIEW PROCESS OBJECTIVE

The Risk Assessment and Management (RAM) Committee was initiated by, and is under the auspices of, the U.S. Aquatic Nuisance Species Task Force (hereafter, Task Force). The Task Force was created for the purpose of developing a strategy in which the appropriate government agencies could meet the goals of the Aquatic Nuisance Prevention and Control Act of 1990. The Task Force was "established to coordinate governmental efforts related to nonindigenous aquatic species in the United States with those of the private sector and other North American interests" (Anonymous 1994). The Task Force is co-chaired by the U.S. Fish and Wildlife Service and the National Oceanic and Atmospheric Administration.

The Generic Nonindigenous Aquatic Organisms Risk Analysis Review Process (Anonymous 1996) is the risk process developed through the RAM committee to help meet the requirements of the Aquatic Nuisance Prevention and Control Act.

The objective of the Review Process is to provide a standardized process for evaluating the risk of introducing nonindigenous organisms into a new environment and, if needed, determining the correct risk management steps needed to mitigate that risk. The Review Process provides a framework where scientific, technical, and other relevant information can be organized into a format that is understandable and useful to managers and decision makers. The Review Process was developed to function as an open process with early and continuous input from all identified interested parties.

The Review Process was designed to be flexible and dynamic enough to accommodate a variety of approaches to nonindigenous organism risk depending on the available resources, the accessibility of biological information, and the risk assessment methods available at the time of the assessment. The Review Process may be used as a purely subjective evaluation or quantified to the extent possible or necessary depending on the needs of the analysis. Therefore, the process will accommodate a full range of methodologies from a simple and quick judgmental process to an analysis requiring extensive research and sophisticated technologies.

The Review Process has two specific functions:

1. Risk Assessment—Develop a process that can be used to evaluate recently established nonindigenous organisms; evaluate the risk associated with individual pathways (i.e., ballast water, aquaculture, aquarium trade, fish stocking, etc.).

2. Risk Management—Develop a practical operational approach to maximize a balance between protection and the available resources for reducing the probability of unintentional introductions; reducing the risk associated with intentional introductions.

HISTORY AND DEVELOPMENT OF THE REVIEW PROCESS

The Review Process was modified from the Generic Non-Indigenous Pest Risk Assessment Process (Orr et al. 1993) developed by the U.S. Department of Agriculture's Animal and Plant Health Inspection Service (APHIS) for evaluating the introduction of nonindigenous plant pests. The development of the Review Process was synchronous with and functionally tied to the development of various ecological risk assessment methodologies and nonindigenous organism issues. Foremost was the U.S. National Research Council's workshops and meetings for the development of the "Ecological Paradigm" (NRC 1993). The Review Process's basic approach and philosophy borrow heavily from the NRC's project. In addition, other major projects and reports have influenced the direction of the Review Process, including the Environmental Protection Agency's "Ecological Framework" (EPA 1992a) and associated documents (EPA 1992b, 1992c, Simberloff and Alexander 1994); the U.S. Congress Office of Technology Assessment's nonindigenous species report (OTA 1993); and the U.S. Forest Service's pest risk assessments on nonindigenous timber pests (USDA, FS 1991, 1992, 1993). Finally, the review process was designed to meet several quality criteria, modified from Fischoff et al. 1981. The Review Process was intended to be

• *Comprehensive.* The assessment should review the subject in detail and identify sources of uncertainty in data extrapolation and measurement errors. The assessment should evaluate the quality of its own conclusions. The assessment should be flexible to accommodate new information.

- *Logically Sound.* The risk assessment should be up to date and rational, reliable, justifiable, unbiased, and sensitive to different aspects of the problem.
- *Practical.* A risk assessment should be commensurate with the available resources.
- *Conducive to Learning.* The risk assessment should have a broad enough scope to have carryover value for similar assessments. The risk assessment should serve as a model or template for future assessments.
- *Open to Evaluation.* The risk assessment should be recorded in sufficient detail and be transparent enough in its approach that it can be reviewed and challenged by qualified independent reviewers.

RISK ANALYSIS PHILOSOPHY

The risk assessment process allows for analysis of factors for which the dimension, characteristics, and type of risk can be identified and estimated. By applying analytical methodologies, the process allows the assessors to utilize qualitative and quantitative data in a systematic and consistent fashion.

The ultimate goal of the process is to produce quality risk assessments on specific nonindigenous aquatic organisms or with nonindigenous organisms identified as being associated with specific pathways. The assessments should strive for theoretical accuracy while remaining comprehensible and manageable; and the scientific and other data should be collected, organized, and recorded in a formal and systematic manner.

The assessment should be able to provide a reasonable estimation of the overall risk. All assessments should communicate effectively the relative amount of uncertainty involved and, if appropriate, provide recommendations for mitigation measures that reduce the risk.

Caution is required to ensure that the process clearly explains the uncertainties inherent in the process and to avoid design and implementation of a process that reflects a predetermined result. Quantitative risk assessments can provide valuable insight and understanding; however, such assessments can never capture all the variables. Quantitative and qualitative risk assessments should always be buffered with careful human judgment. The following goals *cannot* be obtained from a risk assessment:

1. A risk assessment cannot determine the acceptable risk level. What risk, or how much risk, is acceptable depends on how a person, or agency, perceives that risk. Risk levels are value judgments that are characterized by variables beyond the systematic evaluation of information.

2. It is not possible to determine precisely whether, when, or how a particular introduced organism will become established. It is equally impossible to determine what specific impact an introduced organism will have. The best that can be achieved is to estimate the likelihood that an organism may be introduced and estimate its potential to do damage under favorable host/environmental conditions.

The ability of an introduced organism to become established involves a mixture of the characteristics of the organism and the environment in which it is being introduced. The level of complexity between the organism and the new environment is such that whether it fails or succeeds can be based on minute idiosyncrasies of the interaction between the organism and the environment. General statements based only on the biology of the organism cannot predict these in advance. In addition, even if there is extensive information on a nonindigenous organism, many scientists believe that the ecological dynamics are so turbulent and chaotic that future ecological events cannot be accurately predicted.

If all outcomes were certain, there would not be a need for risk assessment. Uncertainty, as it relates to the individual risk assessment, can be divided into three distinct types:

• uncertainty of the process—(methodology)
• uncertainty of the assessor(s)—(human error)
• uncertainty about the organism—(biological and environmental unknowns)

Each type of uncertainty presents its own set of problems. All three types of uncertainty will continue to exist regardless of future developments. The goal is to succeed in reducing the uncertainty in each of these groups as much as possible.

The "uncertainty of the process" requires that the risk methodologies involved with the Review Process never become static or routine but continue to be modified when procedural errors are detected and/or new risk methodologies are developed.

"Uncertainty of the assessor(s)" is best handled by having the most qualified and conscientious people available to conduct the assessments. The quality of the risk analysis will, to some extent, always reflect the quality of the individual assessor(s).

The "uncertainty about the organism" is the most challenging. Indeed, it is the biological uncertainty, more than anything else, that initiated the need for developing a nonindigenous risk process. Common sense dictates that the caliber of a risk assessment is related to the quality of data available about the organism and the ecosystem that will be invaded. Those organisms for which copious amounts of high-quality research have been conducted are the most easily assessed. Conversely, an organism for which very little is known cannot be easily assessed.

A high degree of biological uncertainty, in itself, does not demonstrate a significant degree of risk. However, those organisms that demonstrate a high degree of biological uncertainty do represent a real risk. The risk of importing a damaging nonindigenous organism (for which little information is known) is probably small for any single organism, but the risk becomes much higher when one considers the vast number of these organisms that must be considered. It is not possible to identify which of the "unknowns" will create problems, but only to assume that some will. Demonstrating that a pathway has a "heavy" concentration of nonindigenous organisms for which little information is present may, in some cases (based on the type of pathway and the type of organisms), warrant concern. However, great care should be taken by the assessor(s) to explain why a particular nonindigenous organism load poses a significant risk.

This need to balance "demonstrated risks" against "biological uncertainty" can lead assessors to concentrate more on the uncertainty than on known facts. To prohibit or restrict a pathway or specific nonindigenous organism, the reasons or logic should be clearly described.

Risk assessments should concentrate on demonstrated risk. Applying mitigation measures based on well-documented individual nonindigenous pests will frequently result in a degree of mitigation against other organisms demonstrating high biological uncertainty that might be using the same pathway.

Some of the information used in performing a risk assessment is scientifically defensible, some of it is anecdotal or based on experience, and all of it is subject to the filter of perception. However, we must provide an estimation based on the best information available and use that estimation in deciding whether to allow the proposed activity involving the nonindigenous organism and, if so, under what conditions.

The assessment should evaluate risk in order to determine management action. Estimations of risk are used in order to restrict or prohibit high-risk pathways, with the goal of preventing the introduction of non-indigenous pests.

When conducting risk assessments for government agencies, the most serious obstacles to overcome are the forces of historical precedent and the limitations presented by legal parameters, operational procedures, and political pressure. In order to focus the assessment as much as possible on the biological factors of risk, all assessments need to be completed in an atmosphere as free of regulatory and political influences as possible.

The following quote is taken from the National Research Council's 1983 Red Book *Risk Assessment in the Federal Government: Managing the Process*: "We recommend that regulatory agencies take steps to establish and maintain a clear conceptual distinction between assessment of risks and consideration of risk management alternatives; that is, the scientific findings and policy judgments embodied in risk assessments should be explicitly distinguished from the political, economic, and technical considerations that influence the design and choice of regulatory strategies" (p. 151). This can be translated to mean that risk assessments should not be policy-driven. However, the Red Book then proceeded with a caveat: "The importance of distinguishing between risk assessment and risk management does not imply that they should be isolated from each other; in practice they interact, and communication in both directions is desirable and should not be disrupted." This can be translated to mean that the risk assessment, even though it must not be policy-driven, must be policy-relevant. These truths continue to be valid (NRC 1993).

CONDUCTING PATHWAY ANALYSIS AND ORGANISM RISK ASSESSMENTS

The need for a risk assessment can be established in two ways: the request for opening a new pathway, which might harbor nonindigenous aquatic organisms; or the identification of an existing pathway, which may be of significant risk. All pathways showing a potential for nonindigenous organism introduction should receive some degree of risk screening. Pathways that show a high potential for introducing nonindigenous organisms should trigger an in-depth risk assessment.

The Review Process when focusing on the risk of nonindigenous organisms associated with a pathway follows the steps in Box 16.2. If specific organisms needing evaluation are not tied to a pathway assessment, the assessor then would proceed directly to "Organism Risk Assessments" on page 423.

Pathway Data

Specific information about the pathway is required to undertake the risk assessment (step 2b in Box 16.2). This information will vary with the type of pathway (i.e., ballast water, aquaculture, aquarium trade, fish stocking, etc.). The following generalized list of information has proven useful:

1. Determine exact origin(s) of organisms associated with the pathway.
2. Determine the numbers of organisms traveling within the pathway.
3. Determine intended use or disposition of pathway.
4. Determine mechanism and history of pathway.
5. Review history of past experiences and previous risk assessments (including foreign countries) on pathway or related pathways.
6. Review past and present mitigating actions related to the pathway.

Nonindigenous Aquatic Organisms of Concern

A parallel component of step 2b (Box 16.2) is to create a list of organisms of concern. The following generalized process is recommended. First, determine what organisms are associated with the pathway. Second, using Table 16.1, determine which of these organisms qualify for further evalu-

---

Box 16.2. Pathway Analysis Steps

Step 1. INITIATION includes
a. request to evaluate a pathway, or
b. request to evaluate a single organism.

Step 2. The RISK ASSESSMENT would normally proceed by
a. identifying interested parties and soliciting input.
b. creating a list of nonindigenous organisms of concern while also collecting pathway data or information.
c. conducting the Organism Risk Assessments.
d. assembling the pathway assessment (including the Organism Risk Assessments).
e. providing recommendations to the risk managers.

Step 3. RISK MANAGEMENT
a. developing the Risk/Mitigation Matrix
b. developing Operational Procedures

TABLE 16.1. Steps to evaluate potential organisms for concern.

Category	Organism Characteristics	Concern
1a	species nonindigenous, not present in country (United States)	yes
1b	species nonindigenous, in country and capable of further expansion	yes
1c	species nonindigenous, in country and reached probable limits of range, but genetically different enough to warrant concern and/or able to harbor another nonindigenous pest	yes
1d	species nonindigenous, in country and reached probable limits of range and not exhibiting any of the other characteristics of 1c	no
2a	species indigenous, but genetically different enough to warrant concern and/or able to harbor another nonindigenous pest, and/or capable of further expansion	yes
2b	species indigenous and not exhibiting any of the characteristics of 2a	no

ation. Third, produce a list of the organisms of concern from categories 1a, 1b, 1c, and 2a of Table 16.1. Taxonomic confusion or uncertainty should also be noted on the list. Fourth, conduct Organism Risk Assessments on the developed list of organisms.

Based on the number of organisms identified and the available resources, it may be necessary to focus on fewer organisms than those identified using Table 16.1. When this is necessary, it is desirable that the organisms chosen for complete risk assessments be representative of all the organisms identified. This screening has been done using alternative approaches. A standard methodology is not presented to accomplish this task in this chapter, because the risk assessment requirements will often be site or species specific. Therefore, professional judgment by scientists familiar with the aquatic organisms of concern is often the best tool to determine which organisms are necessary for effective screening. Different approaches can be found in each of the three log commodity risk assessments (USDA, Forest Service 1991, 1992, 1993).

ORGANISM RISK ASSESSMENTS

The Organism Risk Assessment element (step 2c in Box 16.2) is the most important component of the Review Process used in evaluating and deter-

Table 16.2. Risk Assessment Model

I. Probability of Establishment (P)	II. Consequences of Establishment (C)
$P = (X_A)(X_E)(X_C)(X_S)$	$C = X + Y + Z$
X_A = Association with Pathway	X = Economic Impact Potential
X_E = Entry Potential	Y = Environmental Impact Potential
X_C = Colonization Potential	Z = Perceived Impact (Social and Political
X_S = Spread Potential	Influences)

mining the risk associated with a pathway. The Organism Risk Assessment can also be independent of a pathway assessment, if a particular non-indigenous organism needs to be evaluated. Table 16.2 represents the Risk Model that drives the Organism Risk Assessment. For model simplification the various elements are depicted as being independent of one another. In addition, the order of the elements in the model does not necessarily reflect the order of calculation.

The Risk Assessment Model is divided into two major components: probability of establishment and consequence of establishment. This division reflects how one can evaluate a nonindigenous organism, providing the flexibility to weigh individual elements; for example, more restrictive measures are used to lower the probability of a particular nonindigenous organism establishing when the consequences of its establishment are greater.

The Risk Assessment Model is a working model that represents a simplified version of the real world. In reality, the specific elements of the Risk Model are not static or constant, but are truly dynamic and show distinct temporal and spatial relationships. In addition, the elements are neither equal in weighing the risk nor necessarily independent. The weight of the various elements will never be static because they are strongly dependent upon the nonindigenous organism and its environment at the time of introduction.

The two major components of the Risk Assessment Model are further divided into seven basic elements (shown in Table 16.2), which serve to focus scientific, technical, and other relevant information into the assessment. Each of these seven basic elements is discussed in further detail in Box 16.3. A Risk Assessment Form (Appendix 16.1) is used to characterize these seven elements as to probability or impact estimates. These may be determined using quantitative or subjective methods; see also Appendix 16.2 for a minimal subjective approach.

Box 16.3. Elements of Risk Assessment Model

PROBABILITY OF ORGANISM ESTABLISHMENT

1. Nonindigenous Aquatic Organisms Associated with Pathway (At Origin)—Estimate probability of the organism being on, with, or in the pathway.

The major characteristic of this element: Does the organism show a convincing temporal and spatial association with the pathway?

[When evaluating an organism not associated with a pathway, or an organism recently introduced, the first two elements under Group 1 would automatically be rated as high because entry into the new environment either is assumed or has already occurred.]

2. Entry Potential—Estimate probability of the organism surviving in transit.

Characteristics of this element include the organism's hitchhiking ability in commerce, its ability to survive during transit, its stage of life cycle during transit, the number of individuals expected to be associated with the pathway; or whether the organism is deliberately introduced (e.g., biocontrol agent or fish stocking).

3. Colonization Potential—Estimate probability of the organism colonizing and maintaining a population.

Characteristics of this element include the organism coming in contact with an adequate food resource, encountering appreciable abiotic and biotic environmental resistance, and having the ability to reproduce in the new environment.

4. Spread Potential—Estimate probability of the organism spreading beyond the colonized area.

Characteristics of this element include the organism's ability for natural dispersal, its ability to use human activity for dispersal, its ability to readily develop races or strains, and the estimated range of probable spread.

CONSEQUENCE OF ESTABLISHMENT

5. Economic Impact Potential—Estimate economic impact if established.

Characteristics of this element include economic importance of hosts, damage to crop or natural resources, effects to subsidiary industries, exports, and control costs.

continues

Box 16.3. Continued

6. Environmental Impact Potential—Estimate environmental impact if established.

Characteristics of this element include ecosystem destabilization, reduction in biodiversity, reduction or elimination of keystone species, reduction or elimination of endangered/threatened species, and effects of control measures. If appropriate, impacts on the human environment (e.g., human parasites or pathogens) would also be captured under this element.

7. Perceived Impact (Social and Political Input)—Estimate impact from social and/or political influences.

Characteristics of this element include aesthetic damage, consumer concerns, and political repercussions.

The strength of the assessment is that the information gathered by the assessor(s) can be organized under the seven elements. The cumulative information under each element provides the data to assess the risk for that element. Whether the methodology used in making the risk judgment for that element is quantitative, qualitative, or a combination of both, the information associated with the element (along with its references) will function as the information source. Placing the information in order of descending risk under each element will further communicate to reviewers the thought process of the assessor(s).

Adequate documentation of the information sources makes the Review Process transparent to reviewers and helps to identify information gaps. This transparency facilitates discussion, if scientific or technical disagreement on an element rating occurs. For example, if a reviewer disagrees with the rating that the assessor assigns an element, the reviewer can point to the information used in determining that specific element rating and show what information is missing, misleading, or in need of further explanation. Focusing on information to resolve disagreements will often reduce the danger of emotion or a preconceived outcome from diluting the quality of the element rating by either the assessors or the reviewers.

Often the assessor feels uncomfortable dealing with the categories of Economic Impact and Perceived Impact. However, information found by an assessor relating to these categories may be helpful in making risk management decisions. The assessor should present information about the organism that would (or could) affect these decisions, and should not be expected to reflect, or second-guess, what an economist or politician would conclude.

The elements considered under Consequences can also be used to record positive impacts that a nonindigenous organism might have—for example, its importance as a biocontrol agent, aquatic pet, sport fish, scientific research organism, or based on its use in aquaculture. The elements in the case of deliberate introductions would record information that will be useful in determining the element rating that would be a balance between the cost, the benefit, and the risk of introducing the nonindigenous organism.

The Risk Assessment Form (Appendix 16.1) should be considered as a flexible guide. Each nonindigenous organism is unique. The assessor needs to have the freedom to modify the form to best represent the risk associated with that particular organism. The seven elements need to be retained to calculate the risk, but other sections may be added or subtracted. If the assessor feels that information, ideas, or recommendations would be useful, they should be included in the assessment. The assessor can combine like organisms into a single assessment if their biology is similar (e.g., tropical aquarium fish destined to temperate North America).

The number of risk assessments to be completed from the list of nonindigenous organisms in a particular pathway depends on several factors. These include the amount of organism information, available resources, and the assessor's judgment concerning whether the completed assessments effectively represent the pathways' nonindigenous organism risk.

The source of the statements and the degree of uncertainty of the assessor associated with each element need to be recorded on the Risk Assessment Form. The use of the Reference Codes at the end of each statement, coupled with the use of the Uncertainty Codes for each element, fulfill these requirements. Both types of codes are described in Appendix 16.1.

If a federal agency uses the Review Process for potential environmental problems, much of the information may contribute to meeting that agency's National Environmental Policy Act (NEPA) requirements. When both NEPA documentation and a risk assessment are warranted, the two should be coordinated so that resources are not duplicated. Although a risk assessment is similar to an Environmental Impact Statement (EIS), the risk assessment differs by focusing on the probability of occurrence and the impact of that occurrence, while an EIS generally places its emphasis on who or what will be impacted. Therefore, a risk assessment is more likely to clarify possible outcomes, determine or estimate their probabilities of occurrence, and succeed in recording the degree of uncertainty involved in making the predictions.

Assembling and Summarizing Organism and Pathway Risk

An estimate of risk is made at three levels in the Review Process. The first places a risk estimate on each of the seven elements within the Risk Assessment (element rating). The second combines the seven risk element estimates into an Organism Risk Potential (ORP), which represents the overall risk of the organism being assessed. The third links the various ORPs into a Pathway Risk Potential (PRP) that will represent the combined risk associated with the pathway.

Assigning either a quantitative or a qualitative estimate to an individual element, determining how the specific elements in the model are related, and determining how the estimates should be combined are the most difficult steps in a risk assessment. There is no "correct" formula for completing these steps. Various methodologies such as geographical information systems, climate and ecological models, decision-making software, expert systems, and graphical displays of uncertainty may potentially increase the precision of one or more elements in the Risk Assessment Model. Indeed, risk assessments should never become so static and routine that new methodologies cannot be tested and incorporated.

When evaluating new technologies and approaches, it is important to keep in mind that the elements of the Risk Assessment Model are dynamic, chaotic, and not equal in value. New technologies or approaches that may be appropriate for assessing one organism may be immaterial or even misleading in evaluating another organism.

The high, medium, and low approach presented in Appendix 16.2 for calculating and combining the various elements is judgmental. This approach is a generic minimum for determining and combining the element estimates and not necessarily "the best way it can be done." The strength of the Review Process is that the biological statements under each of the elements provide the raw material for testing various approaches. Therefore, the risk assessments will not need to be redone to test new methods for calculating or summarizing the ORP and PRP.

On risk issues of high visibility, the examination of the draft assessment should be completed by pertinent reviewers not associated with the outcome of the assessment. This is particularly appropriate when the risk assessments are produced by the same agency, professional society, or organization that is responsible for the management of that risk.

Risk Management and Operational Requirements

Once the risk assessment for a particular pathway or organism is completed, it is the responsibility of risk managers to determine appropriate policy and operational measures. The key elements include the following:

- Quality of data and degree of uncertainty for risk assessments
- Available mitigation safeguards (i.e., permits, industry standards, prohibition, inspection)
- Resource limitations (i.e., money, time, locating qualified experts, needed information)
- Public perceptions/perceived damage
- Social and political consequences
- Benefits and costs should be addressed in the analysis

Operational steps should be accomplished in determining and implementing risk management policy. First, maintain communication and input from interested parties. Second, maintain open communication between risk managers and risk assessors. Third, match the available mitigation options with the identified risks. Fourth, develop an achievable operational approach that balances resource protection and utilization.

Participation of interested parties should be actively solicited as early as possible. All interested parties should be carefully identified, because adding additional interested parties late in the assessment or management process can result in revisiting issues already examined and thought closed. All identified interested parties should be periodically brought up to date on relevant issues.

Continuous open communication between the risk managers and the risk assessors is important throughout the writing of the risk assessment. This is necessary to ensure that the assessment will be policy-relevant when completed. Risk managers should be able to provide detailed questions about the issues that they will need to address to the risk assessors *before* the risk assessment is started. This will allow the assessors to focus the scientific information relevant to the questions (issues) that the risk managers will need to address.

As important as open communication is between risk managers and risk assessors, it is equally important that risk managers do not attempt to drive, or influence, the outcome of the assessment. Risk assessments need to be policy-relevant, not policy-driven.

Matching the available mitigation options with the identified risks can sometimes be done by creating a mitigation matrix placing the organisms, or groups of organisms, identified in a specific pathway along one axis and the available mitigation options along the other. Where a specific organism, or group of organisms, meets a specific mitigation process in the matrix, the efficacy for control is recorded. Using this process it becomes apparent which mitigation or mitigations are needed to reduce the risk to an acceptable level.

REFERENCES

Anonymous. 1994. Aquatic Nuisance Species Task Force (ANSTF), U.S. Aquatic Nuisance Species Program.

Anonymous. 1996. *Nonindigenous Aquatic Organisms Risk Analysis Review Process.* Report to U.S. Aquatic Nuisance Species Task Force

Fischoff, B., S. Lichtenstein, P. Slovic, S. L. Derby, and R. L. Keeney. 1981. *Acceptable Risk.* London: Cambridge University Press.

National Research Council (NRC). 1983. *Risk Assessment in the Federal Government: Managing the Process.* Washington, DC: National Academy Press.

National Research Council (NRC). 1993. *Issues in Risk Management.* Washington, DC: National Academy Press.

Office of Technological Assessment (OTA). 1993. *Harmful Non-Indigenous Species in the United States.* U.S. Congress, Office of Technology Assessment.

Orr, R. L., S. D. Cohen, and R. L. Griffin. 1993. *Generic Non-Indigenous Pest Risk Assessment Process.* Report to the U.S. Department of Agriculture.

Simberloff, D. and M. Alexander. 1994. Biological Stressors. Issue paper prepared for U.S. Environmental Protection Agency, Risk Assessment Forum. 60 pages.

U.S. Department of Agriculture, Forest Service. 1991. *Pest Risk Assessment of the Importation of Larch from Siberia and the Soviet Far East.* Miscellaneous Publication No. 1495.

U.S. Department of Agriculture, Forest Service. 1992. *Pest Risk Assessment of the Importation of Pinus radiata and Douglas-fir Logs from New Zealand.* Miscellaneous Publication No. 1508.

U.S. Department of Agriculture, Forest Service. 1993. *Pest Risk Assessment of the Importation of Pinus radiata, Nothofagus dombeyi and Laurelia philippiana Logs from Chile.* Miscellaneous Publication No. 1517.

U.S. Environmental Protection Agency. 1992a. Framework for Ecological Risk Assessment. Risk Assessment Forum. U.S. Environmental Protection Agency Report 630/R-92/001.

U.S. Environmental Protection Agency. 1992b. Report on the Ecological Risk Assessment Guidelines Strategic Planning Workshop. Risk Assessment Forum. U.S. Environmental Protection Agency Report 630/R-92/002.

U.S. Environmental Protection Agency. 1992c. Peer Review Workshop Report on a Framework for Ecological Risk Assessment. Risk Assessment Forum. U.S. Environmental Protection Agency Report 625/3-91/022.

ORGANISM RISK ASSESSMENT FORM
(with Uncertainty and Reference Codes)

ORGANISM _____   FILE NO. _____

ANALYST _____   DATE _____

PATHWAY _____   ORIGIN _____

I. LITERATURE REVIEW AND BACKGROUND INFORMATION (summary of life cycle, distribution, and natural history):

II. PATHWAY INFORMATION (include references):

III. RATING ELEMENTS: Rate statements as low, medium, or high. Place specific biological information in descending order of risk with reference(s) under each element that relates to your estimation of probability or impact. Use the Reference Codes at the end of the biological statement where appropriate, and the Uncertainty Codes after each element rating.

---

1. PROBABILITY OF ESTABLISHMENT

Element   Uncertainty
Rating   Code
(L,M,H)   (VC—VU)

_____ , _____   Estimate probability of the nonindigenous organism being on, with, or in the pathway. (Supporting Data with reference codes)

_____ , _____   Estimate probability of the organism surviving in transit. (Supporting Data with reference codes)

Element   Uncertainty
Rating   Code
(L,M,H)   (VC—VU)

_____ , _____	Estimate probability of the organism successfully colonizing and maintaining a population where introduced. (Supporting Data with reference codes)
_____ , _____	Estimate probability of the organism to spread beyond the colonized area. (Supporting Data with reference codes)

2. CONSEQUENCE OF ESTABLISHMENT

Element   Uncertainty
Rating   Code
(L,M,H)   (VC—VU)

_____ , _____	Estimate economic impact if established. (Supporting Data with reference codes)
_____ , _____	Estimate environmental impact if established. (Supporting Data with reference codes)
_____ , _____	Estimate impact from social and/or political influences. (Supporting Data with reference codes)

3. ORGANISM/PATHWAY   RISK   POTENTIAL: (ORP/PRP)

_____

Probability of Establishment | Consequence of Establishment = ORP/PRP RISK

4. SPECIFIC MANAGEMENT QUESTIONS:

5. RECOMMENDATIONS:

6. MAJOR REFERENCES:
REFERENCE CODES TO ANSWERED QUESTIONS

Reference Code	Reference Type
(G)	General Knowledge, no specific source
(10)	Judgmental Evaluation
(5)	Extrapolation; information specific to pest not available; however, information available on similar organisms applied.
(Author, Year)	Literature Cited

UNCERTAINTY CODES TO INDIVIDUAL ELEMENTS

Uncertainty Code	Symbol	Description
Very Certain	VC	As certain as I am going to get
Reasonably Certain	RC	Reasonably certain
Moderately Certain	MC	More certain than not
Reasonably Uncertain	RU	Reasonably uncertain
Very Uncertain	VU	A guess

Appendix 16.2

Judgmental Calculation of Organism Risk and Pathway Risk

STEP 1. CALCULATING THE ELEMENTS IN THE RISK ASSESSMENT

The blank spaces located next to the individual elements of the Risk Assessment Form (Appendix 16.1) can be rated using high, medium, or low. The detailed biological statements under each element will drive the judgmental process. Choosing a high, medium, or low rating, while subjective, forces the assessor to use the biological statements as the basis for his/her decision. Thus, the process remains transparent for peer review.

The high, medium, and low ratings of the individual elements cannot be defined or measured—they have to remain judgmental. This is because the value of the elements contained under "Probability of Establishment" is not independent of the rating of the "Consequences of Establishment." *It is important to understand that the strength of the Review Process is not in the element rating but in the detailed biological and other relevant information statements that motivates them.*

STEP 2. CALCULATING THE ORGANISM RISK POTENTIAL

The Organism Risk Potential and the Pathway Risk Potential ratings of high, medium, and low should be defined (unlike the element rating in Step 1, which has to remain undefined). An example is provided of these definitions at the end of this appendix.

The following three steps must be completed in order to calculate the Organism Risk Potential.

STEP 2A. DETERMINE PROBABILITY OF ESTABLISHMENT

$$P = (X_A)(X_E)(X_C)(X_S)$$

P = Probability of Establishment
X_A = Association with Pathway
X_E = Entry Potential
X_C = Colonization Potential
X_S = Spread Potential

****

The probability of establishment is assigned the value of the element with the lowest risk rating (example: a high, low, medium, and medium estimate for the above elements would result in a low rating).

Because each of the elements must occur for the organism to become established, a conservative estimate of probability of establishment is justified. In reality (assuming the individual elements are independent of each other), when combining a series of probabilities (such as medium—medium—medium), the probability will become much lower than the individual element ratings. However, the degree of biological uncertainty within the various elements is so high that a conservative approach is justified.

STEP 2B. DETERMINE CONSEQUENCE OF ESTABLISHMENT

Economic	Environmental	Perceived	Rating
H	L, M, H	L, M, H	H
L, M, H	H	L, M, H	H
M	M	L, M, H	M
M	L	L, M, H	M
L	M	L, M, H	M
L	L	M, H	M
L	L	L	L

Note that the three elements that make up the Consequence of Establishment are not treated as equal. The Consequence of Establishment receives the highest rating given either the Economic or Environmental element.

The Perceived element does not provide input except when Economic and Environmental ratings are low (see "Perceived" column on the above table).

STEP 2C. DETERMINE ORGANISM RISK POTENTIAL (ORP)

Probability of Establishment	*Consequence of Establishment*	*ORP Risk Rank*
High	High	High
Medium	High	High
Low	High	Medium
High	Medium	High
Medium	Medium	Medium
Low	Medium	Medium
High	Low	Medium
Medium	Low	Medium
Low	Low	Low

Here the conservative approach is to err on the side of protection. When a borderline case is encountered, the higher rating is accepted. This approach is necessary to help counteract the high degree of uncertainty usually associated with biological situations.

STEP 3. DETERMINE THE PATHWAY RISK POTENTIAL (PRP)

ORP Rating	*ORP Number*	*PRP*
High	1 or more	High
Medium	5 or more	High
Medium	>0 but <5	Medium
Low	All	Low

The PRP reflects the highest-ranking ORP. The only exception is when the number of medium-risk organisms reaches a level at which the total risk of the pathway becomes high. The number, 5 or more, used in the above table is arbitrary.

* * * * * * * *

DEFINITION OF RATINGS USED FOR ORGANISM RISK POTENTIAL AND PATHWAY RISK POTENTIAL

Low =   acceptable risk—organism(s) of little concern (does not justify mitigation)

Medium =   unacceptable risk—organism(s) of moderate concern (mitigation is justified)

High = unacceptable risk—organism(s) of major concern (mitigation is justified)

When assessing an individual organism, a determination that the ORP is medium or high often becomes irrelevant because both ratings justify mitigation. When evaluating a pathway, the potential "gray area" between a PRP of medium and one of high may not be a concern for the same reason.

Chapter 17

Pathways-Based Risk Assessment of Exotic Species Invasions

David A. Andow

Risk analysis is often divided into two components, risk assessment and risk management (NRC 1983). Risk assessment is the process by which risk is measured; this measurement can be quantitative, qualitative, probabilistic, or deterministic. In risk management, society determines how to address the risk—for example, whether to tolerate, mitigate, or avoid it. Foot-and-mouth disease is considered a large risk to livestock production in the United States, and is being managed by quarantine regulations (Enserink 2001). Similarly, it is also a serious disease in the United Kingdom: it invaded during 2001 and was managed by massive slaughter of diseased animals and those suspected of contracting the disease (Ferguson et al. 2001, Keeling et al. 2001).

Risk assessment may be conducted *ex ante* or *ex post*. *Ex ante* assessments are conducted before the occurrence of any of the events that could cause the risks and are essential if risks are to be managed, whereas *ex post* assessments are conducted after the possibility of risk is incurred. *Ex post* assessments are useful for determining the effectiveness of previous risk management or for suggesting new assessment or management practices. Because *ex post* assessments are based on the historical data of previous invasions, well-controlled scientific analyses of patterns of risk can be conducted (Kolar and Lodge 2001). In contrast, *ex ante* assessments are predictive and more difficult to complete scientifically because their specificity ensures that gaps in knowledge will erode the credibility of their result. This uncertainty associated with incomplete knowledge must be addressed.

Ecology has focused on *ex post* assessments of invasions by exotic species by analyzing past invasions and evaluating the management that either reduced or increased risk. The goal, however, has been to evolve toward a scientifically sound risk analysis strategy based on sound *ex ante* risk assessment. In this chapter, I briefly review and evaluate some of the findings of *ex post* ecological risk assessment, and develop and illustrate a scientific rationale for *ex ante* risk assessment based on invasion pathways. I argue that a large segment of *ex post* assessment is not very helpful because it does little to inform us how to manage risks. There is ample opportunity for more useful *ex post* ecological risk assessment; however, I argue that a pathways-based assessment may provide greater utility, and sketch some of the ways that such an *ex post* assessment might be conducted.

Development of a scientifically justified risk analysis process will be difficult. Risk assessment is an imperfect activity; any risk assessment procedure will eventually make a faulty assessment, and some invasions will occur despite the best of intentions. Careful analysis of these new invasions should reveal the flaws in the past assessments and lead to improvements. Risk management procedures may also be flawed. Even when risks are estimated accurately, proper actions to mitigate those risks might not be taken because of operational difficulties and/or economic constraints. These operational difficulties include an unsuitably lax regulatory culture and organizational structures that inhibit proper response. Finally, because risk analysis (both assessment and management) is a social activity, the cost of the analysis cannot be excessive and should probably not exceed its perceived benefits. In other words, even though invasions can be extraordinarily costly, a society may choose to underinvest in its oversight of invasions because the benefits are perceived as insufficient. Moreover, the risk analysis must prove capable of reducing the costs associated with biological invasions. Such proof is difficult to provide—it is difficult to demonstrate that a preemptive action has prevented an event from occurring. Hence, a scientifically justifiable risk analysis process must use historical information on its failures to justify its effectiveness. Improvements can then be made and the model (or assessment) retested.

Ex post ASSESSMENT

Problems with Ecological Ex post Risk Assessment

There has been a blossoming of *ex post* risk assessment of exotic species invasions in the ecological literature. With the initiation of the SCOPE (Scientific Committee on Problems in the Environment) projects on species invasions during the late 1970s, many authors have analyzed patterns of species invasions worldwide to uncover ecological trends that could improve risk management (Mooney and Drake 1986, Kornberg and Williamson 1986, Groves and Burdon 1986, Drake et al. 1989, di Castri et al. 1990, Groves and di Castri 1991, Ramakrishnan 1991, Yano et al. 1999, Kolar and Lodge 2001). These assessments have generally focused on two questions (but see Williamson 1996 for a broader treatment): What are the species characteristics that enable some species to have a high probability of colonizing, establishing breeding populations, expanding geographic range, and/or having an adverse effect on the environment? What are the characteristics of the invaded habitat that make it susceptible to invasion by these species?

The answers to these questions depend on which species and which regions of the world are under consideration, but some generalities have emerged (Mooney and Drake 1986, Kornberg and Williamson 1986, Groves and Burdon 1986, Drake et al. 1989, di Castri et al. 1990, Groves and di Castri 1991, Ramakrishnan 1991, Yano et al. 1999, Kolar and Lodge 2001). Species with large natural ranges, with high intrinsic population growth rates, and that arrive in the new habitat with large founding populations are more likely to invade. Habitats with few species present, a high degree of habitat disturbance, and an absence of species similar to the invaders are more likely to be invaded. As additional data are analyzed, many of these generalizations will probably become taxon-specific (Kolar and Lodge 2001). This focused, taxon-specific approach may become useful for identifying organisms from within a taxon that have reduced invasion risk (see, for example, Rejmánek and Richardson 1996). Such taxon-specific analyses could, for example, be used by the horticulture, floriculture, or forest industry in choosing plant species or varieties to release.

Generalized invasion patterns, however, all have a statistical basis. It is possible to make predictions about a large group of potential invading species or invaded habitats, but these generalities do little to help us make predictions about individual species or habitats or about particular cases. Hence, they are of little value for *ex ante* risk assessment, which has been conducted on a case-by-case basis (see "Case-by-Case *ex ante* Assessment," p. 445), using dispersal pathways, taxon-specific data, and expert opinion. At the other extreme, general invasion patterns provide valuable information for broad policy decisions, and have helped to elevate the exotic species problem to the highest levels of policy discussion in the world (CBD 1992, IPPC 1996). Ecological analysis has demonstrated that invasion risk is systematic. Although each particular invasion may be accidental, as a group, invasions occur regularly and have large environmental and economic effects.

It is not possible to change species characteristics, and it is difficult to change the ways that habitats are managed to reduce the risks of invasion. Consider the possible generalization that exotic species with high intrinsic reproductive potential are more likely to colonize and establish in a new habitat (alternately, one could consider any of the more numerous taxon-specific patterns). If the intrinsic reproductive potential for all exotic species was known, in principle, it would be possible to order them from high to low reproductive potential and actively manage those exotics that have a high reproductive potential. Putting aside for the moment the relatively low predictive power of this ranking (that is, let us assume that an appropriately powerful ranking did exist), it is not clear how this information by itself would be used to improve risk management.

Below (in the section on *ex ante* assessment) I suggest that information on the dispersal pathway can have considerable influence on risk management; can our ecological generalizations improve a pathways-based management system? Smith et al. (1999) provide a complex answer to this question based on decision theory. They suggest that the utility of an ecological generalization will depend on the rate at which potential invaders become successful invaders (base rate), the accuracy of the generalization, and the cost of allowing a damaging invader to occur relative to the cost of excluding a potentially useful organism. If the base rate is 2 percent, and the relative cost is 8, then the generalization must be at least 85 percent accurate, or it would be more cost-effective to ignore it. To my knowledge, none of the present ecological generalizations approaches this level of accuracy. Indeed, Smith et al. (1999) conclude that in some conditions, it will be more cost-effective to focus on preventing established species from becoming invasive than on preventing establishment at the importation stage.

Could ecological generalizations be used to develop a list of prohibited (or approved) exotic species, around which an interception and quarantine system could be developed? Economic and public health criteria are routinely used to establish such lists. Foot-and-mouth disease and several human diseases are prohibited in the United States because of potential economic and/or human health risks, and expensive quarantine systems have been developed to enforce the prohibition. While some organisms are prohibited from some countries because of their potential environmental risks, these organisms are identified because of their history of risk, not because they meet some ecological criteria that triggered their exclusion. Hence, ecological principles have not played a major role in developing lists of prohibited or approved species.

Changing Conditions and Ecological Ex post *Assessments*

A serious problem with present results from *ex post* ecological analysis is that the generalities are static—that is, they do not take into account habitat and landscape changes or evolutionary changes in the species. The potential of a species to invade can change because of changes in its habitat of origin or evolutionary changes in the species. For example, the recent invasion of the Asian long-horned beetle, *Anoplophora glabripennis* (Motschulsky), to the United States could have been related to the substantial increase in plantings of hybrid poplar in northern China, a favored host plant of the beetle (EPPO 2001). This increase in planting area caused a substantial increase in beetle population size, which increased the probability that exports from northern China would carry the beetle. Many other invasions may be related to increased population sizes of the potential invader in its habitat of origin. Alternatively, the rice water weevil, *Lissorhoptrus oryzae*, occurred on wetland grasses in its native range on several Caribbean Islands (Iwata 1979). It invaded the southern United States early in the twentieth century, but was restricted to native wetland grasses. In Louisiana, it evolved the ability to feed on rice, and somewhat later a parthenogenic strain evolved from that rice-feeding strain. Around 1959, the parthenogenic strain invaded California rice, and in 1976 it was detected in Japan. It subsequently invaded mainland Asia. Thus, the invasion of rice water weevil in Asian rice paddies occurred because of evolution in host range and reproductive strategy. Similarly, the Colorado potato beetle became invasive only after its host range evolved to include potatoes (Hsiao and Fraenkel 1968).

The habitat being invaded can also change in ways that enable more species to invade it. For example, the crop or habitat in an area could increase or decrease, thereby altering the probability of establishment. There is a positive, temporal correlation between the area of greenhouse production in Japan and the history of invasions of greenhouse pests (Fig. 17.1). Large numbers of insect invasions did not occur until greenhouse production increased sufficiently. In addition, these data suggest that the probability of invasion may have increased as the area of greenhouses increased. However, as Kiritani (1999) noted, greenhouse fauna worldwide is now homogeneous, indicating that the exotic species pool has been depleted by recurrent invasions, so the invasion rate into Japanese greenhouses should now be declining.

Finally, the dispersal pathway can change qualitatively or quantitatively to influence the likelihood of invasion. Sailer (1978) documented insect invasions into the United States. He showed that the composition of

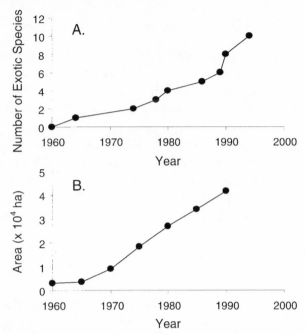

FIGURE 17.1. Insect invasions into greenhouses in Japan. A) Cumulative number of exotic insect species establishing in greenhouses in Japan. Data are cross-referenced between Morimoto and Kiritani (1995) and Kiritani (1999). B) Area of crops produced under greenhouses in Japan. Data are from crop production records published by the Ministry of Agriculture, Forestry and Fisheries in Japan.

these invaders, when classified according to order, changed over time (Fig. 17.2), and that these changes were correlated with changes in shipping practices. During the early 1800s, most of the exotic invading insects were beetles (Coleoptera), many of which were transported to the United States in soil ballast in ships arriving from Europe. By the mid-1800s, the Homoptera began to dominate insect invasions to the United States, including scale insects and other species closely associated with plants. The dominance of the Homoptera is correlated with the rise of fast clipper ships and steamships for transatlantic shipping. These new ships were fast enough that live plant material could be economically shipped from Europe to the United States. Homoptera attached to live plant material became the undesired invaders. The next phase in insect invasions into the United States began in the early 1900s, when deliberate releases of Hymenoptera for use as biological control agents predominated. Thus, changes in dispersal pathways probably have shaped the history of insect invasions into the United States.

Quantitatively, invasions increase when trade increases (e.g., Mills et

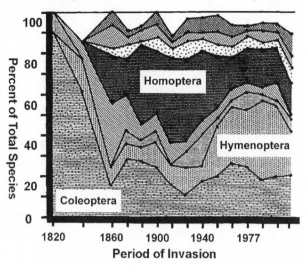

FIGURE 17.2. Change in the composition of insect invasions into continental United States from 1820 to 1977 from Sailer (1978). Unlabeled insect orders from the top of the graph are Thysanoptera (thrips), Heteroptera (true bugs), Diptera (flies). Between the bands designating the Homoptera and Hymenoptera are Lepidoptera (moths and butterflies) and Acarina (mites).

al. 1993, 1994; Chapter 3, this volume), although surprisingly little data are available to evaluate this statement critically. As discussed by Colautti et al. (this volume), aquatic invasions into the Baltic Sea occurred as a result of the building of canals in eastern Europe, which provided a continuous aquatic pathway from the Black and Caspian Seas to the Baltic Sea. These examples show that the invasion process is not static, and that changes in the habitat of origin, the species itself, or the transportation pathways can greatly influence probability of invasion.

EX ANTE ASSESSMENT

Case-by-Case Ex ante *Assessment Based on Dispersal Pathways*

Ex ante risk assessment of species invasions was greatly improved thanks to the analytic framework initiated by Whiteaker and Doren (1989) and developed by Orr et al. (1993). The U.S. Department of Agriculture, Animal and Plant Health Inspection Service (USDA/APHIS) now uses this framework and has conducted risk assessments associated with several commodities. The model focuses on a particular economically important commodity that serves as an importation vector (dispersal pathway) for exotic species. The protocol involves listing all the known exotic species that could become associated with the commodity in the habitat of origin, and for each species qualitatively ranking the probability of establishment and the probability of economic or environmental effect. From these two qualitative measures, a single estimate of unmitigated invasion risk is calculated for each species. Risk management activities are subsequently evaluated by the degree they mitigate invasion risk.

For example, risk of importing pests with raw pine and spruce logs from Mexico into the United States has been evaluated using a panel of experts to assign qualitative risk measures (Thacz et al. 1998). These experts compiled lists of potential pests associated with Mexican species of pine and spruce, and chose twenty-two pine and six spruce species for additional analysis, to be representatives of three ecological niches associated with raw logs. They expected that risk mitigation measures aimed at the selected species would also mitigate risks associated with other species inhabiting a similar niche. Eight species were rated with a high risk potential for either their potential economic effects (tree mortality, wood damage) or their high probability of establishment. The panel concluded that import regulations should address pest importation from Mexico, that the present general permit for unprocessed wood from Mexico should be reassessed, and that USDA/APHIS should review methods to mitigate these risks.

The value of this framework is that by focusing on the dispersal pathway, the risk assessment becomes readily *delimited* and integrates directly with potential risk management activities. In contrast, if a species-based or habitat-based assessment procedure were used, the analyst would have to consider many different potential invasion events, some of which might be difficult to anticipate. Instead, by focusing on a dispersal pathway, only one type of dispersal event needs to be considered, and the uncertainty lies in characterizing the species that might potentially invade. Even this question is reasonably well defined because potential invading species must become associated with the dispersal pathway. Moreover, unlike species characteristics, which are impossible to manage, and habitat characteristics, which society is reluctant to manage prior to an actual invasion, it is possible to manage dispersal pathways, whether by altering their flow or monitoring their content.

Australia has aggressively implemented pathways-based risk assessment. As a part of this effort, Stanaway et al. (2001) evaluated the risk that sea cargo containers would introduce insect pests. About twenty-five years ago, Australia required that all such containers, which were made of wood, either be registered as permanently treated against potential pests, or be inspected every time the container landed in Australia. Containers are now made of steel with plywood floors, and Australia now requires that all containers be treated but no longer registers them. Verification is carried out by random inspection and analysis of the wood components. Only one of 3,001 containers had insect damage to the wooden floor, suggesting that the containers were not a major risk. Timber pests, however, were detected in 3.5 percent of the containers, implying that timber dunnage (low-grade timber used for packing and stabilizing goods) is a risk pathway (Ciesla 1993). About 0.5 percent of the containers had aggressive agricultural or nuisance pests that probably aggregated in containers, and 10 percent had stored-products pests that were transported with the grain inside the containers. Thus, the wood in the containers was not an important dispersal pathway for timber pests, but wood inside the containers could be an important pathway. Instead, the containers could be an important pathway for pests that find shelter in the container or for pests associated with the contents of the containers. Stanaway et al.'s (2001) findings provide clear guidance for quarantine and other risk management activities.

This pathways-centered risk assessment does not rely on any of the *ex post* ecological analysis conducted to date. Ecological theory has focused on the two ends of the invasion process, the organism in its native habitat and the habitat at risk of invasion, while a focus on pathways connects these two. Hence, although a focus on pathways does not invalidate previous *ex post* analysis, it reframes the issues. Consequently, if this approach is to advance based on science, it will be necessary to conduct appropriate *ex post* ecological analyses of invasion risk. Important questions include the following: How are different pathways distinguished from each other? What kinds of species are associated with particular pathways? To what kinds of habitats do the pathways deliver potential invaders? What pathways result in the greatest risks of invasiveness?

Changing Conditions and Ex ante *Assessments*

A pathways-based *ex ante* assessment can be responsive to changing conditions for invasion. This response system could be centered on monitoring changes in the probability of invasion by looking at factors that might affect these probabilities. Invasion probabilities are a product of three probabilities: that a species will leave its habitat of origin, that it will be transported to a new habitat, and that it will be able to establish in the new habitat. As a hypothetical example, the first probability might be proportional to the area of habitat occupied by the species in its habitat of origin; thus, fluctuations in this probability might be monitored by measuring land use changes in habitats of origin. The transport probability might be proportional to the volume of material that is shipped between two locations, so its fluctuations might be monitored by shipping volume. The third probability might be proportional to the area of suitable habitat for the organism in the new habitat, so monitoring land use in the potentially invaded habitat might detect fluctuations in the probability of establishment. It might be possible to construct a pathways approach to assess risk in changing environments associated with two scenarios, large shipping volume and small shipping volume. Oversight associated with large shipping volume should be greater than that associated with small shipping volume, because, all things being equal, invasion risk is higher through dominant pathways.

Changes associated with a large-volume pathway, such as a major trade route, could enable a new suite of invasions to occur. If shipping practices change, then it is possible that a different suite of species from the originating habitat will become associated with the shipment and new invasions will occur. For example, the dramatic increase in containerized shipping has altered invasion risk significantly (OTA 1993, p. 80). In containerized shipping, goods are loaded into a container at the factory or other point of origin; the entire container is then taken to the port, loaded aboard ship, shipped and unloaded intact, and then transported to the point of use (recipient factory or store), where the container is opened. Through the use of containers, potential invaders can enter the container far from the port of origin, and they can be delivered to habitats far removed from the destination port (OTA 1993). Any such change in shipping practice is likely to influence the probability of invasions. Hence, *ex ante* risk assessment could be triggered when a change in shipping practice occurs in some portion of the shipping route. In addition, major changes in the landscape or use patterns in either the country of origin or the receiving country could trigger additional risk assessment. For example, as the dominance of greenhouse production increased in Japan from the early 1960s (Fig. 17.1), additional regulatory oversight could have been triggered in Japan to guard against establishment of the many greenhouse pests. Even if this oversight had delayed establishment of some pests by only a few years, the benefits might have been substantial.

Increases in small-volume pathways, such as minor trade routes, could increase risk of invasions. For example, the shipping route from northern China to the United States has a large volume, but twenty years ago, this volume was much smaller. The increase in shipping volume could have triggered regulatory oversight to assess how this increase would change invasion risks. If this had been done, it might have been possible to recognize the risks associated with the wood shipping material originating from this region, and perhaps the establishment of Asian long-horned beetles could have been avoided or at least significantly delayed.

These examples should illustrate that pathways-based risk assessments can adapt to changing patterns of invasion risk. The scientific basis for relating changes in pathways to changes in invasion risk needs to be developed, but these examples provide an outline for adaptive risk assessment. To begin this process, it is necessary to develop a classification of different kinds of dispersal pathways.

TYPES OF DISPERSAL PATHWAYS AND INVASION RISK

Several types of dispersal pathways can be distinguished (Table 17.1). The purpose of distinguishing different pathways is to simplify risk analysis, either by reducing the complexity of risk assessment or by focusing risk management on a narrower set of possibilities. In some circumstances it may be possible to trade off uncertainty in risk assessment with targeted risk management (Andow et al. 1987).

Deliberate release of exotic organisms is a pathway that focuses risk management, simply because there is a decision point at which an introduction would be allowed or prohibited. This is the largest pathway of invasion for plants and vertebrates (see Chapters 1, 4, and 6, this volume). For example, muskrats were introduced to Europe in the hope that muskrat fur farming would be profitable, gypsy moths were introduced into North America in the hope that an indigenous silk industry could be established, apple snails were introduced in many parts of Asia in the hope that they could be used as human food, and purple loosestrife was introduced into North America as an ornamental plant. It may be possible to distinguish among different types of deliberate releases in a way that differentiates invasion risk, but this has not yet been done except in special cases. For example, a wide taxonomic range of organisms was introduced for biological control (Claussen 1978), and species with a wide host range have the greatest potential to have broader ecological impacts (Howarth 1991). It remains a challenge to incorporate risk assessment into decisions to release potentially beneficial exotic plants and animals.

Another large category of dispersal pathways involves accidental releases (Table 17.1). It may be useful to distinguish several accidental pathways to simplify the risk assessment and focus risk management. An entire community of exotic organisms might be moved when the dispersal pathway moves an intact piece of habitat. For example, dry ballast introduced entire soil communities around the world, and water ballast has

TABLE 17.1. Pathways by which various taxa have invaded and established in new habitats, classified by the kind of pathway. Known pathways are recorded in each cell. (P = plant, I = invertebrate, S = snail, V = vertebrate, M = microorganism)

Pathway	Terrestrial					Aquatic			Marine	
	P	I	S	V	M	P	I	V	P	I
Deliberate	Plant introductions	Biological control, Commercial products	Biological control, Food	Restock game animals, Biological control	Biological control	Aquatic botanical trade		Restock lakes and rivers		Biological control, Ornamental
Accidental Habitat moved	Dry ballast, Trees				Host, Wood		Water ballast		Aquaculture, Water ballast	Water ballast
Organism moved	Hay, etc.	Plants, Logs, Bait	Plants	Pet release	Cut plants	Aquatic trade, Packing material	Fouling, Bait	Pet release, Bait	Fouling	Fouling, Fisheries
Organism aggregates to vector		Logs	Tile	Cargo						
Facilitated movement						Canal building				

introduced entire aquatic communities to new locations (Carlton and Geller 1993). It is not possible to know all of the organisms that could be transported by these pathways. This creates considerable uncertainty in any risk assessment, which might, by necessity, rely on a few broad ecological generalizations, such as climate matching. It may be possible to compensate for this uncertainty by focusing risk management on reducing the probability that the community will survive the transport process.

Another type of accidental dispersal pathway involves the transport of an exotic organism with only a part or none of its associated habitat. Vertebrates (including fish) have been introduced when pets were released after their owners tired of caring for them. Earthworms have been spread when leftover fishing bait was dumped. Fouling organisms have been moved on ship hulls. Snails aggregate on ceramic tile pallets, certain insects aggregate on logs, and rats are attracted to or can be concealed within various forms of cargo. The risks associated with these pathways can be assessed and managed using the commodity-specific, case-by-case procedure discussed earlier. A challenge for ecologists will be to provide robust generalizations as case-by-case assessments accumulate, so that each case does not have to be analyzed independently.

A third type of dispersal pathway includes those in which human activities facilitate natural movement of organisms. For example, canal building has connected many watersheds, enabling many aquatic organisms to move more freely. Landscape use patterns can also facilitate movement of some terrestrial species. While these human activities can increase invasion risk, the incidental association between the exotic species and the activity appears to limit management options. In addition, it is presumptive to expect that an invasive species risk assessment will be conducted prior to all such human activity, so risk analysis options are limited.

Another way to classify the types of dispersal pathways is to concentrate on a particular taxon. Again, the goal of such an exercise is to simplify risk assessment or focus risk management. If it is possible to associate a pathway-taxon combination with a historic risk that a species will become invasive, then it may be possible to allocate oversight in proportion to the risk.

For example, two of the major pathways by which terrestrial plants have invaded are by deliberate introductions and by propagules associated with other plant material (Mack 1991, 1997, 1999, Mack and Lonsdale 2001, Reichard and White 2001). Analyses suggest that more invasive plant species have been established by deliberate introduction than by accidental introduction, yet there is little regulatory oversight in the United

States for these deliberate introductions. This result suggests a need for additional oversight on deliberate plant introductions.

Insects are the major group of invasive terrestrial invertebrates. Insects have invaded along four major pathways: (1) as deliberate releases for biological control or production of commercial products; (2) by association with the movement of their habitat; (3) by aggregating on dispersal vectors; or (4) as dormant stages. Most invasive species have arrived by accidental pathways, which suggests that although deliberate releases of insect biological control agents can cause environmental harm (Howarth 1991, Simberloff and Stiling 1996), greater concern should be given to the accidental pathways. Within the deliberate releases, insect species introduced for the production of commercial products (e.g., gypsy moth and Africanized honeybees) may have had greater risks than those insects introduced for biological control.

Clearly, it will take a systematic analysis of many possible dispersal pathways before a predictive risk analysis framework can be developed. Taxon-specific analyses could lead to a general understanding of the way each taxon interacts with the various possible dispersal pathways that it encounters, which in turn could streamline a pathways-based risk assessment process. I hope that the types of dispersal pathways proposed here might be a useful starting point for improving risk analysis.

CONCLUSION

By focusing research on the occurrence of and changes in dispersal pathways, ecological science can provide the information needed to implement a pathways-based *ex ante* risk assessment. Although there have been many suggestions for a pathways-based risk assessment (e.g., OTA 1993), it is empirically driven scientific principles that are used to address general procedures (Orr et al. 1993). Clearly, many scientific questions need to be evaluated. It is hoped that this brief discussion will help motivate others to provide the needed scientific analysis.

REFERENCES

Andow, D. A., S. A. Levin, and M. A. Harwell. 1987. Evaluating environmental risks from biotechnology: Contributions of ecology. In *Application of Biotechnology: Environmental and Policy Issues,* J. R. Fowle III, ed., pp. 125–144. Boulder: Westview.

Carlton, J. T. and J. B. Geller. 1993. Ecological roulette: The global transport of nonindigenous marine organisms. *Science* 261: 78–82.

CBD (Convention on Biological Diversity). 1992. Article 8, paragraph (h), Convention text at <http://www.biodiv.org/convention/articles.asp>. Nairobi: Environment Program of the United Nations.

Ciesla, W. M. 1993. Recent introductions of forest insects and their effects: A global overview. *FAO Plant Protection Bulletin* 41: 3–13.

Claussen, C. P. 1978. *Introduced Parasites and Predators of Arthropod Pests and Weeds: A World Review.* USDA Agricultural Handbook 480.

di Castri, F., A. J. Hansen, and M. Debussche. 1990. *Biological Invasions in Europe and the Mediterranean Basin.* Dordrecht: Kluwer.

Drake, J. A., H. A. Mooney, F. di Castri, R. H. Groves, F. J. Kruger, M. Rejmánek, and M. Williamson. 1989. *Biological Invasions: A Global Perspective, SCOPE 37.* New York: Wiley.

Enserink, M. 2001. Barricading U.S. borders against a devastating disease. *Science* 291: 2298–2300.

EPPO (European and Mediterranean Plant Protection Organization). 2001. EPPO data sheets on quarantine pests, *Anoplophora glabripennis*. In *Quarantine Pests for Europe.* 2nd ed. London: CAB International Publishing. Also available at <http://www.eppo.org/QUARANTINE/Data_sheets/dsanolgl.html>.

Ferguson, N. M., C. A. Donnelly, and R. M. Anderson. 2001. The foot-and-mouth epidemic in Great Britain: Pattern of spread and impact of interventions. *Science* 292: 1155–1160.

Groves, R. H. and J. J. Burdon. 1986. *Ecology of Biological Invasions: An Australian Perspective.* Canberra: Australian Academy of Science.

Groves, R. H. and F. di Castri. 1991. *Biogeography of Mediterranean Invasions.* Cambridge: Cambridge University Press.

Howarth, F. G. 1991. Environmental impacts of classical biological control. *Annual Review of Entomology* 36: 485–509.

Hsiao, T. H. and G. Fraenkel. 1968. Selection and specificity of the Colorado potato beetle for solanaceous and nonsolanaceous plants. *Annuals of the Entomological Society of America* 61: 493–503.

IPPC (Secretariat of the International Plant Protection Convention). 1996. *International Standards for Phytosanitary Measures: Guidelines for Pest Risk Analysis.* Rome: Food and Agriculture Organization of the United Nations. Updated revisions exist.

Iwata, T. 1979. Invasion of the rice water weevil, *Lissorhoptrus oryzae* Kuschel, into Japan, spread of its distribution and abstract of the research experiments conducted in Japan. *Japan Pesticide Information* 36: 12–21.

Keeling, M. J., M. E. J. Woolhouse, D. J. Shaw, L. Matthews, M. Chase-Topping, D. T. Haydon, S. J. Cornell, J. Kappey, J. Wilesmith, and B. T. Grenfell. 2001. Dynamics of the 2001 UK foot and mouth epidemic: Stochastic dispersal in a heterogeneous landscape. *Science* 294: 813–817.

Kiritani, K. 1999. Exotic insects in Japan. In *Biological Invasions of Ecosystems by Pests and Beneficial Organisms,* E. Yano, K. Matsuo, M. Shiyomi, and D. A. Andow, eds., pp. 60–72. Tsukuba: National Institute of Agro-environmental Sciences.

Kolar, C. S. and D. M. Lodge. 2001. Progress in invasion biology: Predicting invaders. *Tree* 16: 199–204.

Kornberg, H. and M. H. Williamson. 1986. Quantitative aspects of the ecology of biological invasions. *Philosophical Transactions of the Royal Society of London, Series B,* Vol. 314.

Mack, R. N. 1991. The commercial seed trade: An early disperser of weeds in the United States. *Economic Botany* 45: 257–273.

Mack, R. N. 1997. Plant invasions: Early and continuing expressions of global change. In *Past and Future Rapid Environmental Changes: The Spatial and Evolutionary Responses of Terrestrial Biota,* NATO ASI series. Series 2: Global Environmental Change, Vol. 47. B. Huntley, W. A. Cramer, A. V. Morgan, H. C. Prentice, and J. R. M. Allen, eds., pp. 205–216. Berlin: Springer-Verlag.

Mack, R. N. 1999. The motivation for importing potentially invasive plant species: A primal urge? In *Proceedings of the VI International Rangeland Congress, Vol. 2,* D. Eldridge and D. Freudenberger, eds., pp. 557–562. Aitkenvale, Queensland (Australia): International Rangeland Congress, Inc.

Mack, R. N. and W. M. Lonsdale. 2001. Humans as global plant dispersers: Getting more than we bargained for. *BioScience* 51: 95–102.

Mills, E. L., J. H. Leach, J. T. Carlton, and C. L. Secor. 1993. Exotic species in the Great Lakes: A history of biotic crisis and anthropogenic introductions. *Journal of Great Lakes Research* 19: 1–54.

Mills, E. L., J. H. Leach, J. T. Carlton, and C. L. Secor. 1994. Exotic species and the integrity of the Great Lakes. *BioScience* 44: 666–669.

Mooney, H. A. and J. A. Drake, eds. 1986. *Ecology of Biological Invasions of North America and Hawaii.* New York: Springer-Verlag.

Morimoto, N. and K. Kiritani. 1995. Fauna of exotic insects in Japan. *Bulletin of the National Institute of Agro-Environmental Sciences* 12: 87–120.

NRC (National Research Council). 1983. *Risk Assessment in the Federal Government: Managing the Process.* Washington, DC: National Academy Press.

Orr, R. L., S. D. Cohen, and R. L. Griffin. 1993. Generic non-indigenous pest risk assessment process (for estimating pest risk associated with the introduction of non-indigenous organisms). U.S. Department of Agriculture, Animal and Plant Health Inspection Service, 40 pages.

OTA (Office of Technology Assessment). 1993. Harmful non-indigenous species in the United States. U.S. Congress, OTA-F-565. Washington, DC: U.S. Government Printing Office.

Ramakrishnan, P. S. 1991. Ecology of biological invasion in the tropics. New Delhi: International Scientific Publications.

Reichard, S. H. and P. White. 2001. Horticulture as a pathway of invasive plant introductions in the United States. *BioScience* 51: 103–113.

Rejmánek, M. and D. M. Richardson. 1996. What attributes make some plant species more invasive? *Ecology* 77: 1655–1660.

Sailer, R. I. 1978. Our immigrant insect fauna. *Bulletin of the Entomological Society of America* 24: 3–11.

Simberloff, D. and P. Stiling. 1996. How risky is biological control? *Ecology* 77: 1965–1974.

Smith, C. S., W. M. Lonsdale, and J. Fortune. 1999. When to ignore advice: Invasion predictions and decision theory. *Biological Invasions* 1: 89–96.

Stanaway, M. A., M. P. Zalucki, P. S. Gillespie, and C. M. Rodriguez. 2001. Pest risk assessment of insects in sea cargo containers. *Australian Journal of Entomology* 40: 180–192.

Thacz, B. M., H. H. Burdsall Jr., G. A. DeNitto, A. Eglitis, J. B. Hanson, J. T. Kliejunas, W. E. Wallner, J. G. O'Brian, and E. L. Smith. 1998. *Pest Risk Assessment of the Importation into the United States of Unprocessed* Pinus *and* Abies *Logs from Mexico.* General Technical Report FPL-GTR-104. Madison, WI: USDA, Forest Service, Forest Products Laboratory. 116 pages.

Whiteaker, L. D. and R. F. Doren. 1989. Exotic plant species management strategies and list of exotic species in prioritized categories for Everglades National Park. Southeast Regional Office, National Park Service, U.S. Department of the Interior, Research/Resources Management Report SER-89/04.

Williamson, M. 1996. *Biological Invasions.* Kluwer, Amsterdam.

Yano, E., K. Matsuo, M. Shiyomi, and D. A. Andow, eds. 1999. *Biological Invasions of Ecosystems by Pests and Beneficial Organisms.* Tsukuba: National Institute of Agro-environmental Sciences. 232 pages.

Part III

CONCLUSION

Chapter 18

Invasion Vectors: A Conceptual Framework for Management

Gregory M. Ruiz and James T. Carlton

Biological invasions are changing the structure and function of the earth's ecosystems. Nonnative species now dominate many aspects of marine, freshwater, and terrestrial communities throughout the world. Thousands of nonnative species are known to be established in North America alone, and many additional introduced species remain undetected or unrecognized (OTA 1993, Cohen and Carlton 1995, Carlton 1996a, Ruiz et al. 2000a). Invasions cause or contribute to a wide range of high-impact and high-profile effects, including declines in populations of threatened and endangered species, disease outbreaks in human and nonhuman populations, habitat alteration and loss, increased frequency of fires, shifts in food webs and nutrient cycling, decline in fisheries, loss of agricultural crops and productive lands, and reduced water supplies (OTA 1993, Wilcove et al. 1998, Mack et al. 2000, Wittenberg and Cock 2001). Although the full extent and cumulative impact of nonnative species are only coarsely estimated (Parker et al. 1999, Ruiz et al. 1999), biological invasions are clearly a potent force of change, operating on a global scale and affecting many dimensions of society.

The rate of newly detected invasions is increasing over time across many ecosystems and geographic regions (Mills et al. 1993, Rejmánek and Randall 1994, Fuller et al. 1999, Ruiz et al. 2000a; Chapters 4, 5, 6, 7, and 8, this volume). The observed rate increase is often exponential. Search effort has likely increased in recent years, possibly contributing to the observed increase in invasions within some taxonomic groups. However, the overall pattern is very robust, occurring for conspicuous and well-known taxa and in well-studied systems, indicating that invasion rates have indeed increased dramatically in the last half of the twentieth century.

The strong effects of invasions, combined with an apparent increase in the rate of invasions, have caused widespread public and scientific concern, propelling policy and management actions. Many countries now have policies to limit the risk of future invasions and to reduce the impact and spread of established nonnative species (Shine et al. 2000, NISC 2001; Chapters 10, 11, and 12, this volume). In addition, these policies are evolving quickly, undergoing active review and revisions. For example, within the United States, regulations and guidelines to limit the transfer of organisms by ships' ballast water have been established by laws in six states (California, Maryland, Michigan, Oregon, Virginia, Washington), by two federal laws, and by separate requirements governing U.S. Navy vessels (U.S. Congress 1990, 1996, NEMW 2002). None of these laws and regulations existed prior to 1990. Guidelines and strategies for ships have also been pursued internationally through the International Maritime Organization, which is developing an international treaty to limit species transfers by ships (IMO 2002, Raaymakers and Hilliard 2003).

Prevention of new invasions is a clear priority in emerging policies. Management actions directed at established invasions can also have merit, providing options following colonization, but such efforts are often idiosyncratic to the particular species and are potentially expensive, long-term propositions (Mack et al. 2000, Wittenberg and Cock 2001). Successful eradication of established invasions is an unlikely outcome in many instances, even with vast improvements in capacity for "early detection" of new invasions (see discussion below). Furthermore, a separate effort—whether an eradication program or efforts to control spread and abundance—may be required for each invasion event, even if the same species. In contrast, strategies to prevent new invasions are directed at key transfer mechanisms, or vectors. Such vector management may be used to interrupt transfer of a particular target species, but it can also be designed to prevent simultaneously the wholesale transfer of diverse assemblages (including both target and nontarget species), providing a powerful and efficient management approach.

In this chapter, we review the core components of vector management. It is not our intent to provide a comprehensive review, as much detailed information is presented earlier in this volume and elsewhere. Vector management is being implemented in various forms and ecosystems throughout the world, sharing many components. These components, however, are often viewed in isolation of each other and are not integrated into a broader conceptual framework. One goal in this review is to advance such integration, highlighting functional relationships among these elements and some of the associated biological and ecological assumptions that underlie various management strategies. First, we summarize our current understanding of relationships among vector operation, trade, and invasions. Second, we discuss the rationale for broad-based prevention strategies, which includes a precautionary approach. Third, we outline the core components of vector management and their functional relationships. Fourth, we highlight the present status and implementation of vector management. Fifth, we identify some of the gaps and opportunities in this area that we believe are especially significant. This review expands upon a framework for vector management begun by Carlton and Ruiz (in press).

Trade and Vectors: Drivers of Biological Invasion

Trade of commodities is now the primary driver of biological invasions. Trade among geographic regions creates opportunities for biotic interchange—both intentional and unintentional—on regional to global scales, breaching historical barriers to dispersal. Trade provides a diverse supply of organisms to new territories, creating the necessary precursor or first critical phase in a sequence of events that results in invasions and invasion impacts (Fig. 18.1). This transfer phase consists of entrainment of a subset of organisms from a species pool in the source region, and delivery (or release) of these organisms to a recipient region, by human activities. Because of the upstream position of transfers in the invasion sequence, most contemporary invasions simply would not occur without human-mediated delivery through various modes of transportation—ships, planes, trains, automobiles, and trucks.

The observed increases in the rate of invasions have been associated, if not causally linked, to an expansion and globalization of trade (Carlton 1989, Mack 1991, Jenkins 1996, Ruiz et al. 2000a). Although globalization is a widely used term, with many different meanings, there are particular changes in trade characteristics over time that have likely resulted in more opportunities to transfer more species, at both greater numbers and frequency, and among more regions (Table 18.1). Increases in each aspect of organism supply should promote an increase in invasions, due to a few underlying causes:

- Increased density or abundance, and frequency, of inoculation should increase chances of invasion (MacArthur and Wilson 1967, Simberloff

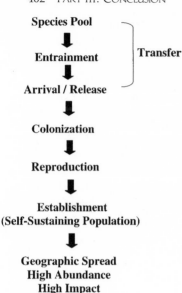

Species Pool

⬇

Entrainment   **Transfer**

⬇

Arrival / Release

⬇

Colonization

⬇

Reproduction

⬇

Establishment
(Self-Sustaining Population)

⬇

Geographic Spread
High Abundance
High Impact

FIGURE 18.1. Stages of the invasion process (after Carlton 1985; see also Richardson et al. 2000 for further discussion). Invasions result from a sequence of events or stages. At each step of the sequence, there is attrition in the total number of organisms and species that transition successfully (survive) to the next stage. There are various ways to characterize the specific steps, which we have outlined here as follows: (a) a subset of the local biota is entrained by a particular vector at a source environment; (b) of entrained organisms, only a fraction survive transit; (c) of surviving organisms, a smaller subset still may be released to a recipient environment; (d) of those released, many will not survive; (e) of those that survive, many will not successfully reproduce and establish self-sustaining populations; (f) of those species that successfully colonize, an unknown fraction will achieve local abundance, spread, and/or have significant impacts. We consider the transfer process to consist of stages a–c.

1989, Schoener and Spiller 1995, Carlton 1996b, Williamson 1996, Kolar and Lodge 2001).

- Transferring more species should also serve to increase the opportunities for invasion.
- The frequency of "sampling" (i.e., entrainment of available organisms by a vector during transfer events) among and within source regions, where the abundance of resident organisms is temporally dynamic, should increase the total number of high-density transfer events.
- The frequency of delivery among and within recipient regions, which also have temporally variable environmental and biological conditions, should increase the chances of encountering favorable conditions.
- Faster transfer should improve the physiological condition of organisms and improve chances of survivorship, since both attributes are often time-dependent (Ruiz et al. 2000a and references therein).

There can be no doubt that biotic exchange by trade has increased dramatically over the past decades to centuries, but changes in the relative contribution—as well as cumulative effect—of the various trade characteristics on organism transfer per unit time or per recipient region remain poorly quantified. Certainly, a wealth of data exists to demonstrate the overwhelming numbers and types of organisms that are transferred throughout the world by most trade-related activities (Carlton and Geller 1993, Mack 1991; Chapter 1, this volume). Further, the number of differ-

TABLE 18.1. Temporal changes in trade characteristics and predicted impact on species transfer.

Trade Characteristic	Change	Predicted Impact of Change
Frequency of Transfer	The amount of traffic for the various modes of transportation (ships, planes, trucks), and therefore frequency of transfer events, has increased over time.	(a) More total organisms are transferred cumulatively. (b) Increased frequency of high-density transfer events. Abundance at source regions is temporally variable. Frequency change results in increased "sampling" through time. (c) Increased range of environmental and biotic conditions experienced at recipient region. Frequency change results in increased "sampling" through time.
Size of Transfer	The capacity of each transfer event, or size of shipments, has increased over time. This is perhaps especially true for ships, which have increased in size through time.	(a) More total organisms are delivered per transfer event. (b) Increased number of species, as well as individuals, per transfer event. Both attributes usually increase with sample effort (number, area, or volume of material examined).
Duration (Speed) of Transfer	The speed with which materials are moved among locations has increased over time.	(a) More organisms are delivered per transfer event. Organisms spend less time in transit. Survivorship of associated organisms is often time-dependent. This may affect both density per species and number of species per transfer event. (b) Organisms arrive in better condition. Physiological condition (food and environmental stress) of associated organisms is often time-dependent.
Number of Commodities	The total number of different commodities traded among locations has increased over time.	(a) More species are entrained and transferred. Each commodity presents a unique set of opportunities for the types of organisms that can be transferred, associated directly with the commodity and transfer process (e.g., packaging materials).
Number of Source Regions	The total number of different geographic sources of suppliers for commodities has increased over time.	(a) More species are entrained and transferred. Each geographic source has a unique species composition. (b) Increased frequency of high-density transfer events. Each geographic source exhibits a unique pattern of abundances for constituent species.
Number of Recipient Regions	The total number of different geographic regions that receive commodities has increased over time.	(a) Increased chance of encountering conditions conducive to invasion. Each geographic recipient region represents a unique environment, including environmental and biotic conditions.

ent transfer mechanisms available for some organisms has increased over time (Carlton and Ruiz, in press). Existing analyses highlight key aspects of transfer mechanisms, but we often lack the data needed to actually estimate changes in the delivery rates of organisms or species over time.

Trade is spatially and temporally very dynamic, creating a complex and shifting pattern. Frequent changes occur in the types, sources, destinations, and volume of cargo, as well as changes in specific methods of cargo transfer (e.g., handling, processing, packaging, and storage), all of which contribute to this complexity. Furthermore, ships and other conveyances can change operating procedures themselves (e.g., type of bottom paints and management of ballast materials on ships; see Chapter 7, this volume), independent of the cargo dynamics, which affect the types and numbers of species transferred. Tracking the multidimensional changes in trade dynamics and their consequences for species transfer is a significant challenge—albeit a core component of vector management (see "Framework for Vector Management").

It is important to note that species distributions are themselves dynamic, resulting from both environmental changes and human transfers, and this dynamic can interact with trade activities to further increase invasions. As new populations of a species become established over time, they create an increased number of source regions for that species that, in turn, may increase the chances for subsequent invasions. New invasions radiate out from an increasing number of "hubs" via trade that may (1) interface with new transfer mechanisms for the species, (2) deliver the species to new destinations, and/or (3) alter the supply characteristics (e.g., frequency, concentration, or timing of delivery) to preexisting destinations. This positive feedback was described by Carlton (1996b) as a "hub-and-spoke model" of invasions; this dynamic could result in more invasions even without an increase in trade, which serves to further accelerate the rate of invasions.

The Dose-Response Relationship

In general, the likelihood of an individual species becoming established, or the number of species that become established, is expected to increase with increasing supply (transfer). This positive relationship likely exists between various aspects of organism supply and the rate of invasion success (establishment). For example, a variety of studies demonstrate an increase in invasion success with increasing frequency and density of inoculation (Roughgarden 1986, Schoener and Spiller 1995, Kolar and Lodge 2001).

In a static environment, this dose-response relationship must be asymptotic, whereby invasion success increases very little or approaches an upper limit above some threshold of supply. Furthermore, if multiple sites received an identical multispecies propagule supply from a single vector, we would expect the number of resulting invasions to differ among sites. This outcome should arise from variation in invasibility or resistance to invasion, due to differences in environmental and biotic conditions among sites (Case 1990, Lonsdale 1999, Ruiz et al. 2000a).

Beyond these general attributes, the specific (quantitative) nature of this dose-response relationship is largely unresolved. We cannot presently predict invasion success even if we knew the supply, either controlling for supply across many sites or varying supply at a single site. Yet, understanding this relationship across species, and vectors, is critical from a management perspective, because it defines the extent to which management-driven changes in supply affect invasion success.

For most vectors, we simply do not yet know whether the operational dose-response relationship (i.e., the range in which it presently operates) is linear, exponential, asymptotic, or some other complex shape (Fig. 18.2). For example, if we observe incremental changes in supply of a single species (along the x-axis), we cannot predict the corresponding change in invasion success in terms of either the type (general shape) of response or

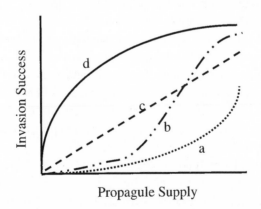

FIGURE 18.2. The dose-response relationship. Shown are hypothetical relationships between organism (propagule) supply and invasion success. Supply refers to concentration or frequency of inoculation. Invasion success is measured as probability of establishment, for species that are physiologically and ecologically capable of colonization. For both single and multi-species inocula, invasion success increases as supply increases, reaching an upper limit or saturation point (dashed horizontal line). The shape of this relationship may vary among taxonomic groups, geographic locations, and time periods.

the specific magnitude (slope) of response. The same is true for a mixed species assemblage, although this response may exhibit a more complex shape, reflecting variation and interactions among species.

We expect that the dose-response relationship will consist of multiple phases and will be asymptotic (e.g., Fig. 18.2b). This means that the number of species that become established, and the likelihood of establishment for any one species (of those that can tolerate the recipient environments), will exhibit thresholds: (1) remaining low up to some threshold, (2) followed by an increase with increasing supply, and (3) approaching saturation at some point. Although such a multiphasic relationship likely exists and has some support (Allee 1931, MacArthur and Wilson 1967, Simberloff 1989), for total density as well as for frequency of inoculation, it remains a challenge to identify the specific shape function (i.e., quantitative equation that describes the relationship) and where the thresholds lie.

Critically, the dose-response relationship is likely to vary tremendously among geographic locations, habitat or ecosystem types, vectors, taxonomic groups, or time periods. A wide variety of characteristics in both the recipient and the source communities influence invasion outcome, independent of transfer mechanisms (Elton 1958, Robinson and Dickerson 1984, Pimm 1991, Carlton 1996a, Vermeij 1991, 1996, Lonsdale 1999, Tilman 1999). For example, many forms of disturbance—from nutrient loading and fisheries harvesting to habitat alteration and chemical contamination—are thought to increase susceptibility to invasion in recipient communities (Hobbs 1989, Moyle and Light 1996, Horvitz 1997, Ruiz et al. 1999). The establishment of nonnative populations themselves may increase susceptibility, possibly enhancing invasion success of subsequent arrivals from the same native region (Simberloff and von Holle 1999). It is thought that directional change (increase) in the levels of both disturbance and invasions has increased susceptibility to invasion through time.

Global climate change may also influence susceptibility through a diverse suite of environmental and biotic mechanisms (Dukes and Mooney 1999, Mooney and Hobbs 2000, Stachowicz et al. 2002). In addition, and independent of changes in susceptibility, environmental changes that result in broadscale shifts in species distribution (i.e., local species composition) or abundance can serve to increase the cumulative species pool that is transferred among regions through time. Such biotic changes at the source region(s) may increase the opportunity for new invasions, as long as transfer mechanisms remain in place. This may be considered an expanded application of Carlton's hub-and-spoke model (discussed earlier), as the general result of sustained transfer (i.e., trade longevity) against a background of changing biological communities.

Although we predict an asymptotic dose-response relationship, which theoretically can result in saturation, it appears that we remain a long way from this endpoint. Available evidence indicates the rate of detected invasions is continuing to increase and shows no signs of abating (Mills et al. 1993, Fuller et al. 1999, Mack et al. 2000, Ruiz et al. 2000a). We believe invasion rates will continue to increase with supply for multiple reasons. First, it appears that we are far from exhausting the available species pool. Simply put, the earth is far from biologically homogenous, and many species appear tolerant of conditions beyond their present distributions (Carlton 1996b). Second, new source regions for trade are continuously coming into play, expanding the species pool for direct and indirect (e.g., hub-and-spoke) transfers. Third, the environmental conditions at recipient regions are changing, opening new opportunities for species, which may have previously been constrained (or prevented) physiologically or ecologically from colonizing. Fourth, the organisms themselves are changing genetically and phenotypically through time, resulting from a variety of mechanisms (e.g., evolutionary response to changing conditions, gene transfer, hybridization; see Adam 1990, Jiang and Paul 1998; Chapters 2, 6, and 8, this volume), in ways that may also present new opportunities for colonization.

PREEMINENCE OF PREVENTION: RATIONALE FOR BROAD-BASED VECTOR MANAGEMENT

A daunting number of variables may affect establishment, spread, and impact of nonnative species, and predictive ability is limited and largely untested. Invasion ecology is a relatively young science, as a great deal of the work to date has been descriptive (Williamson 1999). A growing body of theoretical work has begun to explore many basic and applied questions in invasion ecology, such as which species will become established, which species will have significant impacts, which recipient regions are most susceptible to invasions, and what characteristics influence invasibility of recipient regions (Crawley 1987, Case 1990, Pimm 1991, Williamson 1996). However, empirical tests to corroborate predictions lag far behind. At the present time, the specific mechanisms and quantitative relationships that underlie particular invasion outcomes and dynamics remain poorly resolved (Williamson 1999, Mack et al. 2000, Kolar and Lodge 2001).

Most contemporary invasions are an unintended consequence of trade, and we often do not know which species are being transferred, making many species-based management strategies unworkable. For example, ships transfer diverse assemblages of organisms in their ballast water, consisting of hundreds to thousands of species, which are discharged at coastal ports throughout the world (see descriptions by Carlton 1985, Carlton and Geller 1993, Gollasch et al. 2002); for protists alone, Hülsmann and Galil (2002) estimated that 250 different taxa are commonly found in a single ballast tank. In a similar way, a large number of species are transferred (unintentionally) in association with intentional introductions, such as agricultural and ornamental species (Elton 1958, Cohen and Carlton 1995, Naylor et al. 2001; Chapters 1, 2, 4, 5, 6, and 8, this volume), and many of these associates are not known at the time of transfer.

The magnitude of transfer, and the lack of taxonomic knowledge about transferred organisms, is most extreme for microorganisms such as bacteria, fungi, protists, and viruses. In general, both diversity and abundance of organisms across ecosystems increase with decreasing size of the organisms. Vectors should therefore result in a high transfer rate of microorganisms, in terms of concentration and species diversity, relative to larger organisms. Again, using ships' ballast water as an example, concentrations of organisms per liter are routinely on the order of 10^0–10^2 for zooplankton, 10^3–10^6 for phytoplankton, 10^8–10^9 for bacteria, and 10^9–10^{10} for viruses (Carlton and Geller 1993, Subba Rao et al. 1994, Smith et al. 1999, Dickman and Zhang 1999, Hines and Ruiz 2000, Ruiz et al. 2000b, Drake et al. 2001). However, the identity of most transferred microorganisms is problematic on multiple fronts. The effort required to detect and identify microorganisms is greater than that for larger organisms, resulting in fewer available data (Ruiz et al. 2000a, Hülsmann and Galil 2002). Moreover, taxonomic knowledge of microorganisms is relatively limited, and many of these organisms have never been described (Chapter 2, this volume).

Even when the species are known, robust predictions about which species will become established, and the ecological effects of these species, remain limited at the present time. This results from two fundamental types of information gaps. First, information is often very sparse about the biology and ecology of target species, making it difficult to characterize environmental requirements in the native range. Second, the extent to which performance at one locality (i.e., source or native region) can be used to predict performance in a new locality—often with a completely novel suite of environmental conditions and resident biota—remains unresolved (Grosholz and Ruiz 1996).

A central assumption that underlies most models of invasion impacts is that biological and ecological characteristics of a species in its current range (native or nonnative) are good predictors of interactions and impacts in a novel ecosystem. The magnitude of a species' impact in a community results from its functional role and population characteristics, including especially demographic features that influence density and size structure—since many impacts are density- and size-dependent. Clearly, there can be strong spatial variation in population characteristics within a species, which can be abundant at some localities and rare at others. Such spatial variation can result from any combination of ecological or genetic differences. Population-level differences in demography—resulting in differences in density or size structure—may be driven by environmental forcing functions (e.g., temperature, rainfall, salinity, currents) as well as by differences in biological characteristics (i.e., predators, competitors, prey, or parasites) among communities. Spatial variation in population characteristics can also have a genetic basis. Finally, shifts in the functional role of a species among locations can result from either genetic differences or ecological context (e.g., if key predators, parasites, or prey are present at one location and not another).

Recent debate about the possible consequences of intentionally introducing the Asian oyster *Crassostrea ariakensis* to Chesapeake Bay underscores some of the current limitations surrounding risk analyses (Box 18.1). Relatively little information exists about *C. ariakensis*'s biology and ecology in the native region, making it difficult to assess the probability of

Box 18.1. Effects of Introducing the Asian Oyster *Crassostrea ariakensis* into Chesapeake Bay, Virginia, USA

The Chesapeake Bay has historically supported a large commercial oyster fishery, based upon the native oyster *Crassostrea virginica*. Since the late nineteenth century, production of the native oysters has declined dramatically. Today, the oyster fishery harvest is approximately 1 percent of its historical peak. The decline in oyster harvest has had a large economic and social impact in the region, and is also thought to have resulted in major shifts in the foodweb, chemical cycling, and nursery habitats from many organisms (see Kennedy et al. 1998 and references therein for discussion).

The decline in the native oyster population (standing stock) is attributed to a combination of factors, including overfishing, habitat alteration, sedimentation, and diseases (Kennedy et al. 1998). A large effort exists within the region to restore the native oyster population and a viable commercial fishery (Leffler 2002). It appears that diseases are a primary deterrent to recovery. The native oyster is infected by two protistan parasites, *Haplosporidium*

continues

Box 18.1. Continued

nelsoni (MSX) and *Perkinsus* spp. (Dermo), which can cause very high (>90 percent) mortality in some years. Both parasites emerged as a major source of mortality in the mid–twentieth century. The reason for the emergence of these diseases was unclear, and both parasites were new to science when first discovered. More recently, genetic analysis indicates that one of the parasites (MSX) is not native to the region, and was introduced with the Asian oyster *Crassostrea gigas* (Burreson et al. 2000), although this nonnative host oyster did not become established.

Simultaneous with current efforts to restore the native oyster, there has been some exploration about the possible use of a nonnative oyster to restore an oyster fishery to the region. In the past few years, this effort has focused on another Asian oyster, *Crassostrea ariakensis*. Various lines of research suggest that this oyster could perform well in waters of the Chesapeake, exhibiting relatively high rates of growth and survivorship under local environmental conditions, even when challenged with the local parasites (Calvo et al. 2000).

Based upon these results, a plan to test commercial feasibility of this oyster in Virginia waters was advanced. The initial plan called for introduction of approximately 1 million juvenile oysters for grow-out at multiple sites in the lower Chesapeake Bay. This plan called primarily for the use of sterile triploids in the field-based trials.

Various concerns exist about the uncertainty associated with this pilot introduction, as well as the potential for a rapidly expanding population (Thompson 2001, Chesapeake Bay Program 2002, Leffler 2002). These concerns are highlighted in a recent review by Leffler (2002) as follows:

- First, the initial plan called for use of triploids created by chemical treatment, whereby a subset of the oysters would remain diploid and others could revert to this state. Reproduction and population establishment of the oyster, using this approach, are possibilities. Should this occur, no one can predict whether this nonnative oyster would affect the native oyster (or other species) through a variety of mechanisms, whether it would have a similar or different functional role compared to the native oyster, and how it may alter the Chesapeake ecosystem. This species appears physiologically capable of spreading outside of the Chesapeake, from New England to the Caribbean. However, little is known about the biology and ecology of *C. ariakensis* in its native range or elsewhere, to provide clues (let alone robust predictions) about the possible consequences of introduction.
- Second, independent of the oyster itself, the possibility of introducing

additional species exists. The proponents have followed recommended protocols for intentional introductions of marine organisms (ICES 1995), including use of at least second-generation organisms that were reared in the laboratory and screened for known pathogens. This approach serves to prevent transfer of many organisms, which are associated with the initial imports. However, the screening is limited to known pathogens, and little is known about this oyster (above). The identity and effects of most microorganisms, some of which are transferred vertically from parent to offspring, remain unknown. Thus, despite these protocols, some microorganisms from the original source are likely to be introduced with *C. ariakensis*. It remains a challenge to assess the potential consequences of such transfers of microorganisms into the Chesapeake.

this species reproducing, and to assess the potential consequences that may result from introduction of this oyster, and possibly associated organisms, to eastern North America (Thompson 2001, Chesapeake Bay Program 2002).

The present level of uncertainty surrounding transfers—including the species composition, likelihood of establishment, and potential impacts—argues strongly for a precautionary approach to policy and management strategies. This is further reinforced by the limited scope for success in many eradication and control efforts, following establishment of unwanted species (Mack et al. 2000; also see "Conclusions," this chapter). For both intentional and unintentional transfers, a precautionary approach strives to prevent the transfer and release of all species—to the maximum extent possible—until they are explicitly evaluated and approved for entry.

A precautionary approach to limit unwanted invasions is conceptually simple. It requires the fewest assumptions about which species will become established and which will have unwanted ecological or economic effects. It assumes that all species are "guilty until proven innocent." From a policy perspective, this approach places an emphasis on prevention, directing management actions early in the invasion sequence to interrupt the transfer process. This approach requires explicit consideration of each vector and its associated species whereby transfers are considered problematic in the absence of evidence to the contrary. Such actions would clearly have the greatest effect in reducing future invasions and associated, unwanted impacts.

Below, we describe the core components of vector management, adopting a precautionary approach. This is aimed at reducing all unauthorized transfers and subsequent invasions.

As predictive capability matures beyond its present status, especially concerning dose-response relationships and impacts, management actions can be further refined and directed more efficiently. Based on present understanding, however, we view all transfers as suspect—especially for unintentional transfers that occur in the absence of explicit evaluation—and thus forming an underlying premise for vector management.

FRAMEWORK FOR VECTOR MANAGEMENT

Vector management is comprised of four core components (Fig. 18.3). First, *vector analysis* describes the supply of organisms associated with particular transfer mechanisms, or vectors, including variables that may influence the supply and the characterization of the organisms themselves.

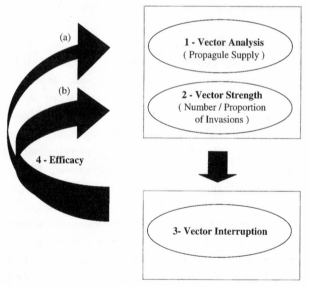

FIGURE 18.3. Framework for vector management. Vector management consists of four components and the interactions among them: (1) Vector analysis—characterization of vector operation, including organisms transferred by the vector; (2) Vector strength—assessment of the number or percent of invasions attributed to a particular vector; (3) Vector interruption—actions taken to reduce the transfer of organisms and likelihood of invasions; (4) Efficacy—measurement of the effect of vector interruption on the transfer of organisms (4a) and number or rate of invasions (4b). See text for further explanation.

This provides an operational understanding of the transfer process. Second, *vector strength* is an assessment of the relative importance or contribution of different vectors to observed invasions (i.e., established populations). Third, *vector interruption* consists of management actions aimed at reducing the chance of transfers and invasions by nonnative species by particular vectors. Fourth, *efficacy* of vector interruption measures the effectiveness of the management actions (i.e., vector interruption) to (4a) prevent or interrupt the transfer of organisms and (4b) reduce the establishment of new invasions.

Implementation of vector management, under this framework, is a stepwise and iterative process. Knowledge of transfer or established invasions (components 1 and 2, respectively) is a precursor and trigger for consideration of management action (component 3). Following implementation, the efficacy of management actions (component 4) is measured by its effect on rates of transfer (4a) and invasion (4b), creating a feedback system using the first two components. This feedback may lead to consideration and implementation of further management actions, providing a scheme for "adaptive management," improving efficacy of management for vectors that are presently targeted and responding to new vectors as they emerge.

These four components are critical underpinnings for vector management. Each is currently a central focus of much applied and basic research for invasion science. However, beyond the importance of each individual component, a key ingredient to vector management is recognition of the interactive nature of relationships among components. Each component is itself dynamic, affecting the other components. Indeed, vector management is truly about tracking the dynamic and interactive associations across the component boxes, to achieve the desired outcome. In essence, this scheme for vector management centers on understanding and measuring the relationships between species transfer (vector operation and vector interruption) and establishment (invasions), or the dose-response relationship.

In the following sections, we describe each of these components, outlining some of the relationships among them that often have consequences for implementation of management programs. This framework for vector management is influenced strongly by our experience with invasions in the marine community, where a focus exists primarily on transfer of entire species assemblages (rather than individual species) associated with particular vectors. For example, management actions are being developed and implemented to reduce ship-mediated transfer and invasion of many species simultaneously, rather than a single species (IMO 2002, NEMW

2002). This strategy reflects both the broadscale transfer associated with this vector as well as the poor predictability about (1) which species are present in any one ship and (2) the outcome (effects) of invasion by any particular species. This marine perspective has therefore shaped our concept of vector management, which we consider to include all species associated with a particular vector. However, we believe that the general principles and approach outlined here are broadly applicable to vector management across ecosystems and can also be directed at transfers of single species. Thus it is that airborne cargoes of flowers or fruit or luggage, shipborne cargoes of containers, truck cargoes of soil, plants, and seeds, and human flow (with their associated biotic assemblage of parasites and symbionts, and the biotic assemblage associated with clothing and paraphernalia [backpacks, shoes, etc.]) are examples of phenomena that also serve to transport multispecies assemblages at a large scale.

Vector Analysis

Vector analysis serves to characterize the delivery of organisms to a recipient region by a particular transfer mechanism. This is a two-step process. First, for each vector, this begins with a general assessment of how the transfer mechanism operates, or may operate, to deliver organisms. This assessment may include a qualitative or quantitative analysis of organisms transferred by a particular vector, providing an indication of the *potential* importance of the vector for transfer. However, since the existence and characteristics of specific vectors are spatially variable, vector analysis is necessarily specific to a recipient (geographic) region. For example, commercial ships deliver organisms in ballast water, which is discharged throughout the world. However, not all bays and estuaries are commercial ports, and the extent of commercial traffic and ballast water discharge varies tremendously among ports (Kerr 1994, Carlton et al. 1995, Smith et al. 1999). Thus, as a second step, the characteristics of transfer associated with a particular vector must be considered for each recipient region of interest.

Characterization of a vector, when applied to a particular recipient region, includes the operational details of how, where, when, and from what geographic sources a vector delivers live organisms (propagules). It also includes biological attributes such as quantity and types of organisms (Carlton and Ruiz, in press). In some cases, the condition or viability of propagules has been considered as well, attempting to further refine the likelihood of invasion. Although this approach can demonstrate when organisms remain viable and capable of reproduction post-transit, particularly when focused upon an organism for which extensive background information exists, it is often difficult to assess viability or performance in a novel environment. For most organisms, data on environmental tolerance or ecological interactions are very limited. Furthermore, viability is very context-specific, resulting from the interactive effects of many different environmental and ecological conditions (see below for further discussion). For this reason, we emphasize characterization of propagule delivery, which does not require assumptions (often untested) about viability; the latter is a subsequent step in the invasion sequence and, although important, is considered separately from delivery (Fig. 18.1).

A strong baseline of information already exists to characterize the operation and propagule delivery for many different vectors, in different geographic regions, and for different time periods (Mack 1991, Carlton and Geller 1993, Carlton et al. 1995, Smith et al. 1999; Chapters 1, 4, 6, and 11, this volume). Vector analysis is most often performed on a vector-by-vector basis, focusing on particular taxonomic groups. Such vector analysis has advanced our understanding of the invasion process and served to structure current management and policy, by identifying transfer mechanisms as well as particular points of the transfer process (within a vector) for management action to interrupt the transfer of species.

The relative contribution of each vector to the overall delivery of organisms, and the dynamic nature of vectors, are important considerations. As discussed earlier, the contribution of various vectors to supply is likely to vary tremendously in space and time. Ideally, supply characteristics could be compared across the entire population of vectors for specific recipient regions or spatial scales, and would be updated on a regular basis to track significant changes in delivery. Such a vector meta-analysis, which integrates across vectors and temporal scales, is not available—even within taxonomic groups or ecosystem types (i.e., terrestrial, freshwater, or marine). For most localities, our ability to compare spatial or temporal differences in propagule supply is limited to qualitative estimates, at best, relying often on trade statistics (e.g., number of arriving ships or amount of a particular commodity) and "snapshot" estimates of organism transfer via vector analysis.

Vector Strength

Vector strength indicates the relative importance of different vectors, in terms of their contribution to the number or proportion of established invasions. This is accomplished by analysis of data on the pattern of invasions for a particular region, identifying which vectors are responsible for the invasions and their relative contribution in space or time. Vector strength is the output of a meta-analysis of invasion pattern across vectors and time periods, analogous to that suggested for transfer or vector analysis (see previous section). Estimates of vector strength are most frequently made for a particular recipient region according to taxonomic group and ecosystem type, such as marine invertebrates (Cohen and Carlton 1995, Hewitt et al. 1999, Reise et al. 1999), freshwater fishes (Fuller et al. 1999), marine plants (Chapter 8, this volume), amphibians (Chapter 4, this volume), and insects (Chapter 3, this volume). A challenge in assembling vector strength data is that multiple vectors may be transporting the same species at the same time, making it difficult to identify the specific vector responsible for an invasion (Carlton and Ruiz, in press; Chapter 7, this volume).

Vector strength is a key measure for invasion science, both alone and in combination with vector analysis. The number of invasions associated with a particular vector (vector strength) may be a more important measure for vector management than propagule supply, since the goal of vector management is to reduce the likelihood of new invasions. Although vector analysis estimates the supply of organisms delivered to a recipient region, many of these organisms will not become established (Williamson 1996, Mack et al. 2000). A poor correlation between supply and established invasions could result from variation in condition or viability of organisms associated with particular vectors, particular source regions, or particular recipient regions. Thus, vector strength is the more direct of these two measures (supply and establishment) and also the more relevant to guide and especially to evaluate vector management.

In combination with propagule supply (vector analysis), measures of established invasions (vector strength) can provide important insights into dose-response relationships. Understanding the nature of this relationship is clearly central to effective vector management to reduce invasions, as it functionally describes the expected outcome of management actions. More specifically, tracking both propagule supply (vector analysis) and invasion pattern (vector strength) across multiple sites can be used to (1) estimate the general shape of dose-response relationships and (2) test for differences in invasibility (Williamson 1996, Lonsdale 1999, Ruiz et al. 2000a). From a management perspective, the vector analysis can help establish goals for reduction of propagule supply, whereas the vector strength may allow fine-tuning of management strategies (for both prevention and possible control, such as "early detection–rapid response" systems) to focus efforts more efficiently on high-risk areas.

Vector Interruption

Vector interruption is the third component of vector management, consisting of those actions designed to disrupt and reduce the flow of propagules to the recipient environment. The goal of vector interruption is to prevent new invasions. Operationally, the implementation of vector interruption seeks to reduce the likelihood of new invasions, thereby causing a decline in the rate, spread, and impact of future invasions. Vector interruption can be accomplished by a wide range of actions, which operate in concert and are directed at specific trade or transport activities, including (1) education and outreach, (2) voluntary guidelines, and (3) regulations and laws, which include various types of enforcement and penalty. Examples of these activities are presented in earlier chapters of this volume and elsewhere (Shine et al. 2000, Wittenberg and Cock 2001, NISC 2001).

The priorities for vector interruption are shaped, in part, by knowledge about patterns of species transfer and invasion. For example, either vector analysis or an assessment of vector strength can be used to target particular vectors for management action. Priority for interruption efforts depends upon the particular management goal(s). Historically, vector interruption has often focused on those vectors that are known to deliver particular high-impact species, operating frequently at the species level (e.g., Chapter 16, this volume). This has been changing in recent years, whereby vector management in some ecosystems targets vectors known to result in relatively high rates of invasions (NISC 2001, IMO 2002; see also "Status and Implementation of Vector Management" for further discussion). In the latter circumstance, measures of vector strength are used to set priorities for vector management, targeting those vectors with high vector strength as a logical entry point (or first step) to minimize the number or rate of all new invasions. This is a substantively different approach than species-level management and requires no a priori knowledge of the individual species associated with a particular vector, at a particular place and time, or its potential impacts.

Efficacy of Vector Interruption

Two different measures for efficacy of vector interruption comprise the fourth critical component of vector management, providing a cause-effect analysis of interruption and a feedback system to evaluate or test the need for further actions (Fig. 18.3). First, the *proximate* effect of vector interruption on propagule supply characteristics is assessed (4a). This assessment can be made by a variety of measures, including direct measurement of organisms associated with the vector (i.e., vector analysis), experimental methods to measure the response of associated biota, or even a theoretical approach. Second, the *ultimate* effect of vector interruption on invasion outcome is assessed (4b). This is accomplished by measuring (monitoring) the rate of new invasions, to test whether those attributed to the target vector show predicted changes associated with vector interruption.

There are good reasons to focus attention on propagule supply as a response variable for vector interruption. First, this measure can provide a relatively quick assessment of performance for vector interruption. Second, using propagule supply, it is possible to implement standards or limits for species transfer (or detection) associated with particular vectors, and these in turn could be used to track or enforce compliance with vector management.

The second measure of efficacy (invasion rate) is especially important, as this evaluates whether the goal of vector management—to reduce the number and rate of, and thereby truly prevent, invasions—is achieved. It seems reasonable to assume that a significant reduction in propagule supply would often result in a reduction in invasions. However, as discussed earlier, the magnitude of this reduction depends greatly on the specific dose-response relationship (Fig. 18.2), which remains largely unresolved. In addition, a variety of other factors may compensate for the per capita propagule delivery with each delivery event. For example, perhaps the number, speed, or sources of shipments is increasing over time, influencing the cumulative number, condition, or types of propagules, respectively (see Table 18.1). The recipient environment may also differ among sites, or change over time, in ways that influence invasion outcome (Carlton 1996b, Williamson 1996, Simberloff and von Holle 1999).

Thus, both measures of efficacy are necessary for effective vector management. Tracking changes in supply is certainly useful but not sufficient to assess efficacy of management actions. We view supply as a valuable short-term proxy, providing a measure for one key factor that should influence invasion outcome. However, at the present time, supply measures cannot be used to predict the subsequent quantitative effect on invasion outcome (as discussed earlier). To rely solely on supply measures would provide an incomplete picture, analogous to assessing the effect of air quality standards by measuring emissions from some individual automobiles without actually measuring changes in air quality—which will depend greatly upon (1) correctly identifying point sources for management action and (2) interactions with local environmental conditions (e.g., atmospheric mixing, temperature, etc.). This issue is highlighted by current efforts to reduce invasions associated with ships, as a major vector (Box 18.2).

Box 18.2. Vector Management for Ships: A Case History

Since the late nineteenth century, commercial ships have used ballast water to maintain trim and stability during voyages, replacing the use of dry ballast (see Carlton 1985 for review). Ballast water is regularly loaded (pumped) from surrounding waters into dedicated ballast tanks and dual-use floodable cargo holds. Ships load ballast water from bays and estuaries (i.e., in the vicinity of ports) as well as the open ocean. This process entrains a taxonomically diverse assemblage of organisms, which are transferred and discharged with ballast water throughout the world. All commercial ships can carry ballast water, and large ships (e.g., oil tankers or bulk carriers) can carry as much as 50,000 metric tons of ballast water at one time. In the United States alone, ports receive approximately 50,000 vessel arrivals per year from foreign countries (U.S. Maritime Administration, unpublished data), and these vessels discharged in excess of 70 million metric tons of ballast water to U.S. coastal waters in 1991 (Carlton et al. 1995).

Vector analyses for several decades have served to characterize the transfer of organisms in ballast tanks. The taxonomic diversity represented in ballast tanks is truly impressive, ranging from viruses and bacteria to crustaceans and fish (e.g., Carlton and Geller 1993, Subba Rao et al. 1994, Smith et al. 1999, McCarthy and Crowder 2000, Drake et al. 2001, Gollasch et al. 2002, Hülsmann and Galil 2002). Any one ballast tank may contain hundreds to thousands of species. Concentrations of organisms per liter of ballast water are routinely on the order of 10^{0-2} zooplankton and 10^{3-6} phytoplankton to 10^{8-9} bacteria (see above references). It is also evident that additional organisms are present in sediments, which accumulate in the bottom of tanks, as well as biofilms on the surfaces of ballast tanks (NRC 1996; Hülsmann and Galil 2002). Although these analyses demonstrate the magnitude of transfer associated with ballast tanks, it is bewildering to attempt prediction of which species (of the thousands available for entrainment) are present in the tanks of any one ship at a particular time, as this is a stochastic process of previous ballast history combined with organisms present at the time (of ballast events) and those surviving through time.

Estimates of vector strength for coastal marine invasions demonstrate the relative importance of shipping at many locations throughout the world (e.g., Hewitt et al. 1999, Reise et al. 1999, Ruiz et al. 2000a). Shipping has often been considered the dominant vector, responsible for most detected marine invasions. Furthermore, the reported increase in rates of invasion for coastal waters have often been associated with changes in the shipping vector (for an exception, see Por 1978).

continues

Box 18.2. Continued

Despite this clear link to shipping, the relative contribution of species transfer by ships' ballast tanks versus ships' hulls is uncertain (Cohen and Carlton 1995; chapter 7, this volume). Many invasions can be linked solely to ballast water or hull fouling as the vector. However, both components of the shipping vector are possible transfer mechanisms for a large proportion of ship-mediated invasions (both historically and currently), reflecting the existence of planktonic (i.e., waterborne) and benthic (bottom-dwelling and attached) life phases for many species. As with ballast water, vector analysis demonstrates the active transfer of many species associated with the ships' hulls (Gollasch 2002).

Over the past two decades, a variety of management actions have been advanced to interrupt the unwanted transfer of organisms in ballast tanks. These began as voluntary guidelines aimed at multiple steps of the transfer process, from entrainment and transfer to discharge, representing a program of Integrated Vector Management or IVM (*sensu* Carlton and Ruiz 2003). Some aspects of this ballast water management are now mandatory (required) in particular geographic regions and are being further considered for international treaty by the International Maritime Organization (IMO 2002, NEMW 2002, Taylor et al. 2002).

Most of the current regulatory efforts for ballast water management are focused on treatments, to reduce organism transfer, and are characterized by a phased approach. The first phase includes the practice of ballast water exchange, or flushing tanks in the open ocean, to reduce transfers of coastal organisms. In essence, exchange replaces coastal organisms with oceanic organisms, and such transfers among systems are thought to result in very few possible invasions. Exchange can be implemented currently on many vessels, as it does not require retrofitting of vessels, but there are some constraints inherent with this approach. First, it may not be safe under some weather conditions, when ships require stability. Second, vessels on coastwise voyages, which do not transit an ocean basin, may not have sufficient time or not move far enough from shore to remove coastal organisms. Third, even if exchange achieves a reduction of coastal organisms >95% (a commonly referenced goal), there is still a significant number of residual organisms, when considering the large volumes of ballast water in play. To overcome these constraints, a second phase is being developed. A large number of treatment technologies are being developed and tested throughout the world, including physical separation methods and physical and chemical biocides (AIRD 2002, Taylor et al. 2002).

Strong support and momentum exist for interruption of transfer by ballast water, but there remain several key issues about the efficacy surrounding efforts. Certainly the effect of management practices on organism transfer can be, and is being, measured. However, the effect of treatment on invasion outcome remains uncertain for two major reasons. As with transfers more generally, the dose-response relationship is not known, resulting in uncertainty about the specific goal or standard for ballast water management.

In addition, the relative importance of ballast water remains unresolved on multiple levels. The vector strength for ballast water versus hull fouling components of shipping is unclear (Fofonoff et al. 2003). It is now thought that the relative importance of hulls may be increasing, due to a ban of Tributyl tin (TBT)—an effective biocide—in bottom paints (Nehring 2001). In addition, even within ballast tanks, the relative importance of residual bottom communities (especially spores and resting stages) and biofilms is not clear (NRC 1996, Hülsmann and Galil 2002, Colautti et al. 2003).

The present uncertainty surrounding the effects of ballast water management practices on invasion rates, as well as ongoing changes in the management of both ballast tanks and hulls, underscores the need for a tracking system to assess associated changes in invasion patterns (Fig. 18.3, component 4b). Some initial steps are under way to develop the baseline information that can provide such feedback to management actions, but these have yet to be considered as an integrated part of vector management. Ideally, a broad-based system for measuring invasion patterns would become an ongoing formal program to both guide and evaluate vector management on an ongoing basis, as discussed above.

STATUS AND IMPLEMENTATION OF VECTOR MANAGEMENT

We consider here various aspects of the status and implementation of vector management. We do not intend this as a comprehensive analysis or synthesis, and much more detail on specific vectors appears in other parts of this volume and elsewhere. Instead, we highlight some of the key issues and constraints associated with implementing vector management.

Vector Analysis and Vector Strength

Vector management has been guided largely by retrospective analyses, relying upon available data for historical propagule supply but more often invasion history. This approach is a reasonable starting point, but some limitations are also evident. First, the information base for analysis of vectors and invasions—to support some form of vector analysis and estimates of vector strength, respectively—is often uneven in quality. As a result, the relative importance of particular vectors may be misrepresented spatially, temporally, or taxonomically (Ruiz et al. 2000a). Second, current analyses may also miss the emergence of new vectors, as well as shifts in vector behavior (including substantive changes in source regions, magnitude, or mode of vector operation), possibly taking years to be reflected by a retrospective approach.

To overcome some of these limitations, management and policy initiatives have begun to emphasize the need for standardized measures of invasion patterns. For example, standardized measures of invasion are now being implemented for marine habitats by individual national initiatives in Australia and the United States (Ruiz and Hewitt 2002). Such baseline data are also being collected for marine invasions, through the International Maritime Organization GloBallast Programme in Brazil, China, India, Iran, South Africa, and Ukraine (IMO 2002). More broadly, there is increasing emphasis on enhancing measures of invasion pattern across many ecosystem types and increasing spatial scales, including both improved field-based measures and effective methods for information sharing (Wittenberg and Cock 2001, NISC 2001, U.S. Congress 2002).

Many of the same initiatives have also emphasized the need for a more comprehensive approach for vector analysis than exists currently. This includes systematic evaluation of vectors as they operate today as well as developing the capacity to forecast emerging vectors and changes in vector activities, estimating possible consequences for invasion outcome. To forecast vector behavior and associated species transfers, Carlton and Ruiz (in press) suggest the need for a formal "Vector Early Warning System" (VEWS). VEWS would combine projections by an interdisciplinary group of specialists (e.g., futurists, economists, trade specialists, biologists) to estimate new patterns of commerce on a regular basis for a time horizon of ten to twenty years. This approach could trigger management actions to reduce transfers associated with new vector patterns, shifting the emphasis of management from a retrospective to a proactive mode.

The direction and growing momentum of recent initiatives in this area are encouraging, and may serve to fill some key information gaps for vector management, but such efforts are still in their infancy and would benefit greatly from further development in several directions. In particular, an expanded effort toward implementation and coordination at the international level is necessary to capture and respond to information on the global scale at which invasions (and trade) truly operate. This includes a well-coordinated effort toward systematic collection, analysis, and sharing of information about vector operations—in the past, present, and future—as well as clear mechanisms to integrate this information into decision making and policy (Shine et al. 2000, Wittenberg and Cock 2001; see "Future Directions and Recommendations" for further discussion).

Vector Interruption

A wide range of management actions and guidelines exist to interrupt or prevent the transfer of species by particular vectors—at the local, regional, national, and international levels (Chapters 10, 11, 12, 13, and 14, this volume)—and have often been triggered by the establishment of a "high-impact" organism, especially those causing significant economic costs or threats to human health. For example, the arrival of zebra mussels to the North American Great Lakes or toxic dinoflagellate blooms in Australia precipitated guidelines and regulations to manage ships' ballast water (U.S. Congress 1990, Jones 1991). In this case, the actions were directed at transfer of all species associated with ballast tanks. Concerns about importation of particular pathogens or insect pests have caused bans (sometimes short-term) on importations of particular fish and shellfish, agricultural products, and livestock. Although such bans target an individual disease or pest organism, they may also function indirectly as vector interruption, preventing the transfer of all biota associated with the target organism.

The threshold or impetus for vector management actions has been shifting in recent years. This threshold depends on the complex interplay of several factors, including (1) the magnitude of invasion impacts associated with a particular vector, (2) assessment of socioeconomic importance of the impacts, (3) costs associated with specific management actions, and (4) political opportunity for management action. Despite a tendency to focus management actions on vectors that are known to have delivered high-impact invasions in the past, transfer of species by any vector is increasingly viewed with concern and as a potential candidate for management. This stems in part from recognition that invasions with significant impacts can result both from vectors that cause relatively few invasions (i.e., low vector strength) and from species that do not have a history of such impacts. Indeed, since the ecology and biology of most species remain poorly understood, predictive ability is low concerning potential impacts upon invasion (Williamson 1999, Wittenberg and Cock 2001). Growing realization about the cumulative scale of invasion impacts—and the capacity of invasions to affect ecological functions and economies significantly—has further served to sensitize the public about this issue, lowering the threshold necessary (and increasing political pressure) for management action.

For example, U.S. policy toward invasions and vector management has undergone a shift in the past ten to fifteen years from addressing only high-impact vectors and target species to addressing potential consequences of any invasion. This policy approach was outlined in a U.S. Presidential Executive Order (1999) and is further amplified in the national management plan that resulted subsequently (NISC 2001). A parallel trajectory in policy is also evident in recent bills that were introduced by the U.S. Congress to reauthorize and revise prior legislation (U.S. Congress 2002). For vector management actions, priority is often placed upon vectors known to deliver the most invasions or species with significant impacts, as a starting point, but the scope for vector management has clearly expanded beyond these immediate priorities.

A broad-based policy of reducing unintentional transfers of all species will clearly be most effective at reducing invasions and their impacts. From a scientific perspective, the precautionary approach of "guilty until proven innocent" is the most defensible. History illustrates that high-impact invasions can result from nearly any (if not all) vectors and taxonomic groups. Following successful invasions, we can be certain that some subset of species will have significant effects. At the present time, we have a very limited capacity to predict (1) which particular species are transferred by a vector, in space and time, and (2) which ones will have significant impacts. A precautionary approach to vector management requires no assumptions about which vectors or species will result in invasions with significant impacts. This is a sound starting position, in the absence of extensive information and robust predictive models about species transferred by each vector, especially given that the species associated with most vectors are both unknown and constantly changing over time and space.

The benefits of such a broad-based approach to vector interruption are often juxtaposed against the costs of implementation. The costs of vector management are certainly real and represent a significant barrier for many actions. In some cases, the resources simply do not exist for implementation. However, the perceived costs of vector management may often be grossly inflated relative to the cost of inaction, for two major reasons. First, it remains difficult to quantify the costs of invasions. Some recent estimates have underscored the known costs associated with some high-impact invasions (OTA 1993, Pimentel et al. 2000). For example, Pimentel et al. (2000) estimated the costs associated with invasions in the United States to exceed $137 billion annually. Although a large number, which has fueled an increased interest in invasion management, this is an underestimate. The direct and indirect impacts of many invasions remain unknown (Parker et al. 1999, Ruiz et al. 1999), and, even when known, valuation of ecological impacts remains difficult and incomplete. Second, the associated costs of invasions are usually not borne by the responsible industry or trade, but instead are externalities—which impact public commons, public goods, or other industries (Hardin 1968, Perrings et al. 2003).

Despite clear evidence of transfer and invasions for particular vectors, and a general shift in attitude increasingly toward broad-based vector management, actions aimed at vector interruption often have not yet been advanced or fully implemented. To some extent, the need for further development of particular technologies and treatment strategies limits the implementation of vector management (NRC 1996, Taylor et al. 2002). In our view, however, the impetus for implementation is a cost-benefit analysis, usually performed indirectly and qualitatively in the public and political arena. At the present time, the actual costs associated with invasions have not been adequately considered and would often support more aggressive, proactive vector management. Thus, necessary ingredients for vector management—in terms of the speed and extent of implementation—are (1) clear analyses of the ecological and economic impacts associated with invasions, conveying the full value of goods and services (including economic costs, ecosystem functions, habitat quality, threatened and endangered species) lost to invasions, and (2) focused outreach and education efforts that are aimed at supplying this information to the public, resource managers, industry groups, and policymakers. In this way policy decisions for vector management can incorporate a balanced view of real costs and benefits of particular actions.

Efficacy of Vector Interruption

Supply is the currency used most commonly for vector management, because it is the most accessible and practical measure for short time frames. Measurement of organism transfer by vector analysis is often used to structure management strategies, and measuring effects of actions (vector interruption) on transfer characteristics provides a proximate and valuable measure of efficacy (Fig. 18.3, component 4a). It is evident that many current management actions serve to reduce the magnitude of organism transfers, as demonstrated across a variety of vectors (e.g., Taylor et al. 2002; Chapters 11 and 12, this volume). However, it is also evident that barriers used for vector interruption are only porous filters, with varying pore sizes, that still allow transfer of some organisms.

It remains challenging to assess the efficacy of management actions to reduce the number of invasions, as the ultimate goal of vector management (Fig. 18.3, component 4b). In general, we expect steps to reduce organism transfers—including number of species and concentrations per transfer event, frequency of transfers, number of source regions—will serve to reduce the chances of successful establishment and the number of new invasions. The extent to which such reduction occurs will depend upon the dose-response relationship, for which we have a limited understanding (as discussed above; Fig. 18.2). Simply put, we do not know how far we must go in reducing supply of organisms (the short-term proximate measure of efficacy) to achieve a desired reduction in invasions, as the ultimate measure of efficacy for vector management.

Although the lack of knowledge about dose-response relationships represents a critical gap for invasion science, this should not be construed as a deterrent to implement vector management. Instead, management actions should proceed with the goal of reducing the transfer of organisms to the maximum extent possible. We cannot afford to wait to further resolve dose-response relationships before implementing management steps, as the underlying science will take many years to advance. Current trends suggest the number of invasions, and their cumulative impacts, will continue to accumulate at an increasing rate—in the absence of actions to interrupt this process. Several lines of evidence indicate that successful establishment of new populations, or invasions, increases with density and frequency of transfer (Simberloff 1989, Williamson 1996, Kolar and Lodge 2001). Thus, we believe management actions to reduce transfers should proceed concurrently with the science necessary to evaluate both dose-response and efficacy.

Multiple approaches are available to understand such dose-response relationships and efficacy for invasion ecology. Theoretical work can provide a useful tool to inform us in this area, exploring the possible outcome of particular biological and ecological processes. Such theoretical work provides working hypotheses that require empirical testing and confirmation. Importantly, the theoretical approach and the output of mathematical models are contingent upon underlying and increasingly refined empirical relationships. Thus, theoretical and empirical approaches must be combined to advance our understanding in this area.

Two general types of empirical measures provide complementary understanding of invasions: (1) manipulative experiments and (2) tracking (monitoring) invasion patterns. Laboratory and field experiments provide a powerful approach to explore dose-response relationships, and thereby estimate efficacy of vector interruption. Experiments can test the relative and interactive effects of multiple factors on invasion outcome, including both delivery patterns (i.e., transfer characteristics such as density and frequency of inoculation) as well as characteristics of the recipient community (e.g., species diversity, disturbance regime, habitat type). The experimental approach has many advantages. First, it can provide the highest quality of quantitative information about specific relationships and interactions, by controlling for other sources of confounding information. Second, experiments provide unparalleled opportunities for rigorous and detailed exploration of many different factors—some of which are difficult to examine in field situations. Third, experiments can address key questions on relatively short time scales, compared to field-based measures (below).

Tracking invasions through monitoring is of paramount importance, however, both to measure vector strength (as above) and to assess the long-term efficacy of vector interruption. Despite the clear value of experiments, they cannot replace the need for such monitoring. Experiments include only a small subset of the physical, chemical, and biological factors, as well as spatial and temporal scales, in each of the source and recipient regions that can influence invasion outcome (as above; see also Diamond 1986). Thus, experiments can be used to test theoretical relationships on a limited scale, providing detailed analyses of underlying assumptions and mechanisms, but the actual patterns of invasion may differ from those predicted by theory or experiments—reflecting the potential importance of complex interactions, which involve a large number of factors and operate on a range of scales that may be missing from experiments. Examination of discrepancies between experimental-based predictions and observed invasion patterns can therefore provide important insights into the processes that influence invasion dynamics.

More generally, we view measurement of invasion patterns and rates as cornerstones of invasion science and vector management. Without this information base, many fundamental questions in invasion ecology will remain unresolved. Theoretical and experimental approaches together are particularly valuable in (a) providing a mechanistic understanding of specific relationships and (b) producing results in relatively short time frames. Although results from these two approaches can be used to guide management and policy, especially in the absence of additional information, it would be a mistake to assume the results—based upon a limited range of conditions (as above)—are robust and applicable broadly. Such confirmation can come only from testing for the existence of predicted associations between invasion pattern (as the dependent variable) and specific independent variables.

Reduction of invasion rate is the most appropriate measure of efficacy (i.e., the primary goal) for vector management, but this must be recognized as a long-term proposition. There are several features that lead to this realization:

- We lack reliable baseline information about the historical rate of invasions for many groups of organisms. In general, the quality of baseline data declines from terrestrial to freshwater to marine habitats, and from large and conspicuous organisms (e.g., birds, mammals, trees, fish, mollusks, and crabs) to small organisms.
- For most habitats and types of organisms, current measurements of invasion are also not sufficient to estimate changes in invasion rates in the future. This stems largely from the lack of standardized, repeated measures across space and time to control for observer bias. Instead, we have a patchwork of measures, collected in different ways, making comparisons tenuous.
- Taxonomic identification of organisms is often unresolved, including not only species identification but also status as either native or nonnative (Carlton 1996a, NISC 2001). This is especially true for small organisms (e.g., many marine and freshwater invertebrate groups) that have received little attention historically. However, more broadly, the number of taxonomists has declined over time, creating significant gaps in knowledge and time lags in identification (NRC 1995).
- Finally, it can take many years for organisms to be detected following an invasion. This results from (a) the lag time that can exist between initial establishment and population growth (Crooks and Soulé 1999, Mack et al. 2000, chapter 3, this volume) and (b) the level of effort required to detect new invasions above a particular threshold in population size.

A high priority for vector management must be a focused effort to track invasion rates over time. The constraints associated with tracking invasion rates provide a compelling case for use of supply as a short-term proxy of efficacy, but they clearly do not obviate the need for measuring invasion patterns (as above). A well-defined and highly coordinated program with this purpose is sorely needed and conspicuously absent for most taxonomic groups and habitats. Without such measures, it is not possible to reliably estimate the efficacy of management actions to reduce invasions, even on decadal timescales.

More broadly, gaps in knowledge about invasion patterns undermine many pursuits of invasion ecology that seek to predict invasion outcomes and measure biological and environmental correlates to invasion success (Ruiz and Hewitt 2002). Such field-based measures are necessary to address the following types of questions:

- How do invasion patterns (rates and extent) vary among regions?
- What factors influence susceptibility and risk of invasion?
- What characteristics are associated with successful invasions?
- Using analysis of vector strength (above), which vectors and geographic regions are responsible for observed invasions? How is this changing over time?
- Is there measurable change in the rate of new invasions that corresponds to management actions (i.e., vector interruption, as above)?
- What is the dose-response relationship between vector-specific supply (or interruption) and invasion rate?

Standardized measures, which are replicated at many sites and repeated regularly over time, are required to effectively address these questions. Such measures are needed across a diverse array of sites and geographic regions, to represent the full spectrum of vector activity, source regions, and recipient regions. Significant variation exists among sites for vector activity and invasion characteristics, such that one or a few sites cannot provide a robust, representative estimate. Measures of such spatial variation among sites are necessary to test for differences in invasibility and the relationship between propagule supply and invasion. Further, repeated measures are necessary to build statistical confidence about the existing assemblage of species (i.e., develop a baseline) with which to measure temporal changes. Toward this end, development of an international network of sites offers a powerful and desirable approach (Ruiz and Hewitt 2002; see below).

CONCLUSIONS

We have outlined core components of vector management, developing a framework that (a) illustrates functional relationships and integration across the various components and (b) highlights current gaps and opportunities. Our focus is exclusively on the prevention of invasions, through vector interruption, but we recognize that this is only one approach used to reduce the risk of unwanted impacts associated with invasions. Prevention falls along a continuum of management tools that include post-colonization actions of eradication, containment (prevention of spread), and population control (Mack et al. 2000, NISC 2001, Wittenberg and Cock 2001). While each of these other tools has its merits, we place the highest premium on prevention through vector management, as the most reliable strategy to minimize unwanted invasions and their effects.

Post-colonization management actions have a more limited scope than prevention, providing a suite of options that can be applied to a particular subset of invasions under some circumstances. These options serve as potentially important "back-ups" or "safety nets" for species that bypass vector interruption. We view post-colonization management options as secondary to prevention, for multiple reasons. First, they are implemented following transfer and are therefore secondary in the sequence of stages at which management can be directed (Fig. 18.1). Second, detection of invasions is not comprehensive, and there can often be significant lag time between colonization and detection (Williamson 1996, Crooks and Soulé 1999; Chapter 3, this volume). This time delay for detection allows species to become abundant and spread, limiting potential management actions. Third, when a new incursion (invasion) is detected, it is often difficult to assess the likely impact and decide on an appropriate trigger for management actions. Given the apparent increase in rate of invasions reported across ecosystems, and the finite resources available, management actions can be directed toward only a fraction of invasions. Uncertainty about impacts for many species further complicates management decisions and especially the efficient use of limited resources, as some high-impact species will initially go unrecognized as such. Finally, even upon detection and a decision to implement management actions, control and eradication efforts may have only mixed success. In many cases, eradication will not be feasible, and sustained efforts are necessary to control population abundance, impacts, and spread (Myers et al. 2000, Wittenberg and Cock 2001).

For example, there is now growing interest in "early detection–rapid response" capability, which is designed to detect particular species of concern and trigger management actions such as eradication, containment, or control (NISC 2001, Wittenberg and Cock 2001, U.S. Congress 2002). A list of target high-impact organisms can certainly be compiled, based upon experience elsewhere in the world, providing the basis for an early detection system. However, this "target list approach" will necessarily include only a subset of future high-impact organisms. Many additional species, which are ecologically or economically potent (i.e., will have significant impacts), will not appear on any such list—simply because they do not have a prior record of high-impact invasions. Thus, although this approach may be well justified for invasions of known pests (i.e., those species with unwanted impacts elsewhere), its effectiveness is inherently limited by both detection capability and the high level of uncertainty in predicting impacts of many species.

We draw a distinction between monitoring efforts, as discussed in previous sections, and detection efforts associated with an "early detection–rapid response" system. The former are designed to measure invasion patterns across different sites, habitats, environmental conditions, taxonomic groups, vectors, source regions, and time periods. This information is used to direct and evaluate management actions, focused largely on vectors and pathways of invasion (Fig. 18.3). Although surveys may provide some early detection capability, this is not the primary goal. In contrast, the goal of "early detection" is to trigger specific post-colonization management actions (e.g., eradication, containment, etc.) for particular (known) species of concern. To provide an early detection capacity would require frequent monitoring of specific habitats for a limited number of organisms. It is simply not feasible to monitor for all organisms on a frequent basis, which allows for rapid response at an early stage of colonization when management options may be most effective (Myers et al. 2000), due to obvious logistical and cost constraints. Thus, early warning detection efforts are focused pragmatically at "sentinel sites" for detection of a subset of future invasions, selecting locations with specific habitat and environmental conditions appropriate for the target species.

Considerable enthusiasm exists for risk assessments and predictions about invasions (CENR 1999, Kolar and Lodge 2001, 2002), but significant limitations currently exist in this area. Despite the clear advantages to developing a predictive capacity about the likelihood of invasion and the associated impacts, confidence about predictions is often low and limited to the context (environmental and biological circumstances) associated with previous specific invasions. In this sense, most of the analyses are "postdictive," describing or explaining patterns from previous invasion events. As Williamson (1999) succinctly states: "Explanation is not prediction." Predictions that may result from such analyses often have not been tested extensively and therefore may not be robust or broadly applicable. Thus, despite the use of and widespread interest in various forms of risk assessment and predictions, particularly surrounding invasion dynamics and impacts, they may often lead to a "false sense of security" (Simberloff as cited in Wittenberg and Cock 2001). A great deal of uncertainty remains for many (if not most) species that is not often reflected or readily evident in risk assessment outputs.

The current level of uncertainty that exists in risk assessments or predictions has significant implications for invasion management and policy. Risk-based models and predictions are clearly a worthwhile and necessary endeavor to advance our understanding, and these should continue to mature. However, policy should be based upon our present understanding and predictive ability. Invasion ecology is still in its infancy, in terms of predictive ability for many dimensions of invasion outcome. The present low predictive ability argues strongly for a precautionary approach, with a goal of preventing transfers unless sufficient evidence exists to demonstrate that (a) colonization is unlikely or (b) significant impacts are unlikely.

For intentional introductions, a precautionary approach is perhaps best implemented by the use of "white lists," whereby the target species must meet some criteria to demonstrate the likelihood of low impact prior to introduction (Wittenberg and Cock 2001; Chapter 10, this volume). This places the burden of proof on an importer or proponent, who ideally must submit a prospectus and obtain approval for importation of the target species (ICES 1995, U.S. Congress 2002). Approval can be withheld for species that do not meet established criteria or when insufficient biological and ecological information exists for evaluation. In the latter case, this provides a feedback mechanism or incentive to fill in critical gaps needed to make informed decisions involving importation of both target species and their associated organisms.

A precautionary approach applied to unintentional introductions focuses on the level of vector instead of species. Here, at its simplest, the primary goal is to minimize all unintentional transfers by known vectors. The framework provided herein outlines an overall approach for implementation. The magnitude of efforts to reduce transfers by any one vector is influenced greatly by the perceived costs of implementation versus those associated with invasions (see earlier discussion); unlike intentional introductions, there are no expected benefits that counterbalance the cost of invasions. As predictive ability is limited about the possible impacts of many individual species, let alone entire assemblages delivered unintentionally by a vector (e.g., ships), the potential costs of unintentional transfers should be considered great unless shown otherwise. In our view, this argues for widespread implementation of vector management.

There now exist several examples of guidelines and regulations that operate at the vector level to prevent unintentional transfers that are not species-specific. We have already discussed ships' ballast water as an example, playing out at state, national, and international levels. Regulations for acceptable vector operation are also seen for the importation of used cars to New Zealand (Chapter 11, this volume), transfers of commercial fisheries stocks (ICES 1995), and elsewhere (Shine et al. 2000; Chapters 10 and 13, this volume). In general, this approach offers a way forward to implement vector management, analogous to the use of "white lists" for intentional introductions, by allowing vectors (trade practices) to operate that meet approved minimum standards to reduce unwanted transfers. Further, these standards may change or evolve through time, in response to increased understanding of invasion processes (as above) and improved capacity to interrupt transfers.

In addition, it is noteworthy that enforcement of regulations for target species operates often at the vector level, to prevent both unintentional transfers as well as those that are intentional but illegal. Such efforts are frequently focused on known modes of importation for target organisms, including a variety of methods to detect, intercept, and prevent transfers (Wittenberg and Cock 2001; Chapters 10, 13, and 14, this volume). Thus, success of measures to prevent transfers of particular target species depends upon knowledge about (a) the species' geographic distribution, (b) availability of possible vectors (i.e., vector analysis), and (c) known vectors of established populations (i.e., vector strength), informing design of both vector interruption and assessment of efficacy. In this sense, the general framework outlined herein for vector management (Fig. 18.3) applies equally to intentional and unintentional transfers as well as target and nontarget species.

Future Directions and Recommendations

Despite the implementation of some aspects of vector management, there is a clear need to advance a more comprehensive and formal approach to vector management, reflecting the fact that "high-impact" invasions can result whenever a vector is actively transferring species. The door to invasions remains open, as evidenced by the increasing rate of newly detected invasions across many habitats and taxonomic groups (Mills et al. 1993, Fuller et al. 1999, Ruiz et al. 2000a; Chapter 4, this volume). At the present time, vector management is often reactive in two respects: (a) developing options for vector interruption after a high-impact invasion occurs and (b) considering possible risk of invasions associated primarily with known high-impact species while discounting (or not incorporating) the potential consequences of poorly known species—a subset of which will surely have significant ecological, economic, and health impacts.

Adoption of a precautionary approach is needed to overcome the current reactive mode to vector management: All species transfers should be considered suspect or unwanted pending evaluation. Such thinking has gained support for intentional introductions, but we are still a long way from this point in considering unintentional introductions. Policies and management are predicated largely on the perceived costs and risks, which are often underestimated in current decision making about invasions. In our view, a precautionary approach would be advanced by a clear, accurate assessment of the actual costs associated with unintentional introductions, as well as acknowledging the limited capacity to predict which species will have significant impacts (as discussed above).

A priority for vector management is to link current national-level efforts through international coordination, enhancing the capacity to track, evaluate, and manage vectors on multiple levels. One obvious area for coordination is the formation of a distributed international network of information on vector operation. Each node in the network could collect data on the transfer and invasion of organisms associated with particular vectors (i.e., vector analysis and vector strength, respectively) for a particular geographic region. This could be further expanded to include changes in vector operation over time, methods for vector interruption, and key information about the biology and ecology of species resident in particular regions. Nodes could collect information in one or many of these topic areas, for a large or small geographic region, or for particular taxonomic groups.

A distributed network approach differs from building a centralized information source, in that it provides both a high degree of flexibility and an opportunity to include existing information and expertise. Many programs already exist in various countries that are collecting information and tracking various aspects of vector operation and invasions. These are distributed across a diverse range of organizations—from universities and museums to state and federal agencies. Although often established independently of one another, many programs are collecting similar types of data in their respective regions. Thus, efforts to link these programs into a distributed network provide an efficient and cost-effective strategy to global-scale information collection and access. Moreover, the best information for any particular region, in terms of contemporary and historical knowledge about transfer mechanisms and invasions, is likely to be found within that region.

The technology currently exists to share this information using queries of distributed databases through the World Wide Web (e.g., Species Analyst 2002; see also Chapter 14, this volume). In a matter of seconds to minutes, a query to a single Web site could simultaneously return data from countless numbers of participating information sources or databases throughout the world. Although some effort would be needed to standardize information characteristics (e.g., nomenclature, categories, etc.) for implementation, this network approach would create a powerful tool for vector management and invasion science that can quickly integrate data across many global regions. For example, such an approach could (a) help track the geographic distribution, impacts, or biological attributes of particular species, (b) provide a strong empirical base to test predictive models of invasion dynamics and impacts; (c) allow quick exchange of new information about emerging vectors or methods of vector treatment; and (d) increase statistical power and generality in measuring dose-response relationships between extent of or changes in transfers and number of invasions, by increasing the number of sites and geographic regions beyond those possible in a single country (see Ruiz and Hewitt 2002 for further discussion).

Another area that deserves significant attention is an expanded international effort to develop and adopt accepted standards for vector operation (i.e., methods to limit species transfers). Such coordination could create an important catalyst toward a more comprehensive, formal, and programmatic approach to vector management. The current management actions directed toward ships' ballast water, using the IMO as an international coordinating body (Box 18.2), provide a useful example. Similar efforts to understand and reduce transfers by particular industry and trade organizations, linking available science to industry knowledge about vector operations, would be a powerful and productive approach.

A range of international organizations and conventions offer important venues for developing and implementing coordinated vector management (Shine et al. 2000; Chapter 14, this volume). The World Trade Organization (WTO) may be the most important platform or venue to develop more comprehensive vector management policy. The WTO, comprised of 145 member countries as of February 2003, was created to establish rules of trade among nations (WTO 2002). The WTO (2002) states its purpose as follows:

> The system's overriding purpose is to help trade flow as freely as possible—so long as there are no undesirable side effects. That partly means removing obstacles. It also means ensuring that individuals, companies and governments know what the trade rules are around the world, and giving them the confidence that there will be no sudden changes of policy. In other words, the rules have to be transparent and predictable.

Current efforts by the WTO are focused primarily on advancing free trade, providing only a very narrow consideration of possible "undesirable side effects." The primary mechanism for considering undesirable effects is the WTO Agreement on the Application of Sanitary and Phytosanitary Measures, or the SPS Agreement (see Chapter 15, this volume, for discussion). This agreement considers explicitly the risk of transfers for known pathogens and pests, which pose clear threats to food safety as well as animal and plant health. The burden of proof is generally placed upon the recipients to demonstrate a significant risk of impacts. In general, the focus of the SPS Agreement is on intentional transfers.

In our view, the WTO should expand its consideration of species transfers, which may lead to invasions, in multiple ways. First, a formal mechanism should exist to evaluate and develop standards governing unintentional transfers through vector operations, which result from trade activities. This would provide a parallel structure to that for consideration of intentional transfers (and organisms associated with those transfers) under the SPS Agreement. In the case of the shipping vector, this would include coordination with the IMO efforts. Second, the approach to considering risks associated with intentional and unintentional transfer requires reexamination. The SPS Agreement largely adopts a "black-list" approach to transfers, limiting restrictions to particular "listed" species in order to maximize trade opportunities, yet the potential impacts of many non-listed species are unknown and remain difficult to predict (as discussed above). The current practice does not adequately reflect actual risks associated with invasions, because the average likelihood of any one species having significant impacts may be low. This bears further formal evaluation, including a different model that more fully incorporates cumulative consequences of transfers, and should be approached with broad-based expert input from ecologists, environmental economists, statisticians, and environmental risk assessment specialists. Such evaluation would lead to a more balanced and precautionary approach to trade activities than presently exists, promoting stronger efforts to minimize unintentional transfers of species.

It is this global scale at which vectors operate and trade rules are established, and improved mechanisms are needed to address vector management at an international level. Policies and research efforts that are created by individual countries have certainly been worthwhile. Nonetheless, such efforts implemented separately, on a country-by-country basis, will result in a piecemeal approach to vector management that is less efficient and effective than a coordinated effort. For these reasons, we emphasize the importance and value of advancing international coordination, for both the research that serves to guide and evaluate vector management (including data collection, analyses, and exchange) and the policies for vector management.

There are various structural models that can be considered to implement coordinated vector management at an international scale. The WTO must play an integral role, given its current position in establishing policies for international trade. One model would have the WTO play a direct role in establishing mechanisms to (a) provide a more comprehensive analysis of intentional and unintentional transfers; (b) incorporate the costs, impacts, and uncertainties associated with invasions into policy decisions; and (c) implement and facilitate vector management. This should include the forecasting of future trade patterns and possible implications for vector operation, or the "Vector Early Warning System" (VEWS) discussed above.

An alternate strategy is the formation of a separate international organization, which is charged with addressing the conspicuous gaps for vector management that exist in the areas of trade policy, research, and management. Along these lines, Jenkins (1996) suggests creation of an "international advisory panel" to advise various international secretariats, trade organizations, and national governments. In addition, Perrings et al. (2003) suggests establishing a "World Environment Organization" to serve this function as part of the Convention on Biological Diversity. Such an approach has some clear advantages, creating an independent body expressly to serve as a catalyst and champion in these areas while avoiding competition with a preexisting mission or purpose in the charter of extant organizations. To a large extent, this is the role that the Global Invasive Species Programme strives to implement (McNeely et al. 2001).

Regardless of the specific organizational structure, many of the elements for concerted, international vector management are best developed as a distributed network, whereby a central organization becomes the focal point for guiding and facilitating the various facets across countries and injecting key findings into WTO processes and other relevant policy venues. This distributed approach is especially useful for tracking vectors and invasions, developing information resources and analyses across countries (as above; see also Ruiz 1998, Ricciardi et al. 2000, Ruiz and Hewitt 2002 for discussion). Furthermore, for multiple reasons, the creation of distributed information networks may be most efficiently implemented according to taxonomic groups, ecosystem type (e.g., freshwater, marine, terrestrial), and vector. First, this approach acquires information closest to the sources, probably increasing the quality and quantity of data (due to the distributed nature of expertise as discussed above). Second, this approach capitalizes on the preexisting networks of scientists and managers, who are usually organized by ecosystem or taxonomic group (and sometimes by vector), and who have already established working relationships. In some cases, such networks have already begun to organize information systems concerning invasions and vector management. For example, the Working Group on Introductions and Transfers of Marine Organisms has been compiling information on marine invasions across multiple countries for many years (ICES 2002); furthermore, this information is shared at annual meetings and compiled into annual reports, with some consideration of developing an online, international information system. Many other groups are working in a similar fashion, organized around other taxonomic groups or ecosystems.

There is a clear value to integrating research and policy efforts across countries, and this represents the next logical step in developing effective vector management. Such an undertaking promises increased understanding of invasion patterns and processes, but also a more coherent and cost-effective approach with the limited resources now available in both developed and developing countries.

ACKNOWLEDGMENTS

We thank the participants in the GISP Pathways workshop for stimulating presentations and discussions, which have contributed to development of this chapter. Many of the ideas presented and the overall framework have emerged from our work on transfer of organisms by ships; this work has been supported by grants from the Maryland Sea Grant Program, National Sea Grant Program, Regional Citizens Advisory Council of Prince William Sound, and U.S. Fish and Wildlife Service. The chapter was improved by comments from Rich Everett and Whitman Miller.

REFERENCES

Adam, P. 1990. *Saltmarsh Ecology.* Cambridge: Cambridge University Press.

AIRD (Aquatic Invasions Research Directory). 2002. Aquatic Invasions Research Directory. Smithsonian Environmental Research Center Web site <http://invasions.si.edu/>.

Allee, W. C. 1931. *Animal Aggregations. A Study in General Sociology.* Chicago: University of Chicago Press.

Burreson, E. M., N. A. Stokes, and C. S. Friedman. 2000. Increased virulence in an introduced pathogen: *Haplosporidium nelsoni* (MSX) in the eastern oysters *Crassostrea virginica. Journal of Aquatic Animal Health* 12: 1–8.

Calvo, G. W., M. W. Luckenback, S. K. Allen, Jr., and E. M. Burreson. 2000. A comparative field study of *Crassostrea ariakensis* and *Crassostrea virginica* in relation to salinity in Virginia. Special report in Applied Marine Science and Ocean Engineering No. 360, Virginia Institute of Marine Science, Gloucester Point, Virginia.

Carlton, J. T. 1985. Transoceanic and interoceanic dispersal of coastal marine organisms: The biology of ballast water. *Oceanography and Marine Biology Annual Review* 23: 313–374.

Carlton, J. T. 1989. Man's role in changing the face of the ocean: Biological invasions and implications for conservation of nearshore environments. *Conservation Biology* 3: 265–273.

Carlton, J. T. 1996a. Biological invasions and cryptogenic species. *Ecology* 77(6): 1653–1655.

Carlton, J. T. 1996b. Pattern, process, and prediction in marine invasion ecology. *Biological Conservation* 78: 97–106.

Carlton, J. T. and J. B. Geller. 1993. Ecological roulette: The global transport of nonindigenous marine organisms. *Science* 261: 78–82.

Carlton, J. T. and G. M. Ruiz. In press. Vector science and integrated vector management in bioinvasion ecology: Conceptual frameworks. In *Invasive Alien Species*, H. A. Mooney, ed. Washington, DC: Island Press.

Carlton, J. T., D. M. Reid, and H. van Leeuwen. 1995. The Role of Shipping in the Introduction of Nonindigenous Aquatic Organisms to the Coastal Waters of the United States other than the Great Lakes and an Analysis of Control Options. Report No. CG-D-11-95, National Technical Information Service, Springfield, Virginia 22161.

Case, T. J. 1990. Invasion resistance arises in strongly interacting species-rich model competition communities. *Proceedings of the National Academy of Science* 87: 9610–9614.

CENR (Committee on Environment and Natural Resources). 1999. Ecological risk assessment in the federal government. Committee on Environment and Natural Resources, National Science and Technology Council. CENR/5-99/001. Washington, DC.

Chesapeake Bay Program. 2002. Report of the ad-hoc panel on the industry trials of triploid non-indigenous oyster species in waters of the Chesapeake Bay basin. Report to the Chesapeake Bay Program, Annapolis, Maryland.

Cohen, A. N., and J. T. Carlton. 1995. Biological Study: Non-Indigenous Aquatic Species in a United States Estuary: A Case Study of the Biological Invasions of the San Francisco Bay and Delta. U.S. Fisheries and Wildlife and National Sea Grant College Program Report, NTIS Number PB96-166525, Springfield, Virginia.

Crawley, M. J. 1987. What makes a community invasible? In *Colonization, Succession, and Stability*, A. J. Gray, M. J. Crawley, and P. J. Edwards, eds., pp. 429–453. Oxford: Blackwell Scientific.

Crooks, J. A., and M. E. Soulé. 1999. Lag times in population explosions of invasive species: Causes and implications. In *Invasive Species and Biodiversity Management*, O. T. Sandlund, P. J. Schei, and Å. Viken, eds., pp. 103–125. The Netherlands: Kluwer Academic Publishers.

Diamond, J. 1986. Overview: Laboratory experiments, field experiments, and natural experiments. In *Community Ecology*, J. Diamond and T. J. Case, eds., pp. 3–23. New York: Harper & Row.

Dickman, M. and F. Zhang. 1999. Mid-ocean exchange of container vessel ballast water. 2. Effects of vessel type in the transport of diatoms and dinoflagellates from Manzanillo, Mexico, to Hong Kong, China. *Marine Ecology Progress Series* 176: 253–262.

Drake, L. A., K-H. Choi, G. M. Ruiz, and F. C. Dobbs. 2001. Global redistribution of bacterioplankton and virioplankton communities. *Biological Invasions* 3: 193–199.

Dukes, J. S. and H. A. Mooney. 1999. Does global change increase the success of biological invaders? *Trends in Ecology and Evolution* 14: 135–139.

Elton, C. S. 1958. *The Ecology of Invasions by Animals and Plants*. London: Methuen & Co. Ltd.

Fuller, P. L., L. G. Nico, and J. D. Williams. 1999. *Nonindigenous Fishes Intro-*

duced into Inland Waters of the United States. American Fisheries Society Special Publication 27. Bethesda, Maryland.

Gollasch, S. 2002. The importance of ship hull fouling as a vector of species introductions into the North Sea. *Biofouling* 18: 105–121.

Gollasch, S., E. Macdonald, S. Belson, H. Botnen, J. T. Christensen, J. P. Hamer, G. Houvenaghel, A. Jelmert, I. Lucas, D. Masson, T. McCollin, S. Olenin, A. Persson, I. Wallentinus, L. P. M. J. Wetsteyn, and T. Wittling. 2002. Life in ballast tanks. In *Invasive Aquatic Species of Europe: Distributions, Impacts, and Management*, E. Leppäkoski, S. Gollasch, and S. Olenin, eds., pp. 217–231. Dordrecht: Kluwer Academic Publishers.

Grosholz, E. D. and G. M. Ruiz. 1996. Predicting the impact of introduced species: Lessons from the multiple invasions of the European green crab. *Biological Conservation* 78: 59–66.

Hardin, G. 1968. The tragedy of the commons. *Science* 162: 1243–1248.

Hewitt, C. L., M. L. Campbell, R. E. Thresher, and R. B. Martin, eds. 1999. *Marine Biological Invasions of Port Phillip Bay, Victoria*. CRIMP Technical Report 20, CSIRO Division of Marine Research, Hobart, Australia.

Hines, A. H., and G. M. Ruiz. 2000. *Biological Invasions of Cold-Water Coastal Ecosystems: Ballast-Mediated Introductions in Port Valdez/Prince William Sound, Alaska*. Final Report, Regional Citizens Advisory Council of Prince William Sound.

Hobbs, R. J. 1989. The nature and effects of disturbance relative to invasions. In *Biological Invasions: A Global Perspective*, J. A. Drake, H. A. Mooney, F. di Castri, R. H. Groves, F. J. Kruger, M. Rejmánek, and M. Williamson, eds., SCOPE 37, pp. 389–405. New York: John Wiley & Sons.

Hobbs, R. J. and L. F. Huenneke. 1992. Disturbance, diversity, and invasion: Implications for conservation. *Conservation Biology* 6: 324–337.

Horvitz, C. C. 1997. The impact of natural disturbances. In *Strangers in Paradise: Impact and Management of Nonindigenous Species in Florida*, D. Simberloff, D. C. Schmitz, and T. C. Brown, eds., pp. 63–74. Washington, DC: Island Press.

Hülsmann, N. and B. S. Galil. 2002. Protists—A dominant component of the ballast transported biota. In *Invasive Aquatic Species of Europe: Distributions, Impacts, and Management*, E. Leppäkoski, S. Gollasch, and S. Olenin, eds., pp. 20–26. Dordrecht: Kluwer Academic Publishers.

ICES (International Council for the Exploration of the Sea). 1995. ICES Code of Practice on the introduction and transfer of marine organisms.

ICES (International Council for the Exploration of the Sea). 2002. Report of the Working Group on Introductions and Transfers of Marine Organisms. [March 2001]. <http://www.ices.uk>.

IMO (International Maritime Organization). 2002. GloBallast Programme Web site, International Maritime Organization, <http://globallast.imo.org>.

Jenkins, P. T. 1996. Free trade and exotic species introductions. *Conservation Biology* 10: 300–302.

Jiang, S. C. and J. H. Paul. 1998. Gene transfer by transduction in the marine environment. *Applied Environmental Microbiology* 64: 2780–2787.

Jones, M. M. 1991. *Marine Organisms Transported in Ballast Water: A Review of the Australian Scientific Position.* Bureau of Rural Resources Bulletin No. 11. Canberra: Australian Government Publishing Service.

Kennedy, V. S., R. I. E. Newell, and A. Eble, Jr., eds. 1998. *The Eastern Oyster: Crassostrea virginica.* College Park: Maryland Sea Grant College Program.

Kerr, S. 1994. Ballast water ports and shipping study. Australian Quarantine Inspection Service, Report Number 5, Canberra.

Kolar, C. S. and D. M. Lodge. 2001. Progress in invasion biology: Predicting invaders. *Trends in Ecology and Evolution* 16: 199–204.

Kolar, C. S. and D. M. Lodge. 2002. Ecological predictions and risk assessment for alien fishes in North America. *Science* 298: 1233–1236.

Leffler, M. 2002. Crisis and controversy. Does Chesapeake Bay need a new oyster? *Chesapeake Bay Quarterly*, Fall 2002. Maryland Sea Grant College Program, College Park.

Lonsdale, W. M. 1999. Global patterns of plant invasions and the concept of invasibility. *Ecology* 80: 1522–1536.

MacArthur, R. H. and E. O. Wilson. 1967. *The Theory of Island Biogeography.* Princeton: Princeton University Press.

Mack, R. N. 1991. The commercial seed trade: An early disperser of weeds in the United States. *Economic Botany* 45: 257–273.

Mack, R. N., D. Simberloff, W. M. Lonsdale, H. Evans, M. Clout, and F. A. Bazzaz. 2000. Biotic invasions: Causes, epidemiology, global consequences, and control. *Ecological Applications* 10: 689–710.

McCarthy, H. P. and L. B. Crowder. 2000. An overlooked scale of global transport: Phytoplankton species richness in ballast water. *Biological Invasions* 2: 321–322.

McNeely, J. A., H. A. Mooney, L. E. Neville, P. Schei, and J. K. Waage, eds. 2001. *A Global Strategy on Invasive Alien Species.* Gland, Switzerland, and Cambridge, UK: IUCN in collaboration with the Global Invasive Species Programme.

Mills, E. L., J. H. Leach, J. T. Carlton, and C. L. Secor. 1993. Exotic species in the Great Lakes: A history of biotic crises and anthropogenic introductions. *Journal of Great Lakes Research* 19: 1–54.

Mooney, H. A. and R. J. Hobbs, eds. 2000. *Invasive Species in a Changing World.* Washington, DC: Island Press.

Moyle, P. B. and T. Light. 1996. Fish invasions in California: Do abiotic factors determine success? *Ecology* 77: 1666–1670.

Myers, J. H., D. Simberloff, A. M. Kuris, and J. R. Carey. 2000. Eradication revisited: Dealing with exotic species. *Trends in Ecology and Evolution* 15: 316–320.

Naylor, R. L., S. L. Williams, and D. R. Strong. 2001. Aquaculture—A gateway for exotic species. *Science* 294: 1655–1656.

Nehring, S. 2001. After the TBT era: Alternative anti-fouling paints and their ecological risks. *Senckenbergiana martima* 31: 341–351.

Nehring, S. 2002. Biological invasions into German waters: An evaluation of the importance of different human-mediated vectors for nonindigenous macrozoobenthic species. In *Invasive Aquatic Species of Europe: Distributions, Impacts, and Management*, E. Leppäkoski, S. Gollasch, and S. Olenin, eds., pp. 373–383. Dordrecht: Kluwer Academic Publishers.

NEMW (Northeast Midwest Institute). 2002. Northeast Midwest Institute, Biological Pollution Web page <http://www.nemw.org/biopollute.htm#laws>.

NISC (National Invasive Species Council). 2001. *Meeting the Invasive Species Challenge: National Invasive Species Management Plan.* Washington, DC: National Invasive Species Council.

NRC (National Research Council). 1995. *Understanding Marine Biodiversity: A Research Agenda for the Nation.* Washington, DC: National Academy Press.

NRC (National Research Council). 1996. *Stemming the Tide: Controlling Introductions of Nonindigenous Species by Ships' Ballast Water.* Washington, DC: National Academy Press.

OTA (Office of Technology Assessment). 1993. *Harmful Non-Indigenous Species in the United States.* Washington, DC: U.S. Congress, Office of Technology Assessment.

Parker, I. M., D. Simberloff, W. M. Lonsdale, K. Goodell, M. Wonham, P. M. Karieva, M. H. Williamson, V. VonHolle, P. B. Moyle, J. E. Byers, and L. Goldwasser. 1999. Impact: Toward a framework for understanding the ecological effects of invaders. *Biological Invasions* 1: 2–19.

Perrings, C., M. Williamson, E. B. Barbier, D. Delfino, S. Dalmazzone, J. Shogren, P. Simmons, and A. Watkinson. 2003. Biological invasion risks and the public good: An economic perspective. *Conservation Ecology* 6: 1–10 online journal <http://www.consecol.org/>.

Pimentel, D., L. Lach, R. Zuniga, and D. Morrison. 2000. Environmental and economic costs of nonindigenous species in the United States. *BioScience* 50: 53–64.

Pimm, S. L. 1991. *The Balance of Nature?* Chicago: University of Chicago Press.

Por, F. D. 1978. *Lessepsian Migration: The Influx of Red Sea Biota into the Mediterranean by Way of the Suez Canal.* Heidelberg: Springer-Verlag.

Raaymakers, S. and R. Hilliard. 2003. Harmful aquatic organisms in ships' ballast water: Ballast water risk assessment. In *Alien marine organisms introduced by ships*, B. Galil and F. Briand, eds., pp. 103–110. CIESM Monography No. 20. Monaco: Commission Internationale pour l'Exploration Scientifique de la mer Méditerranée.

Reise, K., S. Gollasch, and W. J. Wolff. 1999. Introduced marine species of the North Sea coasts. *Helgoländer Meeresunters* 52: 219–234.

Rejmánek, M. and J. M. Randall. 1994. Invasive alien plants in California: 1993 summary and comparison with other areas of North America. *Madrono* 41: 161–177.

Ricciardi, A., W. W. M. Steiner, R. N. Mack, and D. Simberloff. 2000. Toward a global information system for invasive species. *BioScience* 50: 239–244.

Richardson, D. M., P. Pysek, M. Rejmánek, M. G. Barbour, F. D. Paneta, and C. J. West. 2000. Naturalization and invasion of alien plants: Concepts and definitions. *Diversity and Distributions* 6: 93–107.

Robinson, J. V. and J. E. Dickerson, Jr. 1984. Testing the invulnerability of laboratory island communities to invasion. *Oecologia* 61: 169–174.

Roughgarden, J. 1986. Predicting invasions and rates of spread. In *Ecology of Biological Invasions of North America and Hawaii*, H. A. Mooney and J. A. Drake, eds., p. 179–190. New York: Springer-Verlag.

Ruiz, G. M. 1998. Cross-cutting issues. In *Invasive Species Databases, Proceedings of a Workshop*, R. L. Ridgeway, W. P. Gregg, R. E. Stinner, and A. G. Brown, eds., p. 35–37. National Biological Information Infrastructure (NBII), USGS, Reston, Virginia.

Ruiz, G. M., P. W. Fofonoff, J. T. Carlton, M. J. Wonham, and A. H. Hines. 2000a. Invasion of coastal marine communities in North America: Apparent patterns, processes, and biases. *Annual Review of Ecology and Systematics* 31: 481–531.

Ruiz, G. M., P. Fofonoff, A. H. Hines, and E. D. Grosholz. 1999. Non-indigenous species as stressors in estuarine and marine communities: Assessing invasion impacts and interactions. *Limnology and Oceanography* 44: 950–972.

Ruiz, G. M., T. K. Rawlings, F. C. Dobbs, L. A. Drake, T. Mullady, A. Huq, and R. R. Colwell. 2000b. Worldwide transfer of microorganisms by ships. *Nature* 408: 49–50.

Ruiz, G. M. and C. L. Hewitt. 2002. Toward understanding patterns of coastal marine invasions: A prospectus. In *Invasive Aquatic Species of Europe: Distributions, Impacts, and Management*, E. Leppäkoski, S. Gollasch, and S. Olenin, eds., pp. 529–547. Dordrecht: Kluwer Academic Publishers.

Schoener, T. W. and D. A. Spiller. 1995. Effect of predators and area invasion: An experiment with island spiders. *Science* 267: 1811–1813.

Shine, C., N. Williams, and L. Gundling. 2000. A guide to designing legal and institutional frameworks on alien invasive species. Gland: IUCN.

Simberloff, D. S. 1989. Which insect introductions succeed and which fail? In *Biological Invasions: A Global Perspective*, J. A. Drake, H. A. Mooney, F. di Castri, R. H. Groves, F. J. Kruger, M. Rejmánek, and M. Williamson, eds., SCOPE 37, pp. 61–75. New York: John Wiley & Sons.

Simberloff, D. and B. Von Holle. 1999. Positive interactions of nonindigenous species: Invasional meltdown? *Biological Invasions* 1: 21–32.

Smith, L. D., M. J. Wonham, L. D. McCann, G. M. Ruiz, A. H. Hines, and J. T. Carlton. 1999. Invasion pressure to a ballast-flooded estuary and an assessment of inoculant survival. *Biological Invasions* 1: 67–87.

Species Analyst. 2002. <http://tsadev.speciesanalyst.net/>.

Stachowicz, J. J., J. R. Terwin, R. B. Whitlatch, and R. W. Osman. 2002. Linking climate change and biological invasions: Ocean warming facilitates nonindigenous species invasions. *Proceedings of the National Academy of Sciences USA* 99: 15497–15500.

Subba Rao, D. V., W. G. Sprules, A. Locke, and J. T. Carlton. 1994. Exotic phytoplankton from ships' ballast waters: Risk of potential spread to mariculture sites on Canada's east coast. Canadian Data Report of Fisheries and Aquatic Sciences 937.

Taylor, A., G. Rigby, S. Gollasch, M. Voigt, G. Hallegraeff, T. McCollin, and A. Jelmert. 2002. Preventive treatment and control techniques for ballast water.

In *Invasive Aquatic Species of Europe: Distributions, Impacts, and Management,* E. Leppäkoski, S. Gollasch, and S. Olenin, eds., pp. 484–507. Dordrecht: Kluwer Academic Publishers.

Thompson, J. A. 2001. Introduction of *Crassostrea ariakensis* formerly *rivularis* to Chesapeake Bay: The solution to restoring an oyster fishery and water quality in the bay? Report by U.S. Fish and Wildlife Service, Chesapeake Bay Field Office, Annapolis.

Tilman, D. 1999. The ecological consequences of changes in biodiversity: A search for general principals. *Ecology* 80: 1455–1474.

U.S. Congress. 1990. Nonindigenous Aquatic Nuisance Prevention and Control Act (NANPCA).

U.S. Congress. 1996. National Invasive Species Act (NISA).

U.S. Congress. 2002. National Aquatic Invasive Species Act (NAISA), Senate Bill 2964 introduced in U.S. Congress, September 2002.

U.S. Presidential Executive Order. 1999. Executive Order 13112, President Clinton, February 1999.

Vermeij, G. J. 1991. When biotas meet: Understanding biotic interchange. *Science* 253: 1099–1104.

Vermeij, G. J. 1996. An agenda for invasion biology. *Biological Conservation* 78: 3–9.

Wilcove, D. S., D. Rothstein, J. Dubow, A. Phillips, and E. Losos. 1998. Quantifying threats to imperiled species in the United States. *BioScience* 48: 607–615.

Williamson, M. 1996. *Biological Invasions.* London: Chapman & Hall.

Williamson, M. 1999. Invasions. *Ecography* 22: 5–12.

Wittenberg, R. and M. J. W. Cock, eds. 2001. *Invasive Alien Species: A Toolkit of Best Prevention and Management Practices.* Wallingford, Oxon: CAB International.

WTO. 2002. World Trade Organization Web site <http://www.wto.org>.

Contributing Authors

DAVID A. ANDOW
Department of Entomology
University of Minnesota
St. Paul, Minnesota 55108
USA

JAMES A. CAMBRAY
Albany Museum
Somerset Street
Grahamstown 6139
South Africa

JAMES T. CARLTON
Williams College–Mystic Seaport
Maritime Studies Program
75 Greenmanville Avenue
P.O. Box 6000
Mystic, Connecticut 06355-0990
USA

JOSEPH F. CAVEY
USDA, APHIS, PPQ
4700 River Road, Unit 133
Riverdale, Maryland 20737
USA

R. ARTHUR CHAPMAN
CSIR, Division of Water,
Environment and Forestry
Technology
P.O. Box 320
Stellenbosch 7599
South Africa

ROBERT I. COLAUTTI
Great Lakes Institute for
Environmental Research
University of Windsor
Windsor, Ontario N9B3P4
Canada

ROBERT H. COWIE
Center for Conservation Research
and Training
University of Hawaii
3050 Maile Way, Gilmore 408
Honolulu, Hawaii 96822
USA

W. RICHARD J. DEAN
Percy FitzPatrick Institute of
African Ornithology
University of Cape Town
Rondebosch 7701
South Africa

PAUL W. FOFONOFF
Smithsonian Environmental
Research Center
647 Contees Wharf Road
P.O. Box 28
Edgewater, Maryland 21037
USA

PAM L. FULLER
United States Geological Survey
Center for Aquatic Resources
Studies
7920 NW 71st Street
Gainesville, Florida 32654
USA

CHARLES L. GRIFFITHS
Zoology Department, University
of Cape Town
Rondebosch 7701
South Africa

ALEXIS T. GUTIERREZ
Johns Hopkins University
School for Advanced International
Studies
1740 Massachusetts Avenue, NW
Washington, D.C. 20036
USA

ALICIA K. HANCOCK
453 South Venice Blvd.
Venice, California 90291
USA

BARBARA J. HAYDEN
National Institute of Water &
Atmospheric Research Ltd.
P O Box 8602
Christchurch
New Zealand

KEITH R. HAYES
Centre for Research on Introduced
Marine Pests
CSIRO Division of Marine
Research
GPO Box 1538
Hobart, 7001
Tasmania
Australia

KRISTEN HOLECK
Department of Natural Resources
Cornell University Biological Field
Station
Bridgeport, New York 13030
USA

KEIZI KIRITANI
Laboratory of Population Ecology
National Institute of
Agro-Environmental Sciences
Tsukuba 305-8604
Japan

FRED KRAUS
Bishop Museum
1525 Bernice Street
Honolulu, Hawaii 96817
USA

DAVID C. LE MAITRE
CSIR, Division of Water,
Environment and Forestry
Technology
P.O. Box 320
Stellenbosch 7599
South Africa

HUGH J. MACISAAC
Great Lakes Institute for
Environmental Research
University of Windsor
Windsor, Ontario N9B3P4
Canada

RICHARD N. MACK
School of Biological Sciences
Washington State University
Pullman, Washington 99164
USA

EDWARD L. MILLS
Department of Natural Resources
Cornell University Biological Field
Station
Bridgeport, New York 13030
USA

DAVID J. NEWTON
National Representative
TRAFFIC East/Southern Africa:
South Africa
Private Bag X11
Parkview 2122
South Africa

ARTHUR J. NIIMI
Department of Fisheries and
Oceans
Canada Centre for Inland Waters
P.O. Box 5050
Burlington, Ontario L7R4A6
Canada

RICHARD ORR
USDA, APHIS, PPD
4700 River Road, Unit 117
Riverdale, Maryland 20737
USA

MARY E. PALM
APHIS
Rm. 304, B-011A, BARC-West
Beltsville, Maryland 20705
USA

PAUL PHELOUNG
Office of the Chief Plant
Protection Office
Agriculture, Fisheries and
Forestry–Australia
GPO Box 858
Canberra ACT 2611
Australia

PAUL R. PHIFER
1841 20th Avenue, SE
Portland, Oregon 97214
USA

JAMIE K. REASER
Ecos Systems Institute
6210 Julian Street
Springfield, Virginia 22150
USA

MARIA ANTONIA RIBERA SIGUAN
Lab. Botánica, Facultat Farmácia
Universitat Barcelona, 08028
Barcelona, Spain

DAVID M. RICHARDSON
Institute for Plant Conservation
Botany Department
University of Cape Town
Rondebosch 7701
South Africa

DAVID G. ROBINSON
USDA, APHIS, PPQ and
Department of Malacology
Academy of Natural Sciences
1900 Benjamin Franklin Parkway
Philadelphia, Pennsylvania 19130
USA

AMY Y. ROSSMAN
USDA, APHIS
Rm. 304, B-011A, BARC-West
Beltsville, Maryland 20705
USA

GREGORY M. RUIZ
Smithsonian Environmental
Research Center
647 Contees Wharf Road
P.O. Box 28
Edgewater, Maryland 21037
USA

BRIAN STEVES
Smithsonian Environmental
Research Center
647 Contees Wharf Road
P.O. Box 28
Edgewater, Maryland 21037
USA

COLIN D. A. VAN OVERDIJK
Great Lakes Institute for
Environmental Research
University of Windsor
Windsor, Ontario N9B3P4
Canada

CAROLYN F. WHYTE
Border Risk Management
Ministry of Agriculture and
Forestry
P.O. Box 106231
Auckland
New Zealand

TERRY WINSTANLEY
Winstanley & Smith Attorneys
P.O. Box 619
Cape Town 8000
South Africa

KOHJI YAMAMURA
Laboratory of Population Ecology
National Institute of
Agro-Environmental Sciences
Tsukuba 305-8604
Japan

BROOKS B. YEAGER
World Wildlife Fund
1250 25th Street, NW
Washington, D.C. 20037
USA

Index